AF443535

Current Perspectives in

High Energy Physics

Lectures from SERC Schools

Texts and Readings in Physical Sciences

Managing Editors

H. S. Mani, Institute of Mathematical Sciences, Chennai.
h_s_mani@hotmail.com

Ram Ramaswamy, Jawaharlal Nehru University, New Delhi.
r.ramaswamy@mail.jnu.ac.in

Editors

V Balakrishnan, Indian Institute of Technology, Madras, Chennai.
vbalki@chaos.iitm.ernet.in

Jayanta Bhattacharjee, Indian Assoc. for the Cultivation of Science, Kolkata.
tpjkb@mahendra.iacs.res.in

Deepak Dhar, Tata Institute of Fundamental Research, Mumbai.
ddhar@theory.tifr.res.in

Rohini Godbole, Indian Institute of Science, Bangalore.
rohini@cts.iisc.ernet.in

Avinash Khare, Institute of Physics, Bhubaneswar.
khare@iopb.res.in.

Already Published Volumes

Current Perspectives in
High Energy Physics

Lectures from SERC Schools

Edited by
Debashis Ghoshal
Harish-Chandra Research Institute

Published by

Hindustan Book Agency (India)
P 19 Green Park Extension
New Delhi 110 016
India

email: hba@vsnl.com
http://www.hindbook.com

ISBN 81-85931-49-6

Texts and Readings in the Physical Sciences

As subjects evolve, and the teaching and study of a subject evolves, new texts are needed to provide material and to define areas of research. The **TRiPS** series of books is an effort to document these frontiers in the Physical Sciences.

One of the principal aims of the series is to make expositions of current topics accessible to the graduate student and researcher through Textbooks and Monographs. In addition, we also feel that publication of lecture notes emanating from a thematic School or Workshop, and topical volumes of contributed articles can go a long way in providing an insight into a rapidly developing field, or an introduction to a new area.

The pedagogical value of all these forms of exposition is inestimable. We thus hope, in this series of books, to both provide a forum for the physical scientist to give a personal account and definition of his field, and a source of valuable learning material for the student wishing to gain insight and knowledge

H.S. Mani
Chennai

R.Ramaswamy
New Delhi

Contents

2 Higgs Physics   39

Saurabh D. Rindani

9   Brane-World Phenomenology   449
Sreerup Raychaudhuri

Preface

The SERC Schools in Theoretical High Energy Physics, sponsored by the *Science & Engineering Research Council* under the auspices of the *Department of Science & Technology*, Government of India have been held every year since 1985. These schools have served an extremely useful purpose in training students working for their Ph.D's in various institutes and universities across the country. Current research interests have been the principal guide dictating the choice of courses, which have varied over the entire range from 'formal theory' to 'phenomenology'. The courses are designed so as to teach the basics in a thorough and systematic way, but at the same time give a flavour and convey the excitement of the latest developments. More information is available on the homepage of the schools at: `http://www.mri.ernet.in/~sercthep`.

Lecture notes of previous SERC schools have only sporadically been published. However, since we do recognize the value of having the lectures available in print, which also makes them accessible to a larger audience, we have collected, in this volume, the lectures from the XV and XVI SERC Main Schools in Theoretical High Energy Physics held at the Saha Institute of Nuclear Physics, Kolkata and Harish-Chandra Research Institute, Allahabad in 2000 and 2001 respectively. Although it has been over three years since, the topics covered in these schools remain timely: this year's Nobel prize in Physics recognizes the development of Quantum Chromodynamics. Regrettably, though, one set of lectures given at the HRI school could not eventually be included in the present collection.

A number of people have helped, directly and indirectly, in the preparation of this volume. Among them I would particularly like to thank Rohini Godbole, Dileep Jatkar, Sunil Mukhi, Sharmistha Mukhopadhyay, Sumathi Rao, V. Ravindran and all the contributors to this volume. It is a pleasure to acknowledge the help extended by my colleagues at the Harish-Chandra Research Institute in the organization of the XVI SERC Main School in THEP. Finally, I would like to thank Rohini Godbole, H. S. Mani and Ram Ramaswamy, the editors of TRiPS, and HBA for patiently putting up with several failed deadlines.

Debashis Ghoshal
Harish-Chandra Research Institute
Allahabad

Chapter 1

Physics of Massive Neutrinos

Anjan S. Joshipura

1.1   Introduction

1.1.1   History

The existence of an almost massless neutral particle (later on called neutrino by Fermi) was postulated by Pauli in 1932 to account for the continuous energy spectrum of the electrons emitted in nuclear β decay. This particle was required to be a fermion in order to conserve angular momentum. Fermi incorporated this particle into a detailed theory of nuclear beta decay which could account for the observed shape of the electron energy distribution found in many nuclear beta decays. With availability of more experimental results, the original Fermi theory underwent many changes and finally culminated into a simple and elegant $V - A$ theory [1, 2] which universally describes all the known (charged) weak interaction processes at low energy [3, 4, 5, 6]. The $V - A$ theory is basically an effective theory which allows reliable calculations of weak interaction processes at energies $\ll O(100)\,\mathrm{GeV}$. The basic structure of this theory was later on generalized into a full fledged quantum theory based on ideas of spontaneously broken local gauge invariance [7]. It became possible to unify the weak and electromagnetic interactions within this framework.

The resulting theory is now known as the standard electroweak model. The neutrinos have played a very important role in ultimate formulation of this standard model (SM).

On the experimental side, neutrino was detected [8] in 1956, twenty-four years after its postulated existence. The detection became possible due to the availability of intense neutrino beam from reactors and due to advances in electronics which made possible an unambiguous detection of neutrino. Subsequent progress in neutrino physics was made after the development of intense beams of muon neutrinos[1] at Brookhaven Laboratory. These neutrinos are produced in the decays of high energy pions and kaons. It was found that neutrinos produced in these decays gave rise to muons rather than to electrons showing that neutrinos produced in meson decays were different from the ones produced in beta decay.

After the discovery of b quark as a part of an $SU(2)$-doublet, and discovery of τ lepton, the existence of the third type of neutrino, ν_τ, became imperative for consistency of the standard model. The lack of intense ν_τ beam did not allow its direct detection for many years. It was finally discovered in 2001 at Fermilab[9].

Neutrino physics has progressed a great deal in last ten years. The nuclear reactions taking place deep inside the Sun produce intense beam of neutrinos with a flux of $10^{10}/\mathrm{cm}^2/\mathrm{sec}$. It has become possible to precisely determine this flux over a wide energy range ~ 0.2–15 MeV. These experiments have helped in verification of the standard solar model and have also given significant information on neutrino masses and mixings. Intense beam of neutrinos is also produced in our atmosphere by the incoming cosmic rays. These neutrinos have also been detected. It is commonly believed that detection of these neutrinos leads to a conclusion that at least one neutrino is massive with a mass around 0.1 eV. In addition to the solar physics, neutrinos also influence supernova dynamics, the dynamics of expanding universe and also provide

---

[1]At many places we shall use the conventional terminology of collectively denoting neutrino and its antiparticle by the common name "neutrino".

at least a part of the dark matter in the universe. Conversely, one can use astrophysics and cosmology to learn about properties of neutrinos. Neutrino mass plays a very important role in all these.

These lectures notes are intended as a short introduction to the subject of neutrino masses. We discuss experimental evidence showing that neutrinos are massive and also theories which can lead to the masses required on experimental grounds. Considering the available time limit, our main emphasis is on basic concepts involved rather than the details. These details can be found in excellent books [3, 5, 6, 10, 11] and review articles [12, 13, 14, 15]. We proceed according to the following plan. This section contains a very brief historical introduction to weak interactions and the standard electroweak model. The basic formalism of neutrino masses is discussed in the next section. Here we discuss two component neutrino hypothesis. Then we discuss the concept of the Majorana and Dirac neutrino masses. The nature of physical neutrinos based on these mass terms is discussed in detail. Section three and four present evidence for neutrino masses. The direct mass determination (Kurie plot and neutrinoless double beta decay) is discussed in Sec. 3 and phenomena of neutrino oscillations and related experiments are discussed in Sec. 4. This contains discussion of the solar and atmospheric neutrino anomalies. In Sec. 5, we summarize the possible patterns of masses and mixing required experimentally. The final section Sec. 6 contains a brief introduction on how can one theoretically obtain such mass patterns.

I thank the SERC school organizers for giving me an opportunity to give these lectures and Palash Pal for all the arrangements. I am grateful to Nimai Singh for helpful suggestions on this manuscript.

1.1.2 Fermi theory and beyond

Fermi built a theoretical framework to describe nuclear beta decays. This was based on ideas in quantum electrodynamics. All the electromagnetic processes involving fermions result from the following basic interaction Lagrangian in quantum field theory:

$$-\mathcal{L}_{int} = -ie\bar{\Psi}(x)\gamma_\mu\Psi(x). \tag{1.1}$$

Here $\Psi(x)$ describes the basic field of a charged fermion (*e.g.*, electron). By convention $\Psi(\bar{\Psi})$ creates an antiparticle (particle) and destroys particle (antiparticle). $A_\mu(x)$ describes electromagnetic field which creates and destroys photons. The above interaction leads to processes in which a pair of fermions are attached to a spacetime point and photon (or fermions) connect different space time points. The beta decay involves four fermion fields. Fermi dispensed with the photon field and assumed that all the fermions in beta decay are created or destroyed at a single space time point. This is described by the following four fermion Lagrangian:

$$-\mathcal{L}_F = \frac{G_F}{\sqrt{2}}\bar{\Psi}_e(x)\gamma_\mu\Psi_\nu(x)\bar{\Psi}_p(x)\gamma^\mu\Psi_n(x), \tag{1.2}$$

where Ψ_x refers to quantum field corresponding to x=electron (e), proton (p), neutron (n) and neutrino (ν) respectively and $G_F/\sqrt{2}$ is a constant with the dimension[2] of inverse (mass)^2. Its value is approximately given by 1.16×10^{-5} GeV^{-2}. The above interaction can be used to calculate rates and shape of beta decay spectrum and was found to be quite successful in describing a large class of nuclear beta decays. But it could not describe all the beta decays. In the non-relativistic limit (appropriate here since nucleon energy is much smaller than its mass), Eq.(1.2) does not contain nuclear spin operator. As a result, it cannot describe nuclear transitions (called Gamow-Teller transitions) which involve change in nucleon spin. This is accomplished by means of the following interaction:

$$-\mathcal{L}_{GT} = \frac{G_F}{\sqrt{2}}\bar{\Psi}_e(x)\gamma_\mu\gamma_5\Psi_\nu(x)\bar{\Psi}_p(x)\gamma^\mu\gamma_5\Psi_n(x). \tag{1.3}$$

Eqs.(1.2,1.3) are special cases of the following most general four fermion interaction which could be written down using

---

[2]We follow the conventional notations and set $\hbar = c = 1$.

Lorentz invariance:

$$-\mathcal{L}_{gen} = \frac{G_F}{\sqrt{2}} \sum_i \bar{\Psi}_e(x)\Gamma_i(C_i + \gamma_5 C_i')\Psi_\nu(x)\bar{\Psi}_p(x)\Gamma^i\Psi_n(x), \quad (1.4)$$

where the sum over i refers to sum over linearly independent combinations of gamma matrices namely, $\Gamma_i = 1, \gamma_\mu, \sigma_{\mu\nu}, \gamma_\mu\gamma_5, \gamma_5$ referred to respectively as scalar (S), vector (V), tensor (T), axial vector (A) and pseudo-scalar (P). Parity violation found experimentally is already contained in the above interaction. A priori, the above interactions contain 10 independent parameters and it may seem difficult to determine these parameters experimentally. In the early days, it was found that the nuclear beta decays were favouring the S, T and P interactions while the pion and muon decays were clearly showing $V - A$ structure. Very careful analysis [1, 2] of the available information led to the proposition that the signatures of the S, T and P interactions may not be well-founded experimentally and all the available interactions may actually be $V - A$. This proposition was theoretically derived from chiral invariance, *i.e.*, invariance of Eq.(1.4) under independent chiral transformations $\Psi \rightarrow -\gamma_5\Psi$ on each of the fields. This invariance amounts to saying that particles (*e.g.*, electrons, neutrinos) emitted in a decay are left-handed and their antiparticles are right-handed. This was experimentally verified for the electron and also for the neutrino by direct helicity measurements[3]. It can be shown that the chiral invariance reduces the above 10 parameter four fermion interaction to only one parameter $V - A$ structure given by

$$-\mathcal{L}_{V-A} = \frac{G_F}{\sqrt{2}} \sum_i \bar{\Psi}_e(x)\gamma_\mu(1 - \gamma_5)\Psi_\nu(x)\bar{\Psi}_p(x)\gamma^\mu(1 - \gamma_5)\Psi_n(x),$$

$$(1.5)$$

As the evidence gathered, the validity of $V - A$ interaction was proven to be more general. It was found to describe leptonic weak interactions (*e.g.*, μ and τ decay), semi-leptonic weak interactions (*e.g.*, π and K decays to leptons) and non-leptonic decays of baryons and mesons (*e.g.*, K decays to pions). The SM treats quarks as fundamental and thus gives a Lagrangian which contains quark fields instead of the nucleon fields as in Eq.(1.5). The complete four fermion interactions which generalizes Eq.(1.5) is given by

$$-\mathcal{L}_{V-A} = \frac{G_F}{\sqrt{2}} J_\mu^+ J^{-\mu},$$ (1.6)

where

$$J_\mu^- \equiv \sum_{\alpha=e,\mu,\tau} \bar{\nu}_\alpha \gamma_\mu (1-\gamma_5) l_\alpha + \sum_{i,j=1,2,3} (U_{CKM})_{ij}\, \bar{u}_i \gamma_\mu (1-\gamma_5) d_j,$$ (1.7)

with l_α denotes three charged leptonic fields and U_{CKM} is a 3×3 unitary matrix and i,j run over three generations. A remarkable feature of the above equation is its universality. All the weak interactions among leptons are given in terms of a single parameter $G_F/\sqrt{2}$. The same parameter also determines weak interactions of quarks apart from the elements of mixing matrix U_{CKM} which generalizes the original Cabibbo current proposed to describe the strangeness changing weak interactions. Such universality cannot be accidental. It is now known that this is closely linked to the local gauge invariance under an $SU(2)$ group of transformations. This requires the existence of a charged spin 1 bosons $W^\pm$ with the following interaction:

$$-\mathcal{L}_W = \frac{g}{2\sqrt{2}} (J^+ W^- + h.c.).$$ (1.8)

Here g is the fundamental gauge coupling analogous to the electromagnetic charge e. The above interaction leads in the low energy limit $E << M_W$ to the four fermion interactions given in Eq.(1.6) if

$$\frac{G_F}{\sqrt{2}} = \frac{g^2}{8M_W^2}.$$

The basic gauge theoretical framework requires the existence of gauge boson corresponding to each generator of the gauge group. Thus, in addition to the $W^\pm$ associated with the raising and lowering operator of $SU(2)$, there must exist a neutral gauge boson

W_3 corresponding to the third component of the "angular momentum". The current J_3 coupled to W_3 is given by the commutator of $J^\pm$. It is seen from Eq.(1.7), that J_3 determined this way does not coincide with the electromagnetic current. In order to incorporate electromagnetism one needs an additional $U(1)$ group with corresponding gauge boson B and its gauge coupling g'. The physical photon field is a combination of B, W_3 defined as

$$A^\mu = \cos\theta_W B^\mu + \sin\theta_W W_3^\mu,$$

where angle θ_W is determined to be $\tan\theta_W = g'/g$ by requiring that the current coupled to A^μ be the electromagnetic current. The orthogonal combination of B^μ and W_3^μ is conventionally denoted by Z^μ and it leads to additional weak interactions. Structure of this neutral weak interactions is determined by the algebra of the $SU(2) \times U(1)$ group and is given as follows:

$$-\mathcal{L}_Z = \frac{g}{2\cos\theta_W} \sum_f \bar{f}\gamma_\mu[(T_{3f} - 2Q_f \sin^2\theta_W) - T_{3f}\gamma_5]f\, Z^\mu. \quad (1.9)$$

Eqs.(1.8) and (1.9) describe complete weak interactions of all fermions in the standard model. Thus various neutrino production mechanisms which we will be talking about are determined by these equations. The new ingredient which is not part of these equations and of the Standard Model is neutrino mass. We turn to this in the next section.

1.2 Neutrino masses: basic formalism

As we mentioned, neutrinos produced in any experiments are found to be left-handed while antineutrinos are right handed particles. This fact led to a two component hypothesis for neutrinos. According to this hypothesis, neutrinos are described in terms of fields having two (complex) components unlike other spin 1/2 fermions which need four component fields. Two component description of neutrino emerges naturally from the conventional Dirac equation in the massless limit. We discuss this first and use this picture to build up formalism to describe neutrino masses. This formalism is used to describe three basic types of neutrino masses: Majorana, Dirac and pseudo-Dirac.

The evolution of a free massive neutrino is described by the Dirac equation

$$(i\gamma_\mu \partial^\mu - m)\nu = 0. \tag{1.10}$$

The gamma matrices γ^μ ($\mu = 0, 1, 2, 3$) are chosen to satisfy

$$\begin{aligned}
\gamma^{\mu\dagger} &= \gamma^0 \gamma^\mu \gamma^0, \\
\gamma_5 &= i\gamma^0 \gamma^1 \gamma^2 \gamma^3 = \gamma^5 = \gamma^{5\dagger},
\end{aligned} \tag{1.11}$$

in addition to the usual anti-commutation relations. We shall adopt throughout the following specific representation called the chiral representation for the gamma matrices:

$$\gamma^0 = \begin{pmatrix} 0 & -1 \\ -1 & 0 \end{pmatrix}, \quad \vec{\gamma} = \begin{pmatrix} 0 & \vec{\sigma} \\ -\vec{\sigma} & 0 \end{pmatrix}, \quad \gamma^5 = \begin{pmatrix} 1 & 0 \\ 0 & -1 \end{pmatrix}. \tag{1.12}$$

$\vec{\sigma}$ represent here the conventional 2×2 Pauli matrices. The γ_5 is diagonal in this representation and is used to define two chiral components:

$$f_L = \frac{1}{2}(1 - \gamma_5)f, \qquad f_R = \frac{1}{2}(1 + \gamma_5)f, \tag{1.13}$$

for any fermion f. The Dirac equation can be converted to the following equations in the massless limit:

$$\begin{aligned}
i\gamma_\mu \partial^\mu \nu_{L,R} &= 0, \\
\sigma.\hat{\mathbf{p}}\left(\frac{1 \pm \gamma_5}{2}\nu\right) &= \pm\left(\frac{1 \pm \gamma_5}{2}\nu\right),
\end{aligned} \tag{1.14}$$

where $\hat{\mathbf{p}}$ denotes unit vector along the momentum direction and $\sigma.\hat{\mathbf{p}}$ denotes component of spin along the direction of motion. The second of the above equations is known as the Weyl equation. These equations show two important properties of massless fermions. First, the dynamics (*i.e.*, time development) does not mix ν_L and ν_R components. Thus these two components can be regarded as describing two independent particles with no connection between them. Secondly, ν_L (ν_R) carry negative (positive) helicity. As we shall show, if neutrino is left-handed then the anti-neutrino (*i.e.*, corresponding charge conjugate field) is right handed. Both these particles can be describe by a single field ν_L which in field theory, represents creation of antineutrino and destruction of a neutrino. Thus massless neutrinos are completely described in terms of ν_L alone and one does not need the field ν_R. This was the basic idea behind two-component picture of neutrino which was originally proposed to describe massless neutrinos in parity violating theory. But even massive neutrinos can be described in terms of two component fields. To understand this more clearly we now discuss various types of neutrino mass terms.

1.2.1 Neutrino mass terms

Neutrino mass corresponds to a Lorentz invariant renormalizable term in the Lagrangian connecting a left and a right-handed field. There exists possibilities of writing two independent mass terms in case of a neutral fermion. These two different possibilities are termed as Dirac and Majorana masses. In order to introduce these mass terms, we need to discuss the charge conjugation property of the neutrino field.

Charge conjugation symmetry relates particles and antiparticles. A four component neutrino field ν transforms under charge conjugation as follows:

$$\nu \to \nu^c \equiv \mathbf{C}\nu\mathbf{C}^{-1} = C\bar{\nu}^T. \tag{1.15}$$

The Lagrangian for free neutrino field remains invariant if the matrix C is chosen to satisfy

$$C\gamma_\mu C^{-1} = -\gamma_\mu^T .$$

C can be chosen to be $i\gamma^2\gamma^0$ in the specific representation (1.12) for the gamma matrices. In this case one has

$$\nu^c = i\gamma^2\gamma^0\bar{\nu}^T = i\gamma^2\nu^*. \tag{1.16}$$

The matrix C satisfies

$$C^\dagger = C^{-1} = C^T = -C. \tag{1.17}$$

The following relations are easy to prove using properties of gamma matrices:

$$\begin{aligned}
\nu_L^c &\equiv P_L\nu^c = C\bar{\nu}_R^T \equiv \nu_R^c, \\
\nu_R^c &\equiv P_R\nu^c = C\bar{\nu}_L^T \equiv \nu_L^c.
\end{aligned} \tag{1.18}$$

It follows that charge conjugate ν_L^c of a left-handed field ν_L is a right handed object and vice versa. Thus if neutrino emitted in a beta decay is left handed then the corresponding antineutrino would be right handed. The important point to keep in mind is that although the ν_L^c is right handed, it does not coincide with ν_R which as mentioned earlier is an independent field with its own dynamical evolution.

A left handed neutrino field ν_L can form a mass term either with its charge conjugate (and hence right handed) field ν_L^c or it can combine with an independent field ν_R. Moreover, ν_R can also combine with its left-handed charge conjugate ν_R^c to give a mass term.

Let us consider a theory containing two independent fields ν_L' and ν_R'. Here ν_R' is an independent field and thus is not the charge conjugate of ν_L'. The latter would generally represent any of the neutrino fields corresponding to active (*i.e.*, those having weak interactions) neutrinos $\nu_{e,\mu,\tau}$. ν_R can represent a right handed field unrelated to any of these. Such ν_R would transform as a singlet under $SU(2) \times U(1)$. Alternatively, ν_R can be charge conjugate of any of the active neutrinos, *e.g.*, ν_L may represent ν_{eL} and ν_R may be $\nu_{\mu L}^c$. We allow both these possibilities.

We can write the following mass terms between ν_L' and ν_R':

$$-\mathcal{L}_{mass} = \bar{\nu}_L' m_D \nu_R' + \frac{1}{2}m_L\bar{\nu'}_L\nu'_L^c + \frac{1}{2}m_R\bar{\nu'}_R^c\nu_R' + \text{h.c.}, \tag{1.19}$$

where we have used primed fields to distinguish them from the mass eigenstates to be introduced soon. The terms with coefficients $m_{L,R}$ are known as the Majorana mass term and the m_D

term is known as the Dirac mass term. The Majorana mass terms are not invariant under global phase changes of the fields since $\nu'_{L,R}$ and $\bar{\nu}'^{c}_{L,R}$ change by the same amount under this transformation. In contrast, the Dirac mass term can be made invariant if ν'_L and ν'_R are transformed by the same phase. Thus Majorana mass terms violate lepton number while the Dirac mass term respects it.

It is seen from Eq.(1.19) that neither ν'_L nor ν'_R is a mass eigenstate. The nature of physical neutrino is determined by going to the mass basis. To do this, we rewrite Eq.(1.19) as follows:

$$-\mathcal{L}_{\text{mass}} = \frac{1}{2} \left(\begin{array}{cc} \bar{\nu}'_L & \bar{\nu}'^{c}_R \end{array} \right) \left(\begin{array}{cc} m_L & m_D \\ m_D & m_R \end{array} \right) \left(\begin{array}{c} \nu'^{c}_L \\ \nu'_R \end{array} \right) + h.c. \quad (1.20)$$

We have made use of the following relation in writing the above equation.

$$\bar{\nu}'_L \nu'_R = \bar{\nu}'^{c}_R \nu'^{c}_L. \quad (1.21)$$

This relation is a special case of a more general identity which is quite useful in many of the algebra related to charge conjugate fields:

$$\bar{\Psi}\Gamma_i \chi = \bar{\chi}^c C \Gamma_i^T C^{-1} \Psi^c, \quad (1.22)$$

where Ψ, χ are any two Dirac spinors, Γ_i represents products of the Dirac gamma matrices. Let us rewrite Eq.(1.20) as:

$$-\mathcal{L}_{mass} = \frac{1}{2}\bar{n}'_L \mathcal{M}_\nu n'^{c}_L + h.c., \quad (1.23)$$

where $n'_L \equiv (\nu'_L, \nu'^{c}_R)^T$ denotes a column vector for two neutrino states and $\mathcal{M}_\nu$ is a 2×2 matrix defined as:

$$\mathcal{M}_\nu \equiv \left(\begin{array}{cc} m_L & m_D \\ m_D & m_R \end{array} \right). \quad (1.24)$$

It is possible to diagonalize $\mathcal{M}_\nu$ through a unitary matrix U

$$U^T \mathcal{M}_\nu U = \text{diag}\,(m_1, m_2), \quad (1.25)$$

where $m_{1,2}$ are eigenvalues of $\mathcal{M}_\nu$ given by

$$m_{1,2} = \frac{1}{2}\left(m_L + m_R \pm \sqrt{(m_L - m_R)^2 + 4m_D^2}\right). \qquad (1.26)$$

Note that $m_{1,2}$ defined above are not necessarily positive. The physical neutrino mass basis are defined as

$$\begin{pmatrix} \nu'_L \\ \nu'^c_R \end{pmatrix} \equiv U \begin{pmatrix} \nu_{1L} \\ \nu^c_{2R} \end{pmatrix}. \qquad (1.27)$$

The new states ν_{1L} and ν_{2R} represent chiral components of two different neutrino states with masses m_1 and m_2 respectively. If CP conservation is assumed, U can be taken as an orthogonal matrix which is specified in terms of a mixing angle θ giving us

$$\begin{aligned} \nu_{1L} &= \cos\theta\, \nu'_L - \sin\theta\, \nu'^c_R, \\ \nu_{2R} &= \sin\theta\, \nu'^c_L + \cos\theta\, \nu'_R, \end{aligned} \qquad (1.28)$$

with

$$\tan 2\theta = \frac{2m_D}{m_R - m_L}. \qquad (1.29)$$

Since masses $m_{1,2}$ can have either sign, let us define $m_i \equiv |m_i|\eta_i$ ($\eta_i = \pm$) and rewrite Eq. (1.23) as:

$$-\mathcal{L}_m = \frac{1}{2}\left(|m_1|\eta_1 \bar{\nu}_{1L}\nu^c_{1L} + |m_2|\eta_2 \bar{\nu}^c_{2R}\nu_{2R} + h.c.\right), \qquad (1.30)$$

where we made use of Eq.(1.25).

We have been writing all mass terms in terms of chiral projections of the field. We can always define appropriate four component objects and write masses using these new fields. Define

$$\begin{aligned} \chi_1 &= \nu_{1L} + \eta_1 \nu^c_{1L}, \\ \chi_2 &= \nu_{2R} + \eta_2 \nu^c_{2R}. \end{aligned} \qquad (1.31)$$

Eq.(1.30) assumes the following form:

$$-\mathcal{L}_m = \frac{1}{2}\left(|m_1|\bar{\chi}_1\chi_1 + |m_2|\bar{\chi}_2\chi_2\right). \qquad (1.32)$$

From the definition Eq.(1.15), of the charge conjugation, it is obvious that the fields $\chi_{1,2}$ satisfy

$$\chi_{1,2}^{c} = \eta_{1,2}\chi_{1,2}.$$

Thus both the fields $\chi_{1,2}$ are self-conjugate. Neutrinos described by these fields are called Majorana neutrinos.

We started with two independent two component objects ν_L, ν_R with the most general mass term given by Eq.(1.19). This theory could be rewritten in terms of two (four component) objects satisfying Majorana condition. Eq.(1.19) therefore generically defines Majorana neutrinos. However there are special cases which are of considerable theoretical importance. We discuss these cases now.

1.2.2 Dirac neutrino

This case corresponds to the limit $m_L = m_R = 0$. The eigenvalues $m_{1,2}$ in Eq.(1.26) are equal and opposite thus $\eta_1 = -\eta_2$. We can then define

$$\psi = \frac{\chi_1 + \chi_2}{\sqrt{2}}.$$

The mass term can then be rewritten as

$$-\mathcal{L}_{mass} = |m_1|\bar{\psi}\psi. \tag{1.33}$$

By definition, $\psi \neq \psi^c$ and the above mass term describes a four component Dirac fermion. The above mass term is invariant under a phase transformation on ψ. This phase transformation may be identified with the lepton number which is conserved in the limit $m_L = m_R = 0$. In this limit the two original Majorana fields $\chi_{1,2}$ have merged into a Dirac state ψ.

1.2.3 Pseudo-Dirac neutrino

Two eigenvalues turn out to be equal and opposite also in the special case $m_L + m_R = 0$ but $m_{L,R} \neq 0$. Due to equality in the masses, the original mass term can be converted to a Dirac mass term by defining ψ exactly as above. But there is a subtle difference between these two situation. In both cases, the final mass term as given in Eq.(1.33) is invariant under lepton number symmetry. But if we look back at the original mass matrix in Eq.(1.24) then we find that this matrix is invariant under lepton number symmetry if $m_L = m_R = 0$ but is not invariant if only $m_L + m_R = 0$[3]. Due to non-invariance of the mass matrix lepton number is violated [16] by the full Lagrangian although it is respected by Eq.(1.33). As a result, the two different components χ_1 and χ_2 of ψ receive different radiative corrections and the Dirac neutrino gets split. Thus the final theory contains a pair of Majorana neutrinos with almost degenerate masses. Such a pair is referred to as pseudo-Dirac neutrino. Phenomenological importance of this case is discussed in [16, 17].

1.2.4 Seesaw masses

Now let us consider the limit $m_L = 0, m_R \gg m_D$. The m_R is typically assumed to be much larger than the electroweak scale. Since this mass corresponds to the Majorana mass for the right-handed field ν_R, this field has to be identified with some $SU(2) \times U(1)$ singlet field unlike in the previous two cases, where ν_R may represent charge conjugate of the non-singlet neutrinos. The neutrino masses are given in this case by

$$m_1 \sim \frac{m_D^2}{m_R}; \quad m_2 \sim m_R.$$

Thus neutrino masses are hierarchical in this case. This scheme is ideal and one of the most preferred schemes for description of neutrino masses. In this case, very small mass for neutrino is generated through a high scale m_R. The high scale can be naturally introduced in grand unified theory such as $SO(10)$. Moreover,

---

[3]In fact, it is not possible to define any $U(1)$ symmetry under which the neutrino mass matrix remains invariant in this case. The occurrence of Dirac neutrino at tree level in this case is a consequence of the invariance of Eq.(1.24) under a discrete symmetry, see [16] for a detailed discussion on this point.

the angle θ is very small and the light neutrino is predominantly ν_L and heavier is mainly ν_R, see Eq.(1.28). Thus this scheme automatically explains why the conventional neutrinos are much lighter than other fermions. We will talk more about this scheme in our subsequent discussion.

1.3 Neutrino mass measurements: direct detection

We will briefly discuss various experimental methods employed to detect neutrino masses. We shall omit many of the details which can be found, for example, in [5, 6]. The experiments which look for neutrino masses can be divided in two categories. Some experiments are capable of obtaining information on the absolute neutrino masses. This is done through kinematical studies of heavy particle decays which produce a neutrino in the final state. The nuclear beta decay and neutrinoless double beta decays provide information on the electron neutrino mass in this way. The masses of ν_μ and ν_τ can be measured through the pion and tau decays. Direct information on neutrino masses can also be obtained in the neutrino signal from supernova. All these experiments have so far given only upper limits on the neutrino masses.

The second category of experiments study neutrino oscillations and probe neutrino (mass)^2 differences. Three sets of experiments in this category have given us positive information on the neutrino (mass)^2 differences. These correspond to the solar and atmospheric neutrino experiments and the laboratory experiments at Los Alamos. We shall briefly talk about them in this section.

Primary requirement of oscillation experiments is an intense neutrino beam. These beams are either produced in laboratory or arise from some natural sources. We have the following possibilities:

- Nuclear reactors produce intense beam of the electron anti-neutrinos. The average energy of these neutrinos is around few MeV.

- One can produce intense beam of the muon neutrinos and antineutrinos in laboratory experiments. Very energetic protons hitting a target produce large number of pions and kaons. They mainly decay to muons and their neutrinos giving an intense beam of the muon neutrinos. The muons produced this way also decay and lead to large number of electron neutrinos.

- The fusion reactions inside the Sun gives rise to electron neutrinos. These neutrinos carry energies of $\mathcal{O}(1\text{--}10)$ MeV.

- Reactions similar to the ones in item (2) also occur in nature when energetic cosmic rays enter our atmosphere. This gives natural beam of muon and electron neutrinos.

- Intense beams of all types of neutrinos come from supernova explosion. Detection of this however requires explosion to occur not far away from our galaxy and so far there has been only one explosion which has given us detectable neutrino signal.

The neutrino beams obtained through above methods have been used to study neutrino oscillations in laboratory. The oscillating neutrinos are detected through their charged and neutral current interactions. Different experiments use different methods for their detection and we shall describe them as we go along.

1.3.1 Beta decay and m_{ν_e}

The shape of the electron energy distribution in the nuclear beta decay

$$N(A, Z) \rightarrow N(A, Z+1) + e^- + \bar{\nu}_e \tag{1.34}$$

is sensitive to the neutrino mass. The shape function is conventionally defined as

$$K(E) \equiv \left[\frac{d\Gamma/dE}{pE\mathcal{F}(Z, E)} \right]^{1/2}$$

$$\sim \left[(Q + m_e - E)\sqrt{(Q + m_e - E)^2 - m_\nu^2} \right]^{1/2} \tag{1.35}$$

where p, E respectively denote the electron momentum and energy. $d\Gamma/dE$ is the differential probability for the electron emission. Function $\mathcal{F}(Z, E)$ arises because of the coulomb interactions of the departing electron and Q denotes the Q value of the beta decay. It is seen from the above equation that the plot of $K(E)$ versus E (known as Kurie plot) is a straight line in the absence of neutrino mass. The neutrino mass makes this curve bend near $E \sim Q$, the departure being more prominent for lower values of Q. Thus decays with lower Q can give better limit on m_ν. The beta decay of Tritium is used to obtain limit on m_ν due to its low Q value (18.6 KeV). The best limit obtained in this decay is given by[18]:

$$m_\nu \leq 2.2\,\mathrm{eV}. \tag{1.36}$$

The quoted limit is obtained by assuming very small mixing of the electron neutrino with other neutrinos.

1.3.2 Neutrinoless double beta decay

It can happen that ordering of the nuclear energy levels in some nuclei does not allow the ordinary beta decay to take place. For example, the ground state of $^{76}_{32}Ge$ lies lower than that of $^{76}_{33}As$ to which it could have transformed by ordinary beta decay. The former is however higher than the ground state of $^{76}_{34}Se$. Thus $^{76}_{32}Ge$ can decay to $^{76}_{34}Se$ by emitting two electrons. Such type of processes

$$N(A, Z) \rightarrow N(A, Z + 2) + 2e^- + 2\bar{\nu}_e \tag{1.37}$$

are called double beta decay. These processes are second order in Fermi coupling and are very rare but they have been observed.

It turns out that the two neutrinos emitted in the above process can be made virtual if they are massive. This leads to the following process known as neutrinoless double beta decay

$$N(A, Z) \rightarrow N(A, Z + 2) + 2e^- \tag{1.38}$$

Two features of the above process distinguish it from Eq.(1.37). The lepton number is violated here by two units unlike in Eq.(1.37) which conserves lepton number. Secondly, sum of the energies of two electrons is fixed by kinematics. Thus one finds a peak in this energy distribution. This is a distinguishing feature of the neutrinoless double beta decay which is used as a signal for this process.

The lepton number violating neutrino mass probed in this process is given by

$$\langle m_{ee} \rangle = \left| \sum_i U_{ei}^2 m_i \right| . \tag{1.39}$$

Neutrino masses are denoted here by m_i with $i = (1,, 2, 3)$. U is neutrino mixing matrix analogous to U_{CKM} defined in Eq.(1.7).

The presence of a non-zero $\langle m_{ee} \rangle$ has been searched for in number of experiments. The most stringent limit obtained so far comes from the Heidelberg-Moscow collaboration [18]:

$$\langle m_{ee} \rangle \leq 0.38\,\text{eV} , \tag{1.40}$$

This limit is stronger than the limit on absolute neutrino mass obtained from the Kurie plot in Eq.(1.36). But both limits are consistent since there can be cancellations between individual neutrino mass in Eq.(1.39).

Eq.(1.39) not only measures the effective neutrino mass but also provides information on the nature of neutrino. Let us consider two neutrino mixing as in previous section. The effective mass is given in this case by

$$\langle m_{ee} \rangle = m_1 \cos^2 \theta + m_2 \sin^2 \theta . \tag{1.41}$$

If both neutrinos are Majorana particles with unequal masses then $\langle m_{ee} \rangle$ is non-zero and measures the weighted mass as given by the RHS of above equation. When two masses are equal and opposite and mixing angle is $\pi/4$, two neutrinos can be regarded together as a Dirac neutrino with exact lepton number conservation. $\langle m_{ee} \rangle$ in Eq.(1.41) vanishes in this case as expected. If neutrino is pseudo-Dirac then $m_1 = -m_2$ at tree level but θ is not $\pi/4$. $\langle m_{ee} \rangle$ is again non-zero in this case since there is no lepton number conservation in this case. The $\langle m_{ee} \rangle$ can be used together with other evidence for neutrino masses to learn about the nature of neutrino spectrum and there has been extensive studies of this topic [19].

1.4 Neutrino mass measurements: oscillations

Search for neutrino oscillations has proved to be the most powerful way of obtaining information on neutrino masses. Neutrinos are produced through their charged weak interactions given :

$$-\mathcal{L}_W = \frac{g}{2\sqrt{2}} \left(\bar{l}_{\alpha_L} \gamma_\mu \nu_{\alpha L} W^\mu + h.c. \right) . \tag{1.42}$$

$\alpha = e, \mu, \tau$ runs over the physical flavour states corresponding to the mass eigenstates of the charged lepton. The ν_α defined above do not posses definite mass but is given by a combination of neutrinos ν_i having mass m_i.

$$\nu_\alpha = U_{\alpha i} \nu_i \tag{1.43}$$

Neutrinos produced through their charged current interactions are therefore superposition of neutrinos with definite masses each mass states evolving with different phase factor corresponding to its energy. This leads to the phenomena of neutrino oscillations. Let us consider two neutrinos $\nu_{e,\mu}$ which would be produced along with the electrons and the muons respectively. These are given in terms of the mass eigenstates by:

$$\begin{aligned}
\nu_e &= \cos\theta \; \nu_1 + \sin\theta \; \nu_2 \, , \\
\nu_\mu &= -\sin\theta \; \nu_1 + \cos\theta \; \nu_2 \, .
\end{aligned} \tag{1.44}$$

ν_e produced at $t = 0$ evolves to

$$|\nu_e(t)\rangle = \cos\theta \; e^{-iE_1 t} |\nu_1\rangle + \sin\theta \; e^{-iE_2 t} |\nu_2\rangle$$

at time t. The neutrino energies are given for the relativistic neutrinos by

$$E_i \sim p + \frac{m_i^2}{p} \, .$$

As long as the neutrino masses m_i are not equal, relative strengths of ν_1 and ν_2 in original ν_e beam vary with time and the ν_e state does not remain orthogonal to the ν_μ defined by Eq.(1.44). As a result, a non-zero amplitude to find a ν_μ in the original beam of ν_e develops after a time t (or distance L). The probability $P_{e\alpha}$ to find $\alpha = \nu_e, \nu_\mu$ in the original ν_e beam is given by

$$P_{ee} = 1 - \sin^2 2\theta \sin^2 \left(\frac{\Delta m_{12}^2 L}{4E} \right) ,$$

$$P_{e\mu} = \sin^2 2\theta \sin^2 \left(\frac{\Delta m_{12}^2 L}{4E} \right) , \qquad (1.45)$$

where $\Delta m_{12}^2 \equiv m_2^2 - m_1^2$. It is clear that neutrino oscillation probabilities measure their (mass)^2 differences and mixing angle, both of which are required to be non-zero. The conversion probability $P_{e\mu}$ is significant if $\sin^2 2\theta$ is large or $\frac{\Delta m_{12}^2 L}{4E} \sim \pi/2$ or both. In this case $P_{e\mu}$ can be used to obtain information on Δm_{12}^2 and/or $\sin^2 2\theta$. One can increase the sensitivity of Δm_{12}^2 measurement if L is large and/or E is small. The neutrino intensity however drops as $1/L^2$. Thus intense beam, small energy and large path lengths are basic criteria needed to probe smaller (mass)^2 differences through neutrino oscillations. The solar neutrino beam proves to be ideal for this purpose and can measure neutrino (mass)^2 differences as small as 10^{-10} eV^2. We summarize sensitivity of various experiments in Table 1.1 given below.

As seen from Table 1.1, different experiments are sensitive to different values of neutrino masses. The original beam employed is also different either electron or muon type. Thus these experiments search for oscillations of different neutrinos. These searches are made in two ways. One direct way is to start with a neutrino of given flavour and look for the appearance of a different flavour by detecting its charged lepton. This method may not always be feasible. For example, the reactors give electron neutrinos but their energy is not sufficient to produce muons. Thus even if reactors $\bar{\nu}_e$ have oscillated to $\bar{\nu}_\mu$, we cannot easily detect these $\bar{\nu}_\mu$. In such cases, one looks for reduction in fluxes of the original neu-

Type of experiment	Typical energy (MeV)	Typical path length	Δm^2 probed (eV^2)
Reactor	few MeV	10 m-1 km 100 km for KamLand	$10^{-3} - 1$
Accelerator	$10^3 - 10^4$	100-1000m	$1 - 10^2$
Solar	0.1–10	10^8 km	$10^{-10} - 10^{-11}$
Atmospheric	$\sim$ GeV	$10^4 - 10^6$m	$10^{-2} - 10^{-3}$

Table 1.1: Neutrino oscillation experiments and their sensitivity to measurement of the difference in mass-squared between a neutrino pair.

trino flavour which is over and above the normal $1/L^2$ reduction. This type of experiments are known as disappearance experiments. Both these types of experiments have helped us in finding values and in large number of cases restrictions on neutrino mixing angles and (mass)^2 differences. We briefly summarize below positive information that has come from neutrino oscillation experiments.

1.4.1 Atmospheric neutrinos

Cosmic ray interactions with our atmosphere produce very energetic pions and kaons. These particles decay mainly to muons and their neutrinos. Most muons produced in these decays also decay to one electron neutrino and one muon neutrino. Thus one has roughly two muon neutrinos and one electron neutrino for every pion or kaon that decays. Flux of these neutrinos is reasonably large, roughly 100 atmospheric neutrinos pass through our body every second. But due to their weak interactions, only one such neutrino can interact with our body in thousand years! It requires therefore very large mass to detect atmospheric neutrinos.

The detection of atmospheric neutrinos was made in a massive detector called Super Kamiokande[20] in Japan which contains 50,000 tons of pure water. This detector not only detects these neutrinos but can also distinguish between the electron and muon types of neutrinos and thus can detect oscillation of the atmospheric neutrinos. Typical energies of atmospheric neutrinos is 1 GeV or more. Thus electron and muon type neutrinos give rise respectively to energetic electrons and muons after interacting with water. These charged particles travel faster than speed of light in water and give out Cerenkov radiation. These light is detected by photomultiplier tubes surrounding the detector. The neutrino energy in a given event is inferred from the amount of light collected in photomultiplier tubes. The cone generated by the Cerenkov light of muons and electrons produce different types of rings in the detector which is used to differentiate between the electron and muon neutrinos. The charged particles travel almost in the same direction as neutrinos. One therefore has information on flavour, energy and direction of neutrinos which has been used to conclude the presence of neutrino oscillations.

The observations at Super Kamioka detector showed that the ν_μ are oscillating while ν_e do not appear to be oscillating with similar wavelength. These observation suggest that ν_μ are oscillating to ν_τ. It is possible to detect neutrino oscillation in this experiment since different neutrinos travel different distances before entering the detector. Those coming from top travel only in the atmosphere and those coming from below also travel inside the earth -the exact distance traveled is a function of the zenith angle. The observed flux at Superkamioka showed variation with the zenith angle which is in accord with the oscillation probability:

$$P_{\mu\mu} = 1 - \sin^2 2\theta_{\mu\tau} \sin^2 \left(\frac{\Delta m_{23}^2 L}{4E} \right) , \qquad (1.46)$$

where $\Delta m_{23}^2 \equiv m_3^2 - m_2^2$ is the (mass)^2 difference between the mu and the tau neutrinos. The maximum path length correspond to the earth diameter. This distance when translated through Eq.(1.46) for 1 GeV neutrinos correspond to Δm_{23}^2 of about 10^{-3} eV^2. The mixing angle $\theta_{\mu\tau}$ between the mu and tau neutrinos is found to be nearly maximal.

Reaction	E_ν^{max}	Flux ($10^{10}/cm^2 sec$)
$p + p \rightarrow\, ^2H + e^+ + \nu_e$	0.42	5.94
$^7Be + e^- \rightarrow\, ^7Li + \nu_e$	0.86 (90%)	0.48
	0.36 (10%)	
$^8B \rightarrow\, ^8Be^* + e^+ + \nu_e$	14.06	$5.5 \cdot 10^{-4}$

Table 1.2: Important nuclear reactions responsible for the production of the solar neutrinos. The flux of neutrinos produced in each reactions and their highest energies are also shown in the table.

1.4.2 Solar neutrinos

The surface of our atmosphere receives $\sigma = 1.4 \times 10^6$ erg of energy per cm^2 every second. This energy is the result of thermonuclear reactions occurring inside the Sun which converts 4 hydrogen atoms into a Helium atom through complicated chain reactions. Each such conversion is accompanied by 26 MeV of energy and two neutrinos. Thus one neutrino is emitted for every 13 MeV of thermal energy received. The flux of neutrinos coming from the Sun is then given by

$$\phi_{\nu_e} \approx \frac{\sigma}{13 MeV} \approx 6 \times 10^{10} \nu_e/cm^2/sec.$$

This is quite intense source of the electron neutrinos which nevertheless requires very large underground detectors to detect them. By now, these neutrinos are detected in four different types of experiments (1) Cl experiment at Homestake [21] (2) Gallium experiments Gallex [22] and SAGE [23] (3) Elastic scattering experiments Kamioka and Superkamioka [20] and (4) Sudbury detector SNO [24]. We briefly describe results of these experiments and then turn to consequences of experimental results. About 98% of total neutrinos are produced in the reactions shown in Table 1.2.

The neutrino spectrum (*i.e.*, variation of flux with energy) for solar neutrino in these reactions is determined by the standard nuclear physics. The normalization of each of these fluxes is determined in the standard solar model which takes into account

Experiment	Reaction	Threshold (in MeV)	$\frac{\phi_{expt}}{\phi_{SSM}}$
SAGE, GALLEX, GNO	$\nu_e + {}^{71}Ga \to {}^{71}Ge + e^-$	0.233	0.584 ± 0.039
Homestake	$\nu_e + {}^{37}Cl \to {}^{37}Ar + e^-$	0.814	0.335 ± 0.029
Superkamioka	$\nu_x e^- \to \nu_x e^-$	5.5	0.459 ± 0.017
SNO	$\nu_x e^- \to \nu_x e^-$ $\nu_e D \to p + pe^-$ (CC) $\nu_x D \to n + n\nu_x (NC)$	5.0	0.347 ± 0.029 1.0072 ± 0.126

Table 1.3: Characteristics of different solar neutrino experiments and the neutrino flux measured by them. The fluxes are normalized with respect to the standard solar model.

dynamics and conditions existing in the Sun. Experiments mentioned above have different thresholds and thus detect different types of neutrinos. Salient aspects of these experiments are summarized in Table 1.3 below.

Some noteworthy points are:

- Cl and Ga experiments do not have directional and time sensitivity. They only measure the overall conversion rates.

- The Gallium experiments are the only experiments measuring the most dominant pp neutrinos because of their low threshold.

- As shown in the last column, all the experiments find suppression in the neutrino flux compared to the standard solar model. This suppression shows that a part of the solar neutrinos are getting converted to some other neutrino flavour, *e.g.*, muon or tau neutrino. This converted neutrinos cannot produce the corresponding charged lepton because of their low energy and thus produce apparent deficit in the solar flux.

- The amount of deficit seen in different experiments is different. This shows that mechanism which suppresses the solar flux is energy dependent.

- The *Cl* and *Ga* experiments are sensitive only to the charged current process. The elastic scattering rates in Superkamioka and SNO receive a sub-dominant contribution also from the neutral current process ($\nu_\alpha + e \to \nu_\alpha + e$) but it is not possible to separate these two contributions directly.

- SNO experiment has separately measured the charged current process $\nu_e + D \to p + p + e^-$ and the neutral current process $\nu_\alpha + D \to p + n + \nu_\alpha$. This experiment therefore measures the electron neutrinos as well as the neutrinos to which part of the solar neutrinos get converted. The sum of the charged current and neutral current fluxes agrees within errors with the flux predicted in the standard solar model.

1.4.3   Mechanisms for solar neutrino conversion

The above empirical facts can be understood in terms of different mechanisms. We discuss below two of the most popular possibilities.

1.4.4   Vacuum oscillations

It is assumed in this mechanism that the solar neutrinos oscillate to other flavour during their journey to earth from the point of production deep inside the core. The probability P_{ee} that the electron neutrinos retain its flavour after traveling a distance L is given by Eq.(1.45) with $L \sim 10^8$ km. corresponding to the Sun-Earth distance. In case of the MeV neutrinos, the energy dependent factor in P_{ee} averages out if Δm_{12}^2 is $\gg 10^{-10}\,\mathrm{eV}^2$.

This corresponds to energy independent survival probability. The P_{ee} depends on energy if $\Delta m_{12}^2 \sim 10^{-10} - 10^{-11}$ eV^2.

The detailed analysis of the experimental results assuming vacuum oscillation has been made over the years [25]. The latest global analysis using all results from different experiments [26] does not favour this possibility as an explanation for the deficit in the solar flux.

1.4.5 MSW mechanism

This mechanism also requires non-zero mixing and (mass)^2 difference between two neutrinos as in vacuum oscillation but details of conversion to a different flavour are very different. It is possible to explain the solar deficit in this mechanism for a different value of (mass)^2 than in the previous case. Unlike in the vacuum case, the neutrino interactions with the solar matter plays an important role here.

As discussed before, the neutrino produced in some reaction at time $t = 0$ has non-zero amplitude to be found as ν_e or ν_μ at a later time t. Let us denote the neutrino state at time t by

$$|\nu_e(t)\rangle = a_e(t)|\nu_e\rangle + a_\mu(t)|\nu_\mu\rangle \ . \tag{1.47}$$

The evolution of $a_\alpha(t)$ is governed by the free particle Hamiltonian which can be written as

$$i\frac{d}{dt}\begin{pmatrix} a_e \\ a_\mu \end{pmatrix} = \frac{1}{2E}\begin{pmatrix} -\Delta m_{12}^2 \cos 2\theta & \frac{1}{2}\Delta m_{12}^2 \sin 2\theta \\ \frac{1}{2}\Delta m_{12}^2 \sin 2\theta & 0 \end{pmatrix}\begin{pmatrix} a_e \\ a_\mu \end{pmatrix},$$
$$\tag{1.48}$$

where $\Delta m_{12}^2 \equiv m_2^2 - m_1^2$. Derivation of above equation is straightforward and standard [6]. Although it looks different in form, the above equation can be shown to lead to the same oscillation probability as given in Eq.(1.45) when neutrino travels through vacuum. But solar neutrinos do not do that. They interact with solar matter. The effect of these interactions is to modify evolution equation to

$$i\frac{d}{dt}\begin{pmatrix} a_e \\ a_\mu \end{pmatrix} = \frac{1}{2E}\begin{pmatrix} A - \Delta m_{12}^2\cos 2\theta & \frac{1}{2}\Delta m_{12}^2\sin 2\theta \\ \frac{1}{2}\Delta m_{12}^2\sin 2\theta & 0 \end{pmatrix}\begin{pmatrix} a_e \\ a_\mu \end{pmatrix},$$

$$(1.49)$$

where $A = 2\sqrt{2}G_F n_e E$ denotes the effective (mass)^2 difference acquired by neutrino pairs because of their interactions in the Sun. n_e which denotes the electron number density inside the Sun is a function of the distance from the core inside the Sun. This makes A dependent on the distance traveled by the neutrino inside the Sun. The value of A at the solar core is about $10^{-5}\,\mathrm{eV}^2$. This additional (mass)^2 changes as neutrino travels inside the Sun and it also depends upon the energy of neutrinos. If Δm_{12}^2 is of $\mathcal{O}(10^{-5})\,\mathrm{eV}^2$ then it could happen that the matter-dependent (mass)^2 for some of the neutrinos matches exactly with Δm_{12}^2 at some point inside the Sun. When this happens, the diagonal terms in Eq.(1.49) become equal and effective neutrino mixing becomes maximal even if it was very small to start with. This resonance enhancement of the mixing in the presence of matter with varying density is termed as Mikheyev-Smirnov-Wolfenstein (MSW) effect.

The MSW effect can result in strong reduction in the neutrino flux. This can be seen in a simple situation as follows. Assume that the vacuum (mass)^2 difference of neutrino pair is smaller than the value of A at the core In this case, the electron neutrino produced at the core is largely the heavier mass state ν_2. This neutrino goes through resonance and emerges as a ν_2 state in the adiabatic approximation. But the probability to find the electron neutrino in the state ν_2 at the detector in vacuum can be read off from Eq.(1.44) and is given by

$$P_{ee} = \sin^2\theta.$$

This shows that one could obtain sizable suppression in neutrino flux through adiabatic transition in case of large mixing angle. This survival probability appears to be energy independent. But the above expression for P_{ee} holds only for the neutrinos which undergo resonance transformation. Moreover, non-adiabacity in neutrino transition adds another energy dependent factor in P_{ee}. As a result, the MSW effect is found to describe all the available experimental results quite well. We refer the reader to a detailed discussion on P_{ee} to reviews [25] and books [6, 5].

Quite extensive work has been done to test the MSW mechanism against experimental results. The latest analysis [26] implies that the most likely solution for the solar neutrino deficit is the MSW solution with large mixing angle $\tan^2\theta \sim 0.2 - 0.8$ and $\Delta m_{12}^2 \sim 2 \cdot 10^{-5}\,\mathrm{eV}^2 - 3 \cdot 10^{-4}\,\mathrm{eV}^2$ at 3σ level.

Positive result in search of neutrino oscillations also comes from a detector called LSND [27] at the Los Alamos laboratory. This detector uses a $\bar{\nu}_\mu$ beam and looks for its oscillations to ν_e by trying to detect positron and delayed neutron produced by the reaction $\bar{\nu}_e + p \rightarrow e^+ + n$. This experiment reported evidence for the $\nu_\mu \nu_e$ oscillations corresponding to a $(\mathrm{mass})^2$ difference of $\mathcal{O}(\mathrm{eV}^2)$ and mixing $\sin^2 2\theta \sim 10^{-3}$. This value is quite different from the scales found in solar and atmospheric experiments. Most region of parameter space allowed by the LSND experiment is already ruled out by other experiments, *e.g.*, KARMEN[28]. LSND evidence is therefore required to be checked through an independent experiment.

Apart from these positive results, there also exists a strong negative result which provides quite useful information on neutrino mixing. This comes from the reactor experiment at CHOOZ. According to it, probability $(1 - P_{ee})$ for the electron (anti) neutrino to convert to any flavour is required to be less than ~ 0.04 if the corresponding $(\mathrm{mass})^2$ difference is around the atmospheric scale[4].

We now discuss possible neutrino mass patterns implied by these results.

---

[4]In general there is an exclusion curve [29], which gives bound on $1 - P_{ee}$ as a function of $(\mathrm{mass})^2$ difference and mixing angle.

1.5 Neutrino mass spectrum from experiments

We discussed experimental results in previous section assuming mixing between two generations. We found that one needs three non-zero and hierarchical (mass)^2 differences in order to accommodate results from the solar, atmospheric and LSND experiments. Since three neutrino states can have only two independent (mass)^2 differences, all the experimental results cannot be reconciled with the known neutrino flavours and one would need a fourth light neutrino state. Since the LSND results are not as firmly established as other two, we will discuss only three neutrino picture which can explain the solar and atmospheric deficits.

We need neutrino masses to be such that one (mass)^2 difference corresponds to the solar scale $\Delta_\odot \sim 10^{-5}$ eV^2 and the other to the atmospheric scale Δ_{atm}. This implies one of the following three possibilities:

(A) Hierarchical masses: $m_1^2 \le m_2^2 \sim \Delta_\odot \ll m_3^2 \sim \Delta_{atm}$;

(B) Inverted hierarchy: $m_1^2 \sim m_2^2 \sim \Delta_{atm} \gg m_3^2$; $\Delta_\odot \approx m_2^2 - m_1^2$;

(C) Almost degenerate masses: $m_1^2 \sim m_2^2 \sim m_3^2 \gg \Delta_{atm}$.

All these mass patterns are allowed by all the data at present. The first two alternatives give a contribution to effective neutrino mass probed in neutrinoless double beta decay which is at most of the order of the atmospheric scale while the last possibility can give a contribution which is close to the present experimental limit. It is expected [19] that the future neutrinoless double beta decay experiments and direct measurement of the electron neutrino mass will help in distinguishing between the above alternatives.

The mixing pattern among neutrinos is also strongly constrained by the available results. To discuss this, we need to generalize the two generation oscillation picture to the realistic case of three neutrinos. The neutrino production is controlled by the following charged current reaction:

$$-\mathcal{L}_{ch} = \frac{g}{2\sqrt{2}} (\bar{l}_{\alpha L} \gamma_\mu U_{\alpha i} \nu_{iL} W^\mu + h.c.). \tag{1.50}$$

Here U is a unitary matrix describing neutrino mixing. If CP conservation is assumed then U can be parameterized in terms of three mixing angles:

$$U = \begin{pmatrix} c_\phi c_\omega & -s_\phi c_\omega & s_\omega \\ c_\phi s_\theta s_\omega + c_\theta s_\phi & c_\theta c_\phi - s_\phi s_\theta s_\omega & -s_\theta c_\omega \\ -c_\phi c_\theta s_\omega + s_\phi s_\theta & s_\phi s_\omega c_\theta + s_\theta c_\phi & c_\theta c_\omega \end{pmatrix}, \qquad (1.51)$$

A neutrino beam of a definite flavour α evolves in vacuum at time t to

$$|\nu_\alpha(t)\rangle = U_{\alpha i} e^{-iE_i t} |\nu_i\rangle.$$

The probability $P_{\beta\alpha}$ that this beam contains flavour β after traveling distance L can be derived as in Eq.(1.45):

$$P_{\beta\alpha} = \delta_{\alpha\beta} - 4 \sum_{i<j} U_{\beta i} U_{\beta j} U_{\alpha i} U_{\alpha j} \sin^2 \frac{\Delta m_{ij}^2 t}{4E}, \qquad (1.52)$$

with $\Delta m_{ij}^2 \equiv m_{\nu_i}^2 - m_{\nu_j}^2$ and we have assumed a real U. In spite of the presence of two different (mass)^2 difference in Eq.(1.52), the probabilities relevant for the solar and atmospheric experiments are still described in terms of only one mass scale due to the hierarchy in the solar and the atmospheric scales. The effect of the atmospheric scale gets averaged out in solar oscillations and effect of the solar scale is negligible on the oscillations at the atmospheric scale. Eq.(1.52) leads to the following vacuum probabilities:

$$(P_{ee})_{solar} \sim 1 - \sin^2 2\phi \sin^2 \left(\frac{\Delta m_{12}^2 L}{4E} \right) + \mathcal{O}(s_\omega^2),$$

$$(P_{\mu\mu})_{atm} \sim 1 - \sin^2 2\theta \sin^2 \left(\frac{\Delta m_{23}^2 L}{4E} \right) + \mathcal{O}(s_\omega^2),$$

$$(P_{ee})_{chooz} \sim 1 - \sin^2 2\omega \sin^2 \left(\frac{\Delta m_{23}^2 L}{4E} \right). \qquad (1.53)$$

It is seen that the angles θ, ϕ and ω determine the atmospheric, solar and CHOOZ oscillation probabilities respectively. The oscillation results presented earlier implies

$$\tan^2 \phi \sim 0.2 - 0.8,$$

$$\sin^2 2\theta \quad \sim \quad 0.8 - 1,$$
$$\sin^2 \omega \quad \leq \quad 0.01. \tag{1.54}$$

One particular example of the mixing matrix which describes the experimental results is obtained by taking $s_\omega = 0$ and $\theta = \phi = \frac{\pi}{4}$:

$$U = \begin{pmatrix} \frac{1}{\sqrt{2}} & -\frac{1}{\sqrt{2}} & 0 \\ \frac{1}{2} & \frac{1}{2} & -\frac{1}{\sqrt{2}} \\ -\frac{1}{2} & \frac{1}{\sqrt{2}} & \frac{1}{\sqrt{2}} \end{pmatrix}, \tag{1.55}$$

We now discuss theoretical ways which can lead to the required masses and mixing patterns among the three neutrinos discussed above.

1.6 Models of neutrino masses

There is an extensive literature [25] on how to theoretically realize the mass patterns discussed in the last section. We briefly discuss the ideas involved. All the mechanisms for neutrino mass generations finally lead to an effective mass term for the light neutrinos defined as follows:

$$-\mathcal{L}_m = \frac{1}{2}\bar{\nu}_{iL}\mathbf{m}_{ij}\nu_{jL}^c + h.c. \tag{1.56}$$

Here i, j are generation indices and we assumed only three light neutrinos. $\mathbf{m}$ is a complex symmetric 3×3 matrix.

The operator written above is not $SU(2)$ invariant and could arise only after spontaneous breaking of this symmetry. Moreover, it transforms as an $SU(2)$-triplet. Thus one needs to generate an effective $SU(2)$-triplet Higgs field. There are various ways of doing this. All of these need extension in the Higgs or fermion sector of the standard model. The possibilities are:

1. *Add one or more singlet fermions N*: The mixing of the standard neutrino with N along with the lepton number violating mass term for N generates $\mathbf{m}$ in Eq.(1.56). This is the seesaw mechanism already encountered.

2. *Add a triplet Higgs field*: Direct coupling of triplet with leptonic doublet leads to Eq.(1.56). The triplet vacuum expectation value determines neutrino mass scale and is required to be small. This gives so called triplet majoron model[30] if lepton number is spontaneously broken. Consistency of this model with Z width requires explicit violation of lepton number or addition of a singlet Higgs [31] also.

3. *Add a singly or doubly charged Higgs field*: Coupling of this additional Higgs field to the left or the right handed leptons radiatively generate Eq.(1.56). This scenario is realized in the Zee [32] and the Zee-Babu [33] models.

Various neutrino mass models are derived using one or more of the above ingredients. These models have to answer two basic questions. (1) Why neutrino masses are much smaller compared to other fermion masses and (2) why they mix so strongly leading to two large mixing angles. To see how these questions are answered, we discuss two of the above possibilities in some detail.

1.6.1   Seesaw models

The neutrino mass term in this model is analogous to Eq.(1.19) but now for three left-handed and three right handed neutrinos N_{iR}:

$$-\mathcal{L}_{mass} = \bar{\nu}'_{iL}(m_D)_{ij}N'_{jR} + \frac{1}{2}[(\bar{N}'_{iR'})^c(M_R)_{ij}N'_{jR} + \text{h.c.}], \quad (1.57)$$

The above equation leads to the following mass matrix

$$\mathcal{M}_\nu = \begin{pmatrix} 0 & m_D \\ m_D^T & M_R \end{pmatrix}. \quad (1.58)$$

Both m_D and M_R are matrices in generation space. When M_R is nonsingular and is given by a scale M much larger than that in m_D, we get

$$\mathbf{m} \sim -m_D M_R^{-1} m_D^T. \quad (1.59)$$

All the neutrino masses are automatically suppressed due to a large scale M in M_R. A large M is natural in grand unified $SO(10)$ theories which therefore provide a nice framework to understand small neutrino masses. One gets the following mass hierarchy for a diagonal M_R

$$m_1 : m_2 : m_3 \; :: \; m_{D1}^2 : m_{D2}^2 : m_{D3}^2,$$

where m_{Di} are eigenvalues of m_D. As long as these eigenvalues are hierarchical, the neutrino masses also display the hierarchy. There are however ways to obtain non-hierarchical masses as well as large mixing angles [25, 34] within the seesaw models.

1.6.2 Radiative models

It can happen that some symmetry (*e.g.* lepton number) forbids neutrino mass term at the tree level even after extending the Standard Model fields. Soft breaking of this symmetry may however radiatively induce a finite neutrino mass at one or two loop level. This provides an attractive mechanism for understanding the smallness of neutrino masses. The radiatively induced contribution to neutrino masses may be present in addition to the tree level mass. The presence of radiative contribution in this case may explain the hierarchy among neutrino masses. This happens in case of the supersymmetric theories containing explicit lepton number violating interactions.

Radiative mechanism for neutrino mass generation can be implemented in a number of ways. A nice classification of all these possibility can be found in [35]. Many of these models are variations of the basics mechanism proposed by Zee[32]. This mechanism needs an extended Higgs sector containing a charged singlet field h^+ and a doublet field ϕ_2 in addition to the standard Higgs doublet ϕ_1. This allows the following additional terms in the SM Lagrangian:

$$
\begin{aligned}
-\mathcal{L}_{Zee} &= f_{\alpha\beta}\bar{l}_\alpha^c l_\beta h^+ + h.c., \\
V &= \mu\phi_1\phi_2 h^+ + h.c.
\end{aligned}
\tag{1.60}
$$

The first equation by itself cannot generate neutrino mass at tree level since it does not contain a bilinear term in neutrino fields. It cannot give rise to radiative mass also since this term conserves lepton number with appropriately defined lepton number for the field h^+. It is not possible to define a conserved lepton number when both the terms displayed above and the charged lepton Yukawa couplings are simultaneously present. In the presence of all these terms, neutrinos obtain radiative masses.

The general structure of neutrino mass matrix obtained in this model is given by

$$\mathbf{m}_{Zee} = \begin{pmatrix} 0 & m_{e\mu} & m_{e\tau} \\ m_{e\mu} & 0 & m_{\mu\tau} \\ m_{e\tau} & m_{\mu\tau} & 0 \end{pmatrix}, \qquad (1.61)$$

where

$$m_{\alpha\beta} = C f_{\alpha\beta}(m_\beta^2 - m_\alpha^2)$$

and C is a constant which depends upon the Higgs masses and mixing.

The consequences of this mass matrix have been extensively studied [36, 37, 38]. If one of the parameters $m_{\alpha\beta}$ is suppressed compared to the other two then the above mass matrix has approximate symmetry of the type L_α-L_β-L_γ. The above mass matrix then leads to a pseudo-Dirac and an almost massless neutrino desired for phenomenology. But not all such cases can give the correct mixing pattern. For example if $f_{\alpha\beta}$ are of similar magnitudes then $m_{e\mu} \ll m_{\mu\tau} \sim m_{e\tau}$. This case leads to the explanation [37] of the atmospheric neutrino deficit but cannot solve the solar neutrino problem. By allowing hierarchy in $f_{\alpha\beta}$, it is possible to obtain bi-large pattern and simultaneous solutions to both anomalies [39, 36] but solar mixing angle stays very close to maximal [39, 38] in contrast to the non-maximal mixing required in the LMA solution.

A variation of the above model is obtained by replacing h^+ with a doubly charged field κ^{++} [32, 33]. Such model does not need addition field ϕ_2 and lead to neutrino masses at two loop level.

We discussed above two of the most popular models of neutrino masses. While these show typical theoretical ways to obtain neutrino masses, details can substantially change depending upon the phenomenological requirements and theoretical prejudices. A detailed discussion of these topics would take us out of the scope of these lectures.

1.7 Summary

We discussed the subject of neutrino masses in this mini review. Three basic topics were covered: (1) the formalism of neutrino masses (2) experimental methods to detect them and (3) discussion of the presently allowed patterns and models of neutrino masses. We had to completely leave out cosmological implications of neutrino masses. Our main focus was a basic introduction to the subject aimed at graduate students not familiar with it. The material presented is incomplete and one would need to consult existing books and reviews to fill the gaps. Our hope is that the background presented here would be sufficient to motivate and enable fresh students to understand the more elaborate details presented in these books and reviews.

Note Added: After these notes were written, the KamLand experiment in Japan obtained for the first time positive evidence for the oscillations of the man made neutrinos. This detector uses the anti neutrino fluxes from 16 different nuclear reactors located in Japan and Korea. The average distance of these reactors is ~ 180 km and average energy is around 3.0 MeV. This allows the detector to look for the (mass)^2 differences $\sim 1.6 10^{-5}$ eV^2. The results from these detectors have confirmed the large mixing angle solution to the solar neutrino problem. This experiment alongwith the latest results from SNO now narrows down the allowed regions in (mass)^2 difference and mixing angle, for details, see the Plenary talk at PASCOS03 by S. Goswami [40].

Bibliography

[1] E.C.G. Sudarshan and R. Marshak, Phys. Rev. **109** 1860 (1958).

[2] R.P. Feynman and M. Gell-Mann, Phys. Rev. **109** 193 (1958).

[3] E.D. Commins and P.H. Bucksbum, *Weak interactions of leptons and quarks*, Cambridge University Press (1983).

[4] B. Kayser, G. Debu and F. Perrier, *The physics of massive neutrinos*, World Scientific (1989).

[5] R.N. Mohapatra and P.B. Pal, *Massive neutrinos in physics and astrophysics*, World Scientific (1991).

[6] C.W. Kim and A. Pevsner, *Neutrinos in physics and astrophysics*, Harwood Academic Publishers (1993)

[7] E.S. Abers, and B.W. Lee, Phys. Rep. **C9(1)** 1,(1973); S. Weinberg, Rev. Mod. Phys. **46** (1974) 255.

[8] C.L. Cowan *et al*, Science **124** 103 (1956).

[9] K. Kodama *et al*, Phys. Lett. **B504** 218 (2001).

[10] J. Bahcall *et al*, *Solar neutrinos: The first thirty years*, Perseus Publishing (1995).

[11] J. Bahcal, *Neutrino astrophysics*, Cambridge University Press (1989).

[12] P.H. Frampton and P. Vogel, Phys. Rep. **82** 339 (1982).

[13] S.M. Bilenky and S.T. Petcov, Rev. Mod. Phys. **59** 671 (1987).

[14] T. K. Kuo and J. Pantaleone, Rev. Mod. Phys. **61** 937 (1989).

[15] W.C. Hexton and B. Holstein, hep-ph/9905257.

[16] A.S. Joshipura and S.D. Rindani, Phys. Lett. **B494** (2000) 114.

[17] A.S. Joshipura, hep-ph/0204305.

[18] Particle Data Group, K. Hagiwara *et al*, Phys. Rev. **D66** (2002) 010001.

[19] An incomplete list of reference is given by
H. V. Klapdor-Kleingrothaus, H. Pas and A.Yu. Smirnov, Phys. Rev. **D63** (2001) 73005;
H. Minakata and O. Yasuda, hep-ph/9609276;
W. Rodejohann, Nucl. Phys. **B597** (2001) 110 and hep-ph/0203214;
F. Vissani, hep-ph/9904349;
S.M. Bilenky *et al*, Phys. Lett. **B465** (1999) 193;
S.M. Bilenky, S. Pascoli and S.T. Petcov, Phys. Rev. **D64** (2001) 053010;
H. Minakata and H. Sugiyama, hep-ph/0111269 and hep-ph/0202003;
S. Pascoli and S.T. Petcov, hep-ph/0205022.

[20] The latest and earlier results and details of the experiments can be found at www-sk.icrr.u-tokyo.ac.jp

[21] B.T. Cleveland *et al*, Astrophys. J. **496** (1998) 505.

[22] The Gallex collaboration, Phys. Lett. **447** (1999) 127.

[23] The SAGE collaboration, astro-ph/0204245.

[24] The SNO collaboration, nucl-ex/0204008 and nucl-ex/0204009. See also www.sno.phy.queensu.ca/sno.

[25] For reviews of various models, see,
A.Yu. Smirnov, hep-ph/9901208;
R.N. Mohapatra, hep-ph/9910365;
A.S. Joshipura, Pramana **54** (2000) 119.

[26] A. Bandyopadhyay *et. al*, hep-ph/00204286;
V. Barger *et al*, hep-ph/0204253;
J.N. Bahcall *et. al*, hep-ph/0204314;
P.C. de-Hollanda and A.Yu. Smirnov, hep-ph/0205241;
C.V.K. Baba, D. Indumathi and M.V.N. Murthy, Phys. Rev.
D65 (2002) 073033 .

[27] The LSND collaboration, hep-ex/0104049 and Phys. Rev.
Lett. **81** (1998) 1774.

[28] The Karmen Collaboration, hep-ex/0203021 and E.D.
Church *et al*, hep-ex/0203023.

[29] CHOOZ collaboration, M. Apollonio *et al*, Phys. Lett. **B466**
(1999) 415.

[30] G.B. Gelmini and M. Roncadelli, Phys. Lett. **193** (1981) 297.

[31] A.S. Joshipura, Int. J. Mod. Phys. **7** (1982) 2021.

[32] A. Zee, Phys. Lett. **B93** (1980) 389.

[33] K.S. Babu, Phys. Lett. **B203** (1988) 132.

[34] I. Dorsner and S.M. Barr, Nucl. Phys. **B617** (2001) 493.

[35] K.S. Babu and E. Ma, Mod. Phys. Lett. **A4** (1989) 1975.

[36] Y. Koide, Nucl. Phys. Proc. Suppl. **111**, 294 (2002) [hep-
ph/0201250].

[37] A.Y. Smirnov and M. Tanimoto, Phys. Rev. D **55**, 1665
(1997) [hep-ph/9604370].

[38] B. Brahmachari and A.S. Choubey, *Phys. Lett.* **B531**
(192002) 99;

[39] A.S. Joshipura and P. Krastev, *Phys. Rev.* **D50** (191994) 31.

[40] S. Goswami, Plenary talk at the International Symposium,
PASCOS03, Bombay, India (2003), hep-ph/0307224.

Chapter 2

Higgs Physics

Saurabh D. Rindani

2.1 Introduction

Higgs physics is at present poised at an interesting juncture, when a light Higgs boson of the Standard Model (henceforth to be referred to as SM), a spin-zero particle which would signal spontaneous gauge symmetry breaking in the simplest form, has not been seen until the conclusion of experiments at LEP and LEP2 electron-positron collider at CERN, Geneva. It is possible that on-going experiments at the $p\bar{p}$ collider Tevatron at FNAL in U.S.A. may discover the SM Higgs boson if the mass is not too large. If it is not seen at Tevatron, one will have to wait until results come out of the LHC (Large Hadron Collider) which is being built at CERN for a heavier Higgs. From a theoretical point of view, the developments until the present time are complex and interesting. While some of the basic principles underlying spontaneous symmetry breaking of gauge symmetry and the Higgs mechanism are now commonly known, the actual realization of this mechanism in nature is still a subject of investigation. The mass of the SM Higgs boson is an unknown parameter and the pheonemonology is sensitively dependent on the mass. Thus the properties and discovery strategies for the Higgs vary greatly depending on the supposed mass, and the phenomenology rapidly gets complex as the range of the Higgs mass is increased.

This work, based on four lectures delivered mainly for Ph.D. students, aims at describing aspects of the SM Higgs. The scope is limited in extent and depth because of the limited time available for the lectures.

Note that in the following, the SM Higgs is usually denoted by H, but at places, the Higgs field will be denoted by ϕ, or, in the context of the unitarity gauge, by ρ.

Some useful general references are [1, 2, 3, 4, 5].

2.2   The Standard Model Higgs boson

A Higgs boson is a massive spin-0 particle appearing in a local gauge theory, where the local gauge invariance is broken completely, or at least partially, by the mechanism of spontaneous symmetry breaking.   The simplest way in which spontaneous breaking of a symmetry is achieved is by the introduction of elementary scalar fields in a theory. The ideas in the context are best illustrated by means of the Standard Model of electromagnetic, weak and strong interactions. Since SM is currently a theory which explains most experimentally observed phenomena with fairly high accuracy, it is an interesting starting point.

2.2.1   Introduction to the Standard Model

The Standard Model (SM) of electromagnetic, weak and strong interactions is based on a gauge theory with an $SU(2)_L \times U(1) \times SU(3)_C$ gauge group. The quarks and leptons transform according to left-handed (LH) doublet and right-handed (RH) singlet representations of $SU(2)_L$ to account for the $V - A$ nature of the charged-current weak interactions. The quarks (both LH and RH) transform as triplets of $SU(3)_C$ of colour, to account for the strong interactions of the quarks. The leptons are singlets under $SU(3)_C$. The assignment of weak hypercharge corresponding to the $U(1)$ group to the various $SU(2)_L$ and $SU(3)_C$ multiplets is according to the charge formula:

$$Q = T_{3L} + Y, \tag{2.1}$$

where Q is the electric charge, T_{3L} the third component of weak isospin corresponding to $SU(2)_L$, and Y the weak hypercharge. Since no generator of $SU(3)_C$ appears in eq. (2.1), electric charge is independent of colour.

Quarks and leptons come in three generations:

$$L_i \equiv \begin{pmatrix} \nu_i \\ e_i \end{pmatrix}_L ; \quad Q_{i\alpha} \equiv \begin{pmatrix} u_{i\alpha} \\ d_{i\alpha} \end{pmatrix}_L ; \quad \begin{array}{l} i = 1,2,3 \\ \alpha = 1,2,3 \end{array} \tag{2.2}$$
$$e_{iR}; \qquad\qquad u_{i\alpha R}; \qquad\qquad d_{i\alpha R}.$$

Here i is the generation index, and α the colour index.

The fermions interact with the gauge bosons, which correspond to the adjoint representation of the gauge group, through the minimal coupling:

$$\mathcal{L}_{\text{fermion}} = \sum \overline{\psi}_L i \not{D} \psi_L + \sum \overline{\psi}_R i \not{D} \psi_R, \tag{2.3}$$

where the sum is over all fermion multiplets, quarks as well as leptons. D_μ is the covariant derivative appropriate to the representation:

$$D_\mu \equiv \partial_\mu + i\frac{g}{2}\tau^a W_\mu^a + i\frac{g'}{6} B_\mu + i\frac{g_s}{2}\lambda^a G_\mu^a \tag{2.4}$$

for left-handed quarks,

$$D_\mu \equiv \partial_\mu + i\frac{g}{2}\tau^a W_\mu^a - i\frac{g'}{2} B_\mu \tag{2.5}$$

for left-handed leptons,

$$D_\mu \equiv \partial_\mu + ig' Q_q B_\mu + i\frac{g_s}{2}\lambda^a G_\mu^a \tag{2.6}$$

for a right-handed quark q of charge Q_q, and

$$D_\mu \equiv \partial_\mu - ig' B_\mu \tag{2.7}$$

for a right-handed electron with charge -1. The notation is that D_μ is a differential operator and a matrix in $SU(2)_L$ space (2×2) as well as in $SU(3)_C$ colour space (3×3).

The covariant derivative guarantees that $\mathcal{L}_{\text{fermion}}$ is invariant under local gauge transformations

$$\psi \to U\psi, \tag{2.8}$$

where, for LH quarks, for example,

$$U = U_1 U_2 U_3, \tag{2.9}$$

with

$$U_1 = e^{-ig'Y_Q\alpha_1(x)}; \ U_2 = e^{-i\frac{g}{2}\tau^a\alpha_2^a(x)}; \ U_3 = e^{-i\frac{g_s}{2}\lambda^a\alpha_3^a(x)}, \tag{2.10}$$

and similarly for other fields. The gauge fields transform as

$$i\frac{g}{2}\tau^a W_\mu^a \to U_2 \partial_\mu U_2^{-1} + U_2\, i\frac{g}{2}\tau^a W_\mu^a\, U_2^{-1}, \tag{2.11}$$

$$i\frac{g_s}{2}\lambda^a G_\mu^a \to U_3 \partial_\mu U_3^{-1} + U_3\, i\frac{g_s}{2}\lambda^a G_\mu^a\, U_3^{-1}, \tag{2.12}$$

$$ig' B_\mu \to ig' \partial_\mu \alpha_1 + ig' B_\mu. \tag{2.13}$$

The Lagrangian for gauge fields may be written as

$$\mathcal{L}_{\text{gauge}} = -\frac{1}{2} B_{\mu\nu} B^{\mu\nu} - \frac{1}{2} W_{\mu\nu}^a W^{a\mu\nu} - \frac{1}{2} G_{\mu\nu}^a G^{a\mu\nu}, \tag{2.14}$$

where

$$B_{\mu\nu} = \partial_\mu B_\nu - \partial_\nu B_\mu, \tag{2.15}$$

$$i\frac{g}{2}\tau^a W_{\mu\nu}^a = i\frac{g}{2}\tau^a \left(\partial_\mu W_\nu^a - \partial_\nu W_\mu^a \right) + [i\frac{g}{2}\tau^a W_\mu^a, i\frac{g}{2}\tau^b W_\nu^b], \tag{2.16}$$

$$i\frac{g_s}{2}\lambda^a G_{\mu\nu}^a = i\frac{g_s}{2}\lambda^a \left(\partial_\mu G_\nu^a - \partial_\nu G_\mu^a \right) + [i\frac{g_s}{2}\lambda^a G_\mu^a, i\frac{g_s}{2}\lambda^b G_\nu^b]. \tag{2.17}$$

The last two may be rewritten as

$$W_{\mu\nu}^a = \left(\partial_\mu W_\nu^a - \partial_\nu W_\mu^a \right) - g\epsilon^{abc} W_\mu^b W_\nu^c, \tag{2.18}$$

$$G_{\mu\nu}^{a} = \left(\partial_\mu G_\nu^a - p_\nu G_\mu^a\right) - g_s f^{abc} G_\mu^b G_\nu^c, \qquad (2.19)$$

where ϵ^{abc} is the Levi-Civita symbol and f^{abc} are the $SU(3)$ structure constants.

In the gauge-invariant theory described above, none of the fields correspond to massive particles. Fermion mass terms are of the the the form $m\overline{\psi}_L \psi_R + \text{h.c.}$, and are forbidden by even global gauge invariance. Gauge boson mass terms are forbidden by local gauge invariance. The way to give masses without sacrificing the renormalizability of the theory is by allowing for spontaneous symmetry breaking.

2.2.2 Spontaneous symmetry breaking

A symmetry is said to be broken spontaneously if the Lagrangian of the theory is invariant under the symmetry, but the ground state (the vacuum state) is not invariant. In this situation, many degenerate vacuua are related to one another by the symmetry of the Lagrangian, and one is singled out to be the correct vacuum. Hence states constructed out of this vacuum reflect this bias, and the dynamics of the theory no longer shows the invariance.

The Lagrangian we described has a unique minimum energy state corresponding to all the fields taking the value zero. This vacuum is gauge invariant. Hence something has to be added to the theory to achieve spontaneous symmetry breaking.

Moreover, if the vacuum expectation value of any but a spin-0 field is nonzero, even Lorentz invariance would be violated. To avoid this, one introduces only scalar fields in the theory. The scalar field(s) may be elementary or composite. The latter case is not impossible in the standard model as such, but requires extra interactions which can generate a "composite" scalar field corresponding to a bound state.

Let us look more closely at the phenomenon of spontaneous symmetry breaking with elementary scalar fields.

A. Abelian global symmetry

We start with the simple example of a theory with a single complex scalar field ϕ with a global $U(1)$ symmetry under $\phi(x) \rightarrow e^{i\alpha}\phi(x)$, where α is independent of space and time. The corresponding Lagrangian may be written as

$$\mathcal{L}(\phi,\phi^\dagger) = \partial_\mu\phi^\dagger\partial^\mu\phi - V(\phi^\dagger\phi). \tag{2.20}$$

We can write the "potential" term as

$$V(\phi^\dagger\phi) = \mu^2\phi^\dagger\phi + \lambda(\phi^\dagger\phi)^2, \tag{2.21}$$

where higher order terms in $\phi^\dagger\phi$ are not included because they would destroy the renormalizability of the theory.

The Hamiltonian density corresponding to $\mathcal{L}$ is

$$\mathcal{H} = \dot{\phi}^\dagger\dot{\phi} + \nabla\phi^\dagger \cdot \nabla\phi + V(\phi^\dagger\phi). \tag{2.22}$$

Since the first two terms are positive (semi-)definite, it is clear that the field configuration which minimizes the Hamiltonian $H = \int \mathcal{H}d^3x$ should be space and time independent, and should minimize $V(\phi^\dagger\phi)$. For $\mu^2, \lambda > 0$, the only minimum of the potential V is $\phi = 0$. This will correspond to a vacuum state which is symmetric, with no $U(1)$ charge. However, $\lambda > 0$, $\mu^2 < 0$ gives rise to an additional set of nontrivial minima, at

$$\phi^\dagger\phi = -\frac{\mu^2}{2\lambda}, \tag{2.23}$$

that is, at $\phi = \sqrt{-\mu^2/(2\lambda)}e^{i\theta}$, for $0 < \theta < 2\pi$. These configurations are not only local minima, but also global minima, with the energy density $-\mu^4/(4\lambda)$. Note that $\lambda > 0$ ensures stability. Since for this state to make sense it has to be unique, θ has to take some fixed value. Any value of θ corresponds to an unsymmetric vacuum.

Let us work with a shifted field which has a zero value for this minimum:

$$\phi'(x) = \phi(x) - \frac{v}{\sqrt{2}}, \tag{2.24}$$

where $v = \sqrt{-\mu^2/\lambda}$, and where we have chosen $\theta = 0$ for convenience. Then the Lagrangian of eq. (2.20) becomes

$$\mathcal{L}(\phi', \phi'^\dagger) = \partial_\mu \phi'^\dagger \partial^\mu \phi' - \mu^2 \left(\phi' + \frac{v}{\sqrt{2}}\right)^\dagger \left(\phi' + \frac{v}{\sqrt{2}}\right)$$
$$- \lambda \left[\left(\phi' + \frac{v}{\sqrt{2}}\right)^\dagger \left(\phi' + \frac{v}{\sqrt{2}}\right)\right]^2. \tag{2.25}$$

On simplification, this gives

$$\mathcal{L}(\phi', \phi'^\dagger) = \partial_\mu \phi'^\dagger \partial^\mu \phi' - \lambda(\phi'^\dagger \phi')^2 + 2\mu^2 (\mathrm{Re}\,\phi')^2$$
$$+ 2\sqrt{2}\lambda v(\phi'^\dagger \phi')\mathrm{Re}\,\phi' + \frac{\mu^4}{4\lambda}. \tag{2.26}$$

If we write the above equation with

$$\phi' = \frac{1}{\sqrt{2}}(\phi_1 + i\phi_2), \tag{2.27}$$

where $\phi_{1,2}$ are real, the kinetic energy terms have the correct normalizations:

$$\mathcal{L} = \frac{1}{2}\partial_\mu \phi_1 \partial^\mu \phi_1 + \frac{1}{2}\partial_\mu \phi_2 \partial^\mu \phi_2 + \mu^2 \phi_1^2 + \lambda v \phi_1^3 + \phi_1 \phi_2^2 - \frac{\lambda}{4}\left(\phi_1^2 + \phi_2^2\right)^2. \tag{2.28}$$

It can be seen that ϕ_1 has a mass term, with mass $\sqrt{-2\mu^2}$, whereas the field ϕ_2 is massless. The latter corresponds to the massless boson predicted by the Goldstone's theorem for spontaneously broken global continuous symmetries [6].

B. Abelian local gauge symmetry

Let us consider the same $U(1)$ symmetry, but now local. We again take our simple scalar model. Then, gauge invariance requires the use of a covariant derivative $D_\mu \equiv \partial_\mu + igA_\mu$, where A_μ is the gauge field. The gauge-invariant Lagrangian now is

$$\mathcal{L} = D_\mu \phi^\dagger D^\mu \phi - V(\phi^\dagger \phi) - \frac{1}{4}F_{\mu\nu}F^{\mu\nu}. \tag{2.29}$$

It is invariant under

$$\phi \rightarrow e^{i\alpha(x)}\phi, \tag{2.30}$$

$$A_\mu \rightarrow A_\mu - \frac{1}{g}\partial_\mu\alpha. \tag{2.31}$$

A trivial vacuum corresponds to $\phi = 0$, $A_\mu = 0$, which is the case for $\mu^2 = 0$. For $\mu^2 < 0$, again, the vacuum corresponds to $\langle\phi\rangle = v/\sqrt{2}$, with $v^2 = -\mu^2/\lambda$. We again make a shift

$$\phi(x) = \frac{1}{\sqrt{2}}\left(v + \phi_1 + i\phi_2\right). \tag{2.32}$$

Then

$$\begin{aligned}
D_\mu\phi^\dagger D^\mu\phi &= \frac{1}{2}\left(\partial_\mu\phi_1\right)^2 + \frac{1}{2}\left(\partial_\mu\phi_2\right)^2 \\
&\quad + g\partial_\mu\phi_2 A^\mu(v+\phi_1) - g\partial_\mu\phi_1 A_\mu\phi_2 \\
&\quad + \frac{1}{2}g^2 A_\mu A^\mu\left(v^2 + 2v\phi_1 + \phi_1^2 + \phi_2^2\right).
\end{aligned} \tag{2.33}$$

The bilinear terms in ϕ_2 and A_μ can be combined into

$$\frac{1}{2}g^2 v^2\left(A_\mu + \frac{1}{gv}\partial_\mu\phi_2\right)^2 \equiv \frac{1}{2}g^2 v^2 A_\mu'^2. \tag{2.34}$$

This is the mass term for A_μ'.

Actually, it is better to work in the Feynman gauge:

$$\partial_\mu A^\mu - gv\phi_2 = 0. \tag{2.35}$$

Then the gauge fixing term is

$$\mathcal{L}_{gf} = -\frac{1}{2}\left(\partial_\mu A^\mu - gv\phi_2\right)^2. \tag{2.36}$$

This provides the usual $-\frac{1}{2}(\partial_\mu A^\mu)^2$ term for gauge fixing. The cross term $gv\partial_\mu A^\mu\phi_2$ cancels an existing term which has the opposite sign (on integration by parts, dropping surface terms as usual). The $-\frac{1}{2}g^2 v^2\phi_2^2$ corresponds to a mass term for the would-be Goldstone boson.

The spectrum can be seen more easily in the unitarity gauge. Write

$$\phi(x) = \frac{1}{\sqrt{2}}\rho(x)e^{i\alpha(x)}. \tag{2.37}$$

Then the Lagrangian of eq. (2.29) becomes

$$\mathcal{L}_{\mathrm{U}} = \frac{1}{2}(\partial_\mu \rho')^2 + \frac{1}{2}g^2(\rho' + v)^2 A'_\mu A'^\mu - \mu^2(\rho' + v)^2 + \frac{1}{4}\lambda((\rho' + v)^4,$$

(2.38)

where

$$\rho' + v = \rho e^{-i\alpha(x)},$$

(2.39)

and

$$A'_\mu = A_\mu + \frac{1}{g}\partial_\mu \alpha(x).$$

(2.40)

Since the original Lagrangian is gauge invariant, we could have obtained the final Lagrangian simply by making the replacement

$$\phi(x) \rightarrow \frac{1}{\sqrt{2}}\left(\rho'(x) + v\right),$$

(2.41)

and

$$A_\mu(x) \rightarrow A'_\mu(x)$$

(2.42)

We can see that the vector field A'_μ gets a mass gv, $\alpha(x)$ disappears from the Lagrangian, and ρ' corresponds to a massive spin-0 particle of mass $\sqrt{-2\mu^2}$. This mechanism, which gives masses to gauge bosons with the appearance of a massive spin-zero particle, is known as the Higgs mechanism[7]. The spin-zero particle is known as the Higgs boson. The mechanism can be generalized to bigger groups and larger representations of the scalars. Such a generalization is used to achieve spontaneous symmetry breaking and to give masses to various particles in SM. The Higgs sector is described in the next section.

2.2.3   Properties of the Standard Model Higgs boson

The gauge symmetry $SU(2)_L \times U(1) \times SU(3)_C$ of SM is broken to $U(1)_{\mathrm{e.m.}} \times SU(3)_C$ via a single, complex scalar $SU(2)_L$ doublet

$$\phi = \begin{pmatrix} \phi^0 \\ \phi^- \end{pmatrix}.$$

The most general scalar potential which is invariant under the gauge group is

$$V(\phi) = \mu^2 \phi^\dagger \phi + \lambda \left(\phi^\dagger \phi\right)^2. \tag{2.43}$$

ϕ is assigned $Y = -\frac{1}{2}$, just like the lepton doublets.

As before, for $\mu^2 < 0$, there is a nontrivial minimum for the potential, which corresponds to $\langle \phi^0 \rangle = v/\sqrt{2}$, $(v = \sqrt{-\mu^2/\lambda})$, and $\langle \phi^\pm \rangle = 0$. We will assume $\langle \phi^0 \rangle = \langle \phi^0 \rangle^* = v/\sqrt{2}$. The details of the of the minimization are given below.

Differentiating the potential

$$V(\phi) = \mu^2 \phi^{0*} \phi^0 + \mu^2 \phi^+ \phi^- + \lambda(\phi^{0*}\phi^0 + \phi^+\phi^-)^2 \tag{2.44}$$

with respect to ϕ^{0*} and equating the result to 0 we get

$$\frac{\partial V}{\partial \phi^{0*}} = \mu^2 \phi^0 + 2\lambda(\phi^{0*}\phi^0 + \phi^+\phi^-)\phi^0 = 0. \tag{2.45}$$

This implies either

$$\phi^{0*}\phi^0 + \phi^+\phi^- = 0, \tag{2.46}$$

or

$$\phi^0 = 0. \tag{2.47}$$

Next, differentiating with respect to ϕ^+ and equating the result to 0,

$$\frac{\partial V}{\partial \phi^+} = \mu^2 \phi^- + 2\lambda(\phi^{0*}\phi^0 + \phi^+\phi^-)\phi^- = 0 \tag{2.48}$$

This implies (2.46) above, or

$$\phi^- = 0. \tag{2.49}$$

One choice, appropriate from the point of view of charge conservation, is

$$\phi^+ = \phi^- = 0, \ \phi^{0*}\phi^0 = -\frac{\mu^2}{2\lambda}, \ \phi^0 = \sqrt{-\frac{\mu^2}{2\lambda}} = \phi^{0*}. \tag{2.50}$$

To check that the above is indeed a minimum, we need to look at the second derivatives of V with respect to the fields.

$$\frac{\partial^2 V}{\partial \phi^{0*} \partial \phi^0} = \mu^2 + 2\lambda(2\phi^{0*}\phi^0 + \phi^+\phi^-) = 2\lambda\phi^{0*}\phi^0 = -\mu^2 > 0,$$
$$(2.51)$$

where the extremum conditions (2.50) have been made use of. Also,

$$\frac{\partial^2 V}{\partial \phi^+ \partial \phi^-} = \mu^2 + 2\lambda(2\phi^{0*}\phi^0 + \phi^+\phi^-) = 2\lambda\phi^+\phi^- = 0, \quad (2.52)$$

$$\frac{\partial^2 V}{\partial \phi^{0*} \partial \phi^\pm} = 2\lambda\phi^{0*}\phi^\mp = 0, \quad (2.53)$$

$$\frac{\partial^2 V}{\partial \phi^{\pm 2}} = 2\lambda\phi^{\mp 2} = 0. \quad (2.54)$$

For the choice of real ϕ^0, the remaining second-order derivatives give zero. It is clear that conditions for the extremum (2.50) correspond to a minimum for $\lambda > 0$ and $\mu^2 < 0$.

The kinetic energy term for ϕ contains minimal interaction with the gauge fields:

$$\begin{aligned}
\mathcal{L}_{\text{K.E.}} &= (D_\mu\phi)^\dagger (D^\mu\phi) \\
&= \left[\left(\partial_\mu + \frac{i}{2}g\tau^a W_\mu^a - \frac{i}{2}g'B_\mu\right)\phi\right]^\dagger \\
&\quad \times \left[\left(\partial^\mu + \frac{i}{2}g\tau^b W^{b\mu} - \frac{i}{2}g'B^\mu\right)\right]. \quad (2.55)
\end{aligned}$$

To see the spectrum, we can again go to the unitarity gauge:

$$\phi = e^{\frac{i}{2}\alpha^a(x)\tau^a + i\alpha(x)} \begin{pmatrix} \frac{1}{\sqrt{2}}(\rho + v) \\ 0 \end{pmatrix}, \quad (2.56)$$

together with a gauge transformation of all other fields with the same parameters α^a and α. Because of the gauge invariance of the Lagrangian, we can replace ϕ everywhere by $\begin{pmatrix} \frac{1}{\sqrt{2}}(\rho + v) \\ 0 \end{pmatrix}$, and its form would be unchanged. Thus, the Higgs Lagrangian becomes

$$\mathcal{L}_{\text{Higgs}} = \frac{1}{2}\left(\partial_\mu\rho\right)^2 + m_W^2 W_\mu^+ W^{-\mu}\left(1 + \frac{\rho}{v}\right)^2 + \frac{1}{2}m_Z^2 Z_\mu^2\left(1 + \frac{\rho}{v}\right)^2$$

$$-\frac{1}{4}\lambda\rho^4 - \lambda v\rho^3 + \frac{1}{2}m_H^2\rho^2, \tag{2.57}$$

apart from an additive constant. Here use has been made of the relations

$$W_\mu^\pm = \frac{W_\mu^1 \pm iW_\mu^2}{\sqrt{2}}; \quad Z_\mu = \frac{gW_\mu^3 - g'B_\mu}{\sqrt{g^2 + g'^2}}; \tag{2.58}$$

and

$$m_W^2 = \frac{1}{4}g^2v^2; \quad m_Z^2 = \frac{1}{2}(g^2 + g'^2) \tag{2.59}$$

and

$$m_H^2 = -2\mu^2. \tag{2.60}$$

2.2.4   Interactions with fermions

These arise from the Yukawa couplings which also generate fermion masses on symmetry breaking:

$$\mathcal{L}_Y = -h_{ij}^l \overline{L}_i' \tilde{\phi} l_{jR}' - h_{ij}^u \overline{Q}_i' \phi u_{jR}' - h_{ij}^d \overline{Q}_i' \tilde{\phi} d_{jR}' + \text{h.c.}, \tag{2.61}$$

where L, and Q denote respectively the left-handed lepton and quark doublets, l, u and d are the right-handed singlet fields of the charged leptons, up quarks and the down quarks, and i, j denote generation indices. The primes are used because these fields are in the weak basis, to be finally rewritten in terms of the unprimed fields in the mass basis. The conjugate doublet Higgs field is given by

$$\tilde{\phi} = i\tau^2\phi^* = \begin{pmatrix} \phi^+ \\ -\phi^0 \end{pmatrix}. \tag{2.62}$$

On symmetry breaking, in the unitarity gauge these become

$$\mathcal{L}_Y = \frac{1}{\sqrt{2}}h_{ij}^l \overline{l}_{iL}' l_{jR}'(\rho + v) - \frac{1}{\sqrt{2}}h_{ij}^u \overline{u}_{iL}' u_{jR}'(\rho + v)$$

$$-\frac{1}{\sqrt{2}}h_{ij}^d \overline{d}_{iL}' d_{jR}'(\rho + v) + \text{h.c.}. \tag{2.63}$$

We thus have the mass matrices for fermions:

$$M_{ij}^l = -\frac{1}{\sqrt{2}}h_{ij}^l v; \; M_{ij}^u = \frac{1}{\sqrt{2}}h_{ij}^u v; \; M_{ij}^d = -\frac{1}{\sqrt{2}}h_{ij}^d v. \qquad (2.64)$$

These mass matrices are diagonalized by left and right unitary transformations on the fermions:

$$U_L^l M^l U_R^{l\dagger} = \begin{pmatrix} m_e & 0 & 0 \\ 0 & m_\mu & 0 \\ 0 & 0 & m_\tau \end{pmatrix}; \qquad (2.65)$$

$$U_L^u M^u U_R^{u\dagger} = \begin{pmatrix} m_u & 0 & 0 \\ 0 & m_c & 0 \\ 0 & 0 & m_t \end{pmatrix}; \qquad (2.66)$$

$$U_L^d M^d U_R^{d\dagger} = \begin{pmatrix} m_d & 0 & 0 \\ 0 & m_s & 0 \\ 0 & 0 & m_b \end{pmatrix}. \qquad (2.67)$$

Since ρ and v come as a combination $(\rho+v)$ for all the matrices, the ρ couplings to the fermions are diagonal when the mass matrices are diagonalized. It is easy to see that they can be written as

$$\mathcal{L}_{f\bar{f}H} = -\sum_f \frac{m_f}{v}\bar{f}f\rho, \qquad (2.68)$$

where the sum is over all massive fermions. Note that the neutrino is massless in SM, and does not couple to the Higgs.

In summary, the Higgs boson couplings to the gauge bosons and fermions can be written as

$$\mathcal{L}_{\text{Higgs}} = \frac{\rho}{v} m_i \frac{\partial}{\partial m_i}\mathcal{L}_{\text{mass}} + \frac{\rho^2}{2v^2}m_i^2 \frac{\partial}{\partial m_i^2}\mathcal{L}_{\text{mass}}. \qquad (2.69)$$

Note that v can be rewritten in terms of G_F, the Fermi weak coupling constant:

$$v = \frac{2m_W}{g} = \frac{2m_W}{(4\sqrt{2}G_F)^{\frac{1}{2}}m_W} = (\sqrt{2}G_F)^{-\frac{1}{2}} \qquad (2.70)$$

For $G_F = 1.166 \cdot 10^{-5}$ GeV^{-2}, $v = (\sqrt{2}G_F)^{-\frac{1}{2}} \approx 246$ GeV.

Problem: Work out the magnitudes of the Higgs couplings g_f^H in $g_f^H \overline{f} f H$ for the various leptons and quarks, and also g_W^H and g_Z^H in $g_W^H W^{+\mu} W_\mu^- H$ and $g_Z^H Z^\mu Z_\mu H$.

2.2.5 Summary of the properties of SM Higgs

1. The mass is a free parameter.

2. Couplings to various particles are proportional to the masses of the particles. Hence it couples dominantly to the heaviest particle relevant to a particular process. As a corollary, there is no coupling to neutrinos.

3. The couplings to fermions are diagonal in flavour.

4. The ρ parameter

$$\rho \equiv \frac{m_W}{m_Z \cos \theta_W}$$

is predicted to be 1.

2.3 Theoretical Constraints on the Mass of the Standard-Model Higgs Boson

As seen earlier, the Standard Model leaves the mass of the Higgs a free parameter. However, the dynamics can put some constraints on the mass. These may be from some general theoretical considerations like vacuum stability, or experimental, coming from measurement of other physical quantities to which the Higgs contributes indirectly.

2.3.1 Bound from vacuum stability

A lower bound on the Higgs mass may be obtained from the stability of the vacuum. This bound is known as the Linde-Weinberg bound after the authors who first obtained it [8, 9]. The condition of vacuum stability is the condition that the non-trivial vacuum remains lower than the trivial vacuum in the presence of one-loop radiative corrections.

The approach is to calculate the one-loop effective potential according to the method of Coleman and Weinberg [10]. This corresponds to the sum of all one-loop diagrams with external Higgs legs of arbitrary number, all evaluated at zero momentum (corresponding in configuration space to no derivatives).

The one-loop effective potential may be obtained by their method to be [10]

$$V = \mu^2 \phi^2 + \lambda \phi^4 + B\phi^4 \ln \frac{\phi^2}{M^2}, \qquad (2.71)$$

where ϕ is the field corresponding to the physical SM Higgs, M is an arbitrary scale (the subtraction point) used to define λ, and B is given by

$$B = \frac{3}{16\pi^2 v^4} \left[2m_W^4 + m_Z^4 - 4m_t^4 \right]. \qquad (2.72)$$

Originally, the effective potential was used by Coleman and Weinberg to make a prediction of the Higgs mass if the bare mass for the scalar is assumed to be zero, and the gauge symmetry is broken by the minimum of the one-loop effective potential.

Writing

$$m_{\phi_0}^2 = \frac{1}{2} \frac{\partial^2 V}{\partial \phi^2} \bigg|_{\phi=v/\sqrt{2}}, \qquad (2.73)$$

we can evaluate

$$m_{\phi_0}^2 = 2v^2 \left[\lambda + B \left\{ \ln \frac{v^2}{2M^2} + \frac{3}{2} \right\} \right]. \qquad (2.74)$$

We can rewrite the potential in terms of the value at the minimum:

$$V(\phi)|_{\phi=v/\sqrt{2}} = -\frac{1}{2}\mu^2 v^2 + \frac{1}{2}\lambda v^4 + \frac{1}{4}Bv^4 \ln \frac{v^2}{2M^2}. \qquad (2.75)$$

Then

$$m_{\phi_0}^2 = -\frac{8}{v^2} V\left(\frac{v}{\sqrt{2}}\right) + Bv^2. \qquad (2.76)$$

Requiring that $V(v/\sqrt{2}) < V(0)$ gives

$$m_{\phi_0}^2 > Bv^2. \tag{2.77}$$

When all fermions are light, this gives $m_{\phi_0} \gtrsim 7$ GeV, which is the original Linde-Weinberg bound [8, 9]. For $m_t \gtrsim 80$ GeV, however, $B < 0$, and so this bound is not useful. On the other hand, for $B < 0$, the potential is not bounded below, because $V \to \text{inf}$ for $\phi \to \text{inf}$. For ϕ large, however, perturbation theory breaks down. So it is not sensible to take ϕ very large. Politzer and Wolfram [11], and independently Hung [12], therefore imposed the constraint that $V(\phi_1) > V(v/\sqrt{2})$, where ϕ_1 is an estimate of where perturbation theory breaks down. They could obtain an upper bound on fermion masses (in the region of 100 GeV) as a weak function of the Higgs mass.

Alternatively, ϕ_1 could be taken to be the cut-off scale, which is larger than either the grand unification scale (10^{15} GeV) or Planck scale (10^{19} GeV). This again leads to an upper bound on fermion mass. Alternatively, if the masses of all the fermions are known, it gives a lower bound on the Higgs mass [13].

It is of course possible that the unsymmetric vacuum we now live in is unstable, but the decay lifetime is larger than the age of the universe. This gives a bound of $m_H < 260$ MeV if fermion masses are assumed zero [14]. For a heavy top, the bound depends on the assumed top mass.

These bounds have been considerably refined by renormalization group methods and also updated [15]. The results, together with those on upper bound on the Higgs mass are presented in Fig. 2.1.

2.3.2   Upper bound on the Higgs mass

An upper bound on the Higgs mass may be obtained from considerations of triviality of the scalar field theory [17].

Consider a pure ϕ^4 theory. We have the evolution equation for the coupling λ,

$$\frac{d\lambda}{dt} = \frac{1}{16\pi^2} \cdot 12\lambda^2, \tag{2.78}$$

where $t \equiv \ln M$. The solution is

$$\frac{1}{\lambda(v)} - \frac{1}{\lambda(\Lambda)} = \frac{3}{4\pi^2} \ln\left(\frac{\Lambda^2}{v^2}\right), \tag{2.79}$$

where Λ is some large scale, and v is the electroweak scale. Suppose we are allowed to take $\Lambda \to \inf$, i.e., the theory remains unchanged for indefinitely large scales. Then, however small $\lambda(v)$ is, $\lambda(\Lambda)$ will blow up at some large Λ, because

$$\lambda(\Lambda) = \frac{\lambda(v)}{1 - \lambda(v)\frac{3}{4\pi^2} \ln \frac{\Lambda^2}{v^2}}. \tag{2.80}$$

This is the Landau pole, $\Lambda_0 = v e^{\frac{2\pi^2}{3\lambda(v)}}$, or $m_H^2 = \frac{8\pi^2 v^2}{3 \ln(\Lambda^2/v^2)}$. Looking at it differently, if $\lambda(\Lambda)$ is finite for all Λ, however large, $\lambda(v)$ can only be zero, leading to trivial theory. Of course, one cannot rely on the perturbation theory beta function, and one must check by non-perturbative methods if the problem exists. This means that the theory is trivial (a free theory) unless it is understood as an effective theory, to be replaced by a more complete theory above a large cut-off scale Λ. The cut-off cannot be taken to infinity. However, if we lower the cut-off, the value of $\lambda(v)$, for fixed $\lambda(\Lambda)$, increases. Now since $m_H = \sqrt{2\lambda(v)}v$, this means that m_H increases. For the theory to continue to make sense, m_H cannot be allowed to be so large that it exceeds the cut-off. This gives an upper bound on m_H of

$$m_H < \Lambda = v \exp\left(\frac{4\pi^2 v^2}{3m_H^2}\right), \tag{2.81}$$

where $1/\lambda(\Lambda)$ is taken to be zero. This is solved by $m_H/v \approx 3.75$, or $m_H = 920$ GeV. Of course, for such a large coupling $m_H^2/(2v^2) \approx 7$, perturbation theory is not valid and one must take into account non-perturbative effects. It turns out that non-perturbative calculations more or less confirm this limit [18]. All these considerations were in the limit of no gauge or Yukawa couplings. However, the effect of these latter couplings can be included, and this changes the limit only weakly.

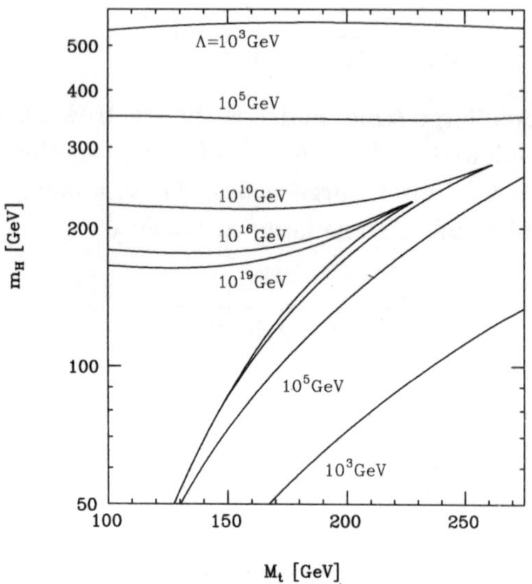

Figure 2.1: Stability bounds on the SM Higgs boson mass. Λ denotes the cut-off scale where the particles become strongly interacting.

The upper limit may also be derived in another way, by requiring that the cut-off can be removed to at least 10^{15} GeV (GUT scale) or 10^{19} GeV (Planck scale) [13, 15]. This is shown in Fig. 2.1, taken from [16]. The limit is about 140 GeV in the latter case.

2.3.3   Unitarity limit

The Equivalence Theorem

We have the relation $m_H^2 = -2\mu^2 = 2\lambda v^2 = \lambda\sqrt{2}/G_F$. Since v and G_F are fixed from experiment, large m_H means large λ. Hence for $m_H \overset{>}{\sim} (G_F/\sqrt{2})^{-1}$, perturbation theory is not valid. (This corresponds to $m_H \approx 300$ GeV). In fact, a limit may be obtained from unitarity, if the tree amplitudes are to be valid at high energies [19, 20]. A convenient way of calculating vector boson scattering amplitudes at high energies is to use the equivalence theorem [20]. This theorem relies on the fact that at high energies, zero-helicity amplitudes dominate. These, in turn, may be calculated in the Feynman-'t Hooft gauge by replacing longitudinally polarized gauge bosons by the spin-0 bosons (the "would-be" Goldstones) responsible for the masses of the gauge bosons.

To study the high-energy scattering amplitudes, it is useful to take a fresh look at the scalar Lagrangian,

$$\mathcal{L} = (D_\mu \phi)^\dagger (D^\mu \phi) - \mu^2 \phi^\dagger \phi - \lambda(\phi^\dagger \phi)^2. \tag{2.82}$$

We write

$$\phi = \begin{pmatrix} \phi^0 \\ \phi^- \end{pmatrix} = \frac{1}{\sqrt{2}} \begin{pmatrix} \phi_1 + i\phi_2 \\ \phi_3 + i\phi_4 \end{pmatrix}. \tag{2.83}$$

Then $\phi^\dagger \phi = \frac{1}{2}(\phi_1^2 + \phi_2^2 + \phi_3^2 + \phi_4^2)$. The Lagrangian thus has a global $O(4)$ symmetry, under which ϕ_i ($i = 1, 2, 3, 4$) transforms as a four-dimensional representation. Explicitly,

$$\mathcal{L} = \frac{1}{2} \sum_{i=1}^{4} \left(\partial_\mu \phi_i \partial^\mu \phi_i - \mu^2 \phi_i \phi_i \right) - \frac{\lambda}{4} \left(\sum_{i=1}^{4} \phi_i \phi_i \right)^2, \tag{2.84}$$

where we have not included the gauge fields, which do not have the full $SO(4) \equiv SU(2) \times SU(2)$ symmetry. To write the scalar Lagrangian in terms of the representations of the two $SU(2)$'s, we use

$$\vec{\pi} \equiv (\phi_2, \phi_3, \phi_4), \quad \sigma = \phi_1. \tag{2.85}$$

The Lagrangian is then

$$\begin{aligned} \mathcal{L} &= \frac{1}{2}\partial_\mu \vec{\pi} \cdot \partial^\mu \vec{\pi} + \frac{1}{2}\partial_\mu \sigma \partial^\mu \sigma - \frac{1}{2}\mu^2 \vec{\pi} \cdot \vec{\pi} \\ &\quad - \frac{1}{2}\mu^2 \sigma^2 - \frac{\lambda}{4}\left(\vec{\pi} \cdot \vec{\pi} + \sigma^2 \right)^2. \end{aligned} \tag{2.86}$$

For one $SU(2)$ rotation, take rotations in the three-dimensional $\vec{\pi}$ space:

$$\delta\vec{\pi} = -i\delta\vec{\alpha} \cdot \vec{T}\vec{\pi}; \quad \delta\sigma = 0, \tag{2.87}$$

where $\vec{T}$ are the 3×3 rotation matrices corresponding to the three-dimensional irreducible representation of $SU(2)$. This is just an isospin symmetry. Next consider another transformation

$$\delta\vec{\pi} = -\delta\vec{\beta}\sigma; \quad \delta\sigma = \delta\vec{\beta}\cdot\vec{\pi}. \tag{2.88}$$

Then

$$\delta(\vec{\pi}^2 + \sigma^2) = -2\vec{\pi}\cdot\delta\vec{\beta}\sigma + 2\sigma\delta\vec{\beta}\cdot\vec{\pi} = 0. \tag{2.89}$$

This is also an invariance of the Lagrangian. There is also a discrete symmetry, *viz.*, parity;

$$\vec{\pi} \to -\vec{\pi}, \tag{2.90}$$

$$\sigma \to \sigma. \tag{2.91}$$

$\delta\vec{\beta}$ interchanges these, so it is axial. Thus the $O(4)$ is equivalent to $SU(2)_V \times SU(2)_A$. We now allow for spontaeous symmetry breaking,

$$\sigma = \phi_1 = \langle\phi_1\rangle + H = \langle\sigma\rangle + H, \tag{2.92}$$

with

$$\langle\sigma\rangle^2 = v^2 = -\mu^2/\lambda. \tag{2.93}$$

Rewrite the Lagrangian in terms of the shifted fields:

$$\begin{aligned}
\mathcal{L} &= \frac{1}{2}\partial_\mu\vec{\pi}\cdot\partial^\mu\vec{\pi} + \frac{1}{2}\partial_\mu H\partial^\mu H - \frac{1}{2}m_H^2 H^2 \\
&\quad - \frac{m_H^2}{2v}(H^2 + \vec{\pi}^2) - \frac{m_H^2}{8v^2}(H^2 + \vec{\pi}^2).
\end{aligned} \tag{2.94}$$

In the above equation, H is the only physical field. The $\vec{\pi}$ fields correspond to the longitudinal degrees of freedom of the massive gauge bosons:

$$\pi_3 = z \, (\equiv Z_L)\pi^\pm = w^\pm \, (\equiv W_L^\pm). \tag{2.95}$$

For example, one can check that the $H \to W_L W_L$ width comes out right:

$$\mathcal{M} \approx -\frac{m_H^2}{v}, \tag{2.96}$$

$$\Gamma = \frac{1}{8\pi}\frac{G_F}{\sqrt{2}}m_H^3. \tag{2.97}$$

2.3.4 $W_L W_L$ scattering

We can use the Lagrangian in eq. (2.94) to calculate amplitudes involving longitudinal gauge bosons. As an illustration of how a bound on the Higgs mass can be obtained from unitarity, we take the process $W_L W_L \to W_L W_L$. A similar analysis can be carried out for other channels like $W_L W_L \to Z_L Z_L$, $Z_L Z_L \to Z_L Z_L$, etc.

There are three contributions to the amplitude for $W_L W_L \to W_L W_L$, one from the quartic coupling,

$$-i\mathcal{M}_1 = -i\frac{2m_H^2}{v^2}, \tag{2.98}$$

one from the diagram with s-channel Higgs exchange,

$$-i\mathcal{M}_2 = \left(\frac{-im_H^2}{2v}\right)^2 \frac{4i}{s - m_H^2} \tag{2.99}$$

and from the diagram with t-channel Higgs exchange,

$$-i\mathcal{M}_3 = \left(\frac{-im_H^2}{2v}\right)^2 \frac{4i}{t - m_H^2}. \tag{2.100}$$

The sum gives

$$-i\mathcal{M} \approx i\frac{(s+t)}{v^2} \approx -i\frac{u}{v^2} \tag{2.101}$$

for $s, t \ll m_H^2$. Note that without the Higgs exchange contributions, the amplitude goes as constant. Hence, Higgs exchange is needed for good high-energy behaviour. Secondly, the Higgs cannot be very heavy, or else the unitarity limit would be crossed too soon. To see this, write the old-fashioned scattering amplitude f_{cm}, so that $|f_{cm}|^2 = \frac{d\sigma}{d\Omega}$. Then we have

$$f_{cm} = \frac{1}{8\pi\sqrt{s}}\mathcal{M}. \tag{2.102}$$

The scattering amplitude has the partial wave expansion

$$f(\theta) = \frac{1}{k}\sum_l (2l+1)P_l(\cos\theta)a_l \tag{2.103}$$

where $a_l = e^{i\delta_l}\sin\delta_l$ is the partial wave amplitude written in terms of the phase shift δ_l. Unitarity is expressed by the optical theorem relation

$$\sigma = \frac{4\pi}{k}\Im f(0).\tag{2.104}$$

For elastic scattering, the phase shift δ_l is real, but has a positive imaginary part if there is inelasticity. A convenient way to express elastic unitarity is

$$\Im\frac{1}{a_l} = -1.\tag{2.105}$$

From this it follows that

$$|a_l| \leq 1; \quad |\Re a_l| \leq \frac{1}{2}.\tag{2.106}$$

Projecting $l = 0$,

$$a_0 = \frac{1}{16\pi}\int d\cos\theta\,\sqrt{2}G_F m_H^2\left(2 + \frac{m_H^2}{s - m_H^2}\right.$$
$$\left. + \frac{m_H^2}{m_W^2 - \frac{s}{2}\left(1 - \sqrt{1 - \frac{4m_W^2}{s}}\cos\theta\right) - m_H^2}\right),\tag{2.107}$$

we get

$$a_0 \approx \frac{1}{8\sqrt{2}\pi}G_F m_H^2\left(2 + \frac{m_H^2}{s - m_H^2} - \frac{m_H^2}{s}\ln\frac{m_H^2 + s}{m_H^2}\right)\tag{2.108}$$

The $l = 0$ partial-wave unitarity for large s then implies,

$$\frac{G_F m_H^2}{4\sqrt{2}\pi} < 1,\tag{2.109}$$

or

$$m_H^2 < \frac{4\sqrt{2}\pi}{G_F}.\tag{2.110}$$

This gives a limit of $m_H < 1.2$ TeV.

2.4 Experimental Constraints on the Mass of the Standard-Model Higgs Boson

2.4.1 Electroweak precision data

The fact that SM is renormalizable only after including the top and Higgs in the loop implies that electroweak observables are sensitive to the masses of these particles.

At lowest orders, the Fermi coupling can be written in terms of the weak coupling and W mass as

$$G_F/\sqrt{2} = g^2/(8m_W^2). \tag{2.111}$$

Putting in the electromagnetic coupling, the weak mixing angle and the W mass, this relation can be rewritten as

$$\frac{G_F}{\sqrt{2}} = \frac{2\pi\alpha}{\sin^2 2\theta_W m_Z^2} \left[1 + \Delta r_\alpha + \Delta r_t + \Delta r_H\right]. \tag{2.112}$$

The Δ's come from radiative corrections. Δr_α is the shift in the electromagnetic coupling if evaluated at m_Z^2 instead of at 0; Δr_t denotes the top (and bottom) contributions to the W and Z masses, which are quadratic in the top mass; Δr_H accounts for the virtual Higgs contributions to the masses. This term depends logarithmically on the Higgs mass at leading order, due to Veltman's screening theorem [21, 22]:

$$\Delta r_H = \frac{G_F m_W^2}{8\sqrt{2}\pi^2} \frac{11}{3} \left[\ln \frac{m_H^2}{m_W^2} - \frac{5}{6}\right]. \quad (m_H^2 \gg m_W^2) \tag{2.113}$$

For a large Higgs mass, the quartic scalar coupling ($\lambda \approx \frac{1}{2}m_H^2/v^2$) is large, and the theory is strongly interacting. Hence it would be expected that radiative corrections grow with m_H^2. However, Veltman [21] showed that the growth is only logarithmic at lowest loop order. At two loops, there is a term $g^4 m_H^2/m_W^2$r:

$$\Delta r_H = g^2 \left[\ln \frac{m_H^2}{m_W^2} + g^2 \frac{m_H^2}{m_W^2} + \cdots\right]. \tag{2.114}$$

Thus terms proportional to m_H^2 are screened by an extra g^2. This is called the "screening" theorem, and explains why low-energy observables are relatively insensitive to the Higgs mass.

Even though the sensitivity on the Higs mass is only logarithmic, the increasing precision in the measurement of the electroweak observables gives the estimates [23]:

$$m_H \;=\; 88 \; {}^{+60}_{-37} \; \text{GeV};$$

$$m_H \;<\; 206\text{GeV}. \quad (95\%\text{C.L.})$$

2.5   Decays of the SM Higgs boson

The mass of the SM Higgs boson is not fixed by theory, and is therefore not known. As a consequence, the decay channels and their relative partial widths are not known—they depend on the mass. Thus, certain channels may be forbidden for lighter Higgs. The dominant decay is into the most massive channel permissible by the Higgs mass, because of the fact that the coupling is proportional to the mass of the particle-antiparticle pair in the decay channel.

We now take up the decays in terms of different categories.

2.5.1   Higgs decays into fermions

The Higgs coupling to quarks and leptons is given by

$$\mathcal{L} = -\frac{m_f}{v}\overline{f}f\phi, \tag{2.115}$$

where m_f is the mass of the fermion f. Since $v \approx 246$ GeV, the couplings of light fermions are quite small. The amplitude for the decay $H \to f\bar{f}$ (related to the S-matrix element S by $S = -iM$) is given by

$$M = \frac{m_f}{v}\bar{u}(p_f)v(p_{\bar{f}}). \tag{2.116}$$

This leads to the matrix element squared, summed over final spins,

$$\sum |M|^2 \;=\; \frac{4m_f^2}{v^2}\left[\frac{1}{2}m_H^2 - 2m_f^2\right]$$

$$= 2\sqrt{2}G_F m_f^2 \left(m_H^2 - 4m_f^2\right). \qquad (2.117)$$

The partial decay width can then be easily worked out to be [24]

$$\Gamma(H \to f\bar{f}) = N_c \frac{G_F}{4\pi\sqrt{2}} m_f^2 m_H \left(1 - \frac{4m_f^2}{m_H^2}\right)^{3/2}. \qquad (2.118)$$

N_c is a colour factor, being 1 for leptons and 3 for quarks. In practice, one has to include QCD corrections to this result. The calculation of these is beyond the scope of these lectures. They were first calculated in Refs.[25]. However, the result is that bulk of the QCD corrections are accounted for by using a running quark mass: $m_f^2(q^2 = m_H^2)$. For $m_H \approx 100$ GeV, the relevant parameters are $m_b(m_H^2) \approx 3$ GeV and $m_c(m_H^2) \approx 0.6$ GeV. The residual QCD corrections $\delta\Gamma$ are small:

$$\frac{\delta\Gamma}{\Gamma} \approx 5.7 \frac{\alpha_s}{\pi}. \qquad (2.119)$$

For a Higgs mass below $2m_W$, the fermionic decays are dominant. For $m_H > 2m_b$, the $b\bar{b}$ mode dominates.

2.5.2 Higgs decays into WW and ZZ

The coupling of the Higgs to W^+W^- and ZZ is given by

$$\mathcal{L}_{HVV} = \frac{\phi}{v} \left(2m_W^2 W_\mu^+ W^{-\mu} + m_Z^2 Z_\mu Z^\mu\right). \qquad (2.120)$$

The amplitudes for $H \to VV$ are given by

$$\begin{aligned}
H \to W^+W^- \quad &: \quad M_{\mu\nu} = (\sqrt{2}G_F)^{1/2} 2m_W^2 g_{\mu\nu}, \\
H \to ZZ \quad &: \quad M_{\mu\nu} = (\sqrt{2}G_F)^{1/2} 2m_Z^2 g_{\mu\nu}.
\end{aligned} \qquad (2.121)$$

Thus, the squared matrix element, summed over final spins and averaged over initial spins, is given by

$$\sum |M|^2 = \sqrt{2}G_F 4m_V^4 g_{\mu\nu} g_{\mu'\nu'} \left(-g_{\mu\mu'} + \frac{q_\mu q_{\mu'}}{m_V^2}\right)$$

$$\times \left(-g_{\nu\nu'} + \frac{q_\nu q_{\nu'}}{m_V^2} \right)$$

$$= \sqrt{2} G_F 4 m_V^4 \left(2 + \frac{(q_1 \cdot q_2)^2}{m_V^4} \right)$$

$$= \sqrt{2} G_F m_H^4 (1 - 4x + 12x^2), \qquad (2.122)$$

where $x = m_V^2/m_H^2$. The phase space is the same as in the case of $H \to f\bar{f}$, giving[26]

$$\Gamma = \frac{2 G_F}{16\sqrt{2}\pi} m_H^3 (1 - 4x + 12x^2)\sqrt{1-x}, \qquad (2.123)$$

for $W^+ W^-$, and without the factor of 2 in the numerator in the case of ZZ, the two final-state particles being identical.

Since longitudinal polarization vectors of massive spin-1 particles go as E_V/m_V for large vector boson energy E_V, the amplitude increases with $E_V = \frac{1}{2} m_H$ as m_H^2. This explains the large power of m_H^3 in the width. Note that

$$\frac{\Gamma(H \to WW)}{\Gamma(H \to ZZ)} \approx 2 \qquad (2.124)$$

for $m_H \gg m_V$.

Below the threshold for two vector bosons, the Higgs can decay into one real and one virtual vector boson V^*. This becomes important for $m_H \gtrsim 140$ GeV.

Above the threshold, the four lepton channel $H \to ZZ \to 4l^\pm$ provides a very clean signal for the Higgs. $H \to WW$ is not so clean, since the final state has either neutrinos or jets.

2.5.3   Higgs decays into gluon and photon pairs

The SM Higgs does not couple to gg or $\gamma\gamma$ at tree level. Hence these decays are suppressed. However, below the $\tau^+\tau^-$ threshold these decays could have large branching ratios. Also, even for larger Higgs masses, these decays are important because the signature is clean. These decays occur at one-loop level, with top

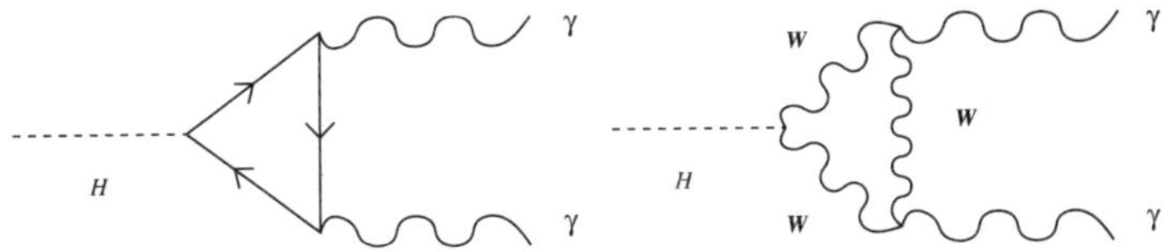

Figure 2.2: Diagrams with fermion and W loops contributing to the decay $H \to \gamma\gamma$.

and bottom quarks in the loop, and in the case of $H \to \gamma\gamma$, even W can contribute, as shown in Fig.2.2.

Approximate expressions or these partial widths are (in the limit of $m_H^2 \ll 4m_W^2, 4m_t^2$) [24]

$$\Gamma(H \to gg) = \frac{G_F \alpha_s^2(m_H^2)}{36\sqrt{2}\pi^3} m_H^3 \left[1 + \left(\frac{95}{4} - \frac{7N_F}{6} \right) \frac{\alpha_s}{\pi} \right]$$

$$\Gamma(H \to \gamma\gamma) = \frac{G_F \alpha^2}{128\sqrt{2}\pi^3} m_H^3 |\frac{4}{3} N_c e_t^2 - 7|^2, \tag{2.125}$$

where e_t, the top charge, is $\frac{2}{3}$, N_F is the number of light flavours, and N_c is 3, the number of colours. QCD corrections to $H \to gg$ should be included. These correspond to the $q\bar{q}g$ and ggg final states, and are important, because they increase the width by about 65%.

Loop mediated Higgs couplings can be calculated in the limit of m_H small compared to the loop mass m by using a low-energy theorem for an amplitude with an external Higgs[27, 28, 29]:

$$\lim_{p_H \to 0} \mathcal{A}(XH) = \frac{1}{v} \frac{\partial \mathcal{A}(X)}{\partial \ln m}. \tag{2.126}$$

This theorem can be derived by observing that if an external zero-mass Higgs line is introduced in the propagator, it is equivalent to the substitution

$$\frac{1}{\not{p} - m} \longrightarrow \frac{1}{\not{p} - m} \cdot \frac{m}{v} \cdot \frac{1}{\not{p} - m}$$
$$= \frac{1}{v} \cdot \frac{\partial}{\partial \ln m} \cdot \frac{1}{\not{p} - m},$$

$$\frac{(-g^{\mu\nu} + Ap^\mu p^\nu)}{p^2 - m_W^2} \rightarrow \frac{(-g^{\mu\alpha} + Ap^\mu p^\alpha)}{p^2 - m_W^2} \times \left(\frac{-2m_W^2 g^{\alpha\beta}}{v}\right)$$

$$\times \frac{\left(-g^{\beta\nu} + Ap^\beta p^\nu\right)}{p^2 - m_W^2}$$

$$= \frac{1}{v} \cdot \frac{\partial}{\partial \ln m_W} \cdot \frac{(-g^{\mu\nu} + Ap^\mu p^\nu)}{p^2 - m_W^2},$$

This works for $A = 0$ (Feynman gauge), or $A = 1/p^2$ (Landau gauge). In either of these gauges, one has to worry about unphysical scalars.

To take an example, for the Hgg coupling, we need the gluon propagator Π_{gg} as a function of the loop masses. With

$$\Pi_{gg} \sim \frac{\alpha_s}{12\pi} \ln m \tag{2.127}$$

for $q^2 = 0$,

$$\mathcal{A}(Hgg) = \frac{\alpha_s}{12\pi} \cdot \frac{1}{v}. \tag{2.128}$$

More precise expressions for m_H not small can be found in [4].

2.5.4   Higgs decay into $Z\gamma$

The calculations are similar to those for $H \rightarrow \gamma\gamma$. Because the Z coupling is proportional to g, whereas the γ coupling is proportional to $e = g\sin\theta_W$, the branching ratio for $H \rightarrow Z\gamma$ is higher than that for $H \rightarrow \gamma\gamma$ by a factor of about $1/\sin^2\theta_W$. However, since only a fraction of Z decays provide a good signature, this mode is not very useful for a search of the Higgs boson.

2.5.5   Conclusion

Branching rations for various channels and the total decay width are shown as a function of the Higgs mass in Fig. 2.3 taken from [30].

We can get the total width of the Higgs by adding up all the decay channels. Upto masses of about 140 GeV, the Higgs is

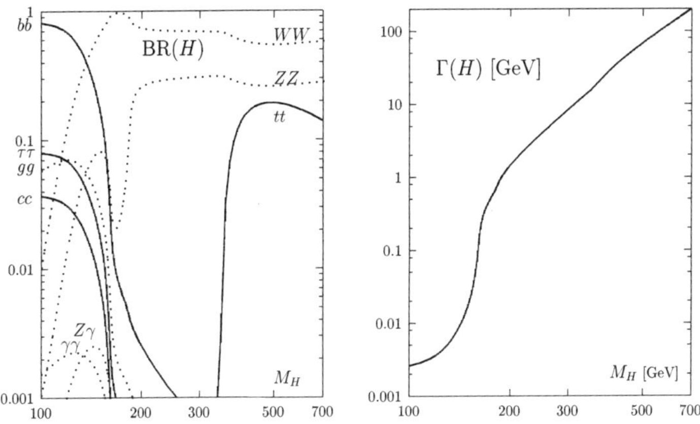

Figure 2.3: The branching ratios (left figure) and the total decay width (right figure) of the SM Higgs shown as a function of the Higgs mass, from [30].

very narrow, $\Gamma(H) \lesssim 10$ Mev. After the opening of the real and virtual gauge-boson channels, the state rapidly becomes wider, reaching a width of about 1 GeV at around $m_H \approx 200$ GeV. The width cannot be measured directly in the intermediate mass region at LHC or e^+e^- colliders. However, it could be measured at $\mu^+\mu^-$ colliders. Above a mass of about 250 GeV, the state is wide enough to be observable, in general. Above the two-vector-boson threshold, the width is $\Gamma(H) \approx \frac{1}{2}m_H^3$ (TeV). For $m_H \approx 1$ TeV, $\Gamma_H \approx \frac{1}{2}$ TeV.

2.6 Production of Higgs at e^+e^- colliders

Since the Higgs couples to particles with strength proportional to the mass of the particle, one should look for processes where a heavy particle is produced. The heaviest particles we know are $W^\pm$, Z and the top quark. So we can look for mechanisms where Higgs is produced from one of these particles, either real or virtual.

The first process studied at LEP ($\sqrt{s} = m_Z$), which ruled out Higgs masses less than 65.4 GeV, was the Bjorken process.

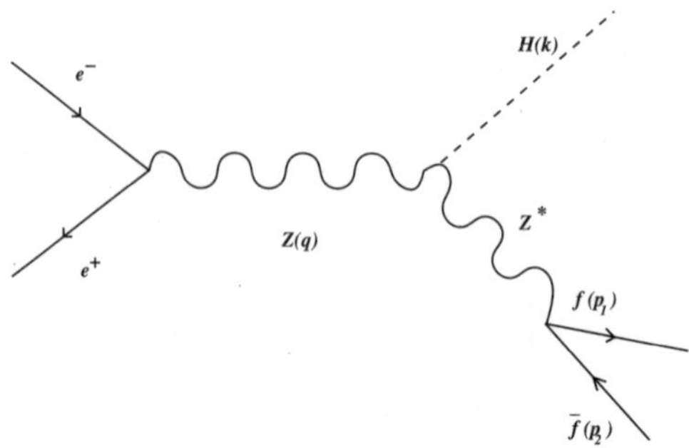

Figure 2.4: The Bjorken process for the production of Higgs

2.6.1   Bjorken process

This process was first studied by Bjorken [31, 32]. The process, shown in Fig.2.4 is $e^+e^- \to Z(q) \to Z^*H(k)$, where the virtual Z decays in to a fermion pair. Since we are considering real Z production in e^+e^- collisions, we can separate the production and decay of Z, and look at $Z(q) \to H(k)f(p_1)\overline{f}(p_2)$. The corresponding amplitude is given by

$$M(Z \to Hf\overline{f}) = \epsilon_\mu^Z \frac{2m_Z^2}{v}\left(-g^{\mu\nu} + \frac{(q-k)^\mu(q-k)^\nu}{m_Z^2}\right)$$
$$\times \frac{\overline{u}(p_1)\gamma_\nu(v_f - a_f\gamma_5)v(p_2)}{q^2 - m_Z^2}, \tag{2.129}$$

where

$$v_f = \frac{g}{\cos\theta_W}\left(\frac{1}{2}t_{3L}^f - q^f \sin^2\theta_W\right),$$
$$a_f = \frac{g}{\cos\theta_W}\left(\frac{1}{2}t_{3L}^f\right) \tag{2.130}$$

are the vector and axial-vector couplings of the fermion f to Z.

The square of the matrix element summed over final spins and averaged over initial spins is given by

$$\overline{\sum}|M|^2 = \overline{\sum}\left(\epsilon_\mu(q)\epsilon^*_{\mu'}(q)\right)\frac{4m_Z^4}{v^2}\left[\frac{1}{(q-k)^2-m_Z^2}\right]^2$$

$$\times\left(-g^{\mu\nu}+\frac{(q-k)^\mu(q-k)^\nu}{m_Z^2}\right)$$

$$\times\left(-g^{\mu'\nu'}+\frac{(q-k)^{\mu'}(q-k)^{\nu'}}{m_Z^2}\right)$$

$$\times \text{Tr}\left[\gamma_\nu(v_f-\gamma_5 a_f)\slashed{p}_2\gamma_{\nu'}(v_f-\gamma_5 a_f)\slashed{p}_1\right], \quad (2.131)$$

with

$$\overline{\sum}\left(\epsilon_\mu(q)\epsilon^*_{\mu'}(q)\right) = \frac{1}{3}\left(-g_{\mu\mu'}+\frac{q_\mu q_{\mu'}}{m_Z^2}\right). \quad (2.132)$$

The trace in eq. (2.131) evaluates to

$$\text{Tr}\left[\gamma_\nu(v_f-\gamma_5 a_f)\slashed{p}_2\gamma_{\nu'}(v_f-\gamma_5 a_f)\slashed{p}_1\right]$$

$$= 4\left(v_f^2+a_f^2\right)[p_{1\nu}p_{2\nu'}+p_{1\nu'}p_{2\nu}-g_{\nu\nu'}p_1\cdot p_2]$$

$$-8iv_f a_f\epsilon_{\nu\nu'\alpha\beta}p_1^\alpha p_2^\beta. \quad (2.133)$$

The last term in the expression for the trace can be dropped because it is antisymmetric in ν and ν', whereas the spin-averaged product of polarization vectors (2.132) is symmetric. Also, the terms in eq.(2.131) with $(q-k)$ factors can be dropped because when they are contracted with the trace, they give zero because of current conservation, and the fact that $q-k = p_1+p_2$. We therefore have the relation, for $q^2 = m_Z^2$,

$$\frac{1}{3}\left(-g_{\mu\mu'}+\frac{q_\mu q_{\mu'}}{m_Z^2}\right)4\left(v_f^2+a_f^2\right)\left[p^{1\nu}p^{2\nu'}+p^{1\nu'}p^{2\nu}-g^{\nu\nu'}p_1\cdot p_2\right]$$

$$= \frac{4}{3}\left(v_f^2+a_f^2\right)\left[p_1\cdot p_2+\frac{2p_1\cdot q p_2\cdot q}{m_Z^2}\right]. \quad (2.134)$$

We can then write the expression for the matrix element squared (2.131) as

$$\overline{\sum}|M|^2 = \frac{4}{3}\frac{4m_Z^4}{v^2}\left[\frac{1}{(q-k)^2-m_Z^2}\right]^2\left(v_f^2+a_f^2\right)$$

$$\times \left[\frac{1}{2}(q-k)^2 + \frac{2p_1 \cdot q\, p_2 \cdot q}{m_Z^2} \right], \qquad (2.135)$$

which becomes, in the rest frame of the initial Z,

$$\overline{\sum}|M|^2 = \frac{4g^2 m_Z^4}{3m_Z^2} \left(\frac{1}{q_{Z*}^2 - m_Z^2} \right)^2 \left(v_f^2 + a_f^2 \right) \left[\frac{1}{2}q_{Z*}^2 + 2p_{10}p_{20} \right], \tag{2.136}$$

where q_{Z*} is the four-momentum of the virtual Z.

The decay width can be obtained by multiplying (2.136) above by the phase space factor and carrying out the momentum integrations. The phase space factor is

$$\frac{1}{2m_Z} \int \frac{d^3p_1}{2p_{10}(2\pi)^3} \int \frac{d^3p_2}{2p_{20}(2\pi)^3} \int \frac{d^3k}{2k_0(2\pi)^3} (2\pi)^4 \delta^{(4)}(p_1+p_2+k-q)$$

which, using the on-mass-shell condition

$$\frac{d^3p_2}{2p_{20}(2\pi)^3} = d^4p_2 \delta((q-p_1-k)^2) \qquad (2.137)$$

can be written, after carrying out integration over d^4p_2 as

$$\frac{1}{2m_Z} \frac{1}{(2\pi)^5} \int \frac{d^3p_1}{2p_{10}} \int \frac{d^3k}{2k_0} \delta((q-p_1-k)^2)$$

Now going to a frame in which $\vec{p}_1 + \vec{p}_2 = 0$ and $\vec{q} = \vec{k}$, and writing

$$p_1 \cdot q = p_{10}q_0 - |\vec{p}_1||\vec{q}| \cos\theta_{pq}$$
$$p_2 \cdot q = p_{20}q_0 + |\vec{p}_1||\vec{q}| \cos\theta_{pq},$$

we can carry out the angular integration in d^3p_1 using θ_{pq} as the integration variable. The p_{10} integral can be carried out using the δ function. One can then write everything in a Lorentz-invariant form and then make a transition to the Z rest frame, to get

$$\Gamma = \frac{1}{2m_Z} \frac{m_Z^4}{(2\pi)^5} \int \frac{d^3k}{2k_0} \frac{g^2\pi}{9m_W^2} \frac{v_f^2 + a_f^2}{\left[(q-k)^2 - m_Z^2\right]^2}$$
$$\times (q-k)^2 \left[4 + \frac{2(m_Z - k_0)^2}{(q-k)^2} \right]. \tag{2.138}$$

Since the integrand has no angular dependence, the angular integral gives 4π. Then using the variables

$$x = 2E_H/m_Z \quad \text{and} \quad y = m_H/m_Z, \tag{2.139}$$

we get

$$\Gamma(Z \to Hf\bar{f}) = \frac{v_f^2 + a_f^2}{(2\pi)^3} \frac{g^2 m_Z^3}{288}$$
$$\times \int_{2y}^{1+y^2} dx \frac{\sqrt{x^2 - 4y^2}}{(x - y^2)^2} \left(12(1 - x) + x^2 8y^2\right). \tag{2.140}$$

This may be compared with the decay width into fermions:

$$\Gamma(Z \to f\bar{f}) = \frac{m_Z}{12\pi} \left(v_f^2 + a_f^2\right). \tag{2.141}$$

The normalized decay distribution is then

$$\frac{1}{\Gamma(Z \to f\bar{f})} \frac{d\Gamma(Z \to Hf\bar{f})}{dx}$$
$$= \frac{g^2}{192\pi^2 \cos^2 \theta_W} \frac{(12 - 12x + x^2 + 8y^2)\sqrt{x^2 - 4y^2}}{(x - y^2)^2} \tag{2.142}$$

Integrating over x,

$$\frac{\text{BR}(Z \to Hf\bar{f})}{\text{BR}(Z \to f\bar{f})} = \frac{g^2}{192\pi^2 \cos^2 \theta_W} \left[\frac{3y(y^4 - 8y^2 + 20)}{\sqrt{4 - y^2}}\right.$$
$$\cos^{-1}\left(\frac{y(3 - y^2)}{2}\right) - 3(y^4 - 6y^2 + 4)\ln y$$
$$\left.-\frac{1}{2}(1 - y^2)(2y^4 - 13y^2 + 47)\right]. \tag{2.143}$$

Because the Z width has been neglected, there is a singularity for $y = 0$. Using a Breit-Wigner form regulates the singularity.

The ratio is 10^{-2} or less. The branching ratio itself is much smaller, $10^{-4} - 10^{-7}$ (for $m_H \approx 0 - 60$ GeV), decreasing with increasing m_H. The signal at LEP is $e^+e^- \to l^+l^-l'^+l'^-$, $e^+e^- \to l^+l^-q\bar{q}$ etc., with a peak in the invariant mass of a pair of fermions. The Higgs could be detected without observing the decay products of the Higgs. One simply measures the four-momenta of the outgoing fermions f, $\bar{f}$ coming from the virtual Z, and uses energy-momentum conservation to obtain the Higgs-boson four-momentum. In addition, the invariant mass distribution $M_{f\bar{f}}$ is strongly peaked at very large values. Roughly, $M_{f\bar{f}} \approx r\,(m_Z - m_H)$, where r ranges from 0.96 at $m_H = 0.1 m_Z$ to 0.73 at $m_H = .9 m_Z$. This is obtained from the earlier expression for $d\Gamma/dx$, by using $M_{f\bar{f}}^2 = m_H^2 + m_Z^2(1-x)$. Physically, the distribution peaks when the Z^* is almost on-shell, but since kinematics allows a maximum of $M_{f\bar{f}}^2 = m_Z^2 - m_H^2$, it peaks there.

The strategy to search for the Higgs boson at LEP was to look for Z^* going into e^+e^- or $\mu^+\mu^-$, and the Higgs decay products depending on the mass of the Higgs

Mass range	Decay products	Signature
$m_H < 2m_\mu$	$e^+e^-,\ \gamma\gamma$	acollinear lepton pair; no calorimetric energy
$2m_\mu < m_H < m_{c\bar{c}}$	$\mu^+\mu^-$	Lepton pair plus a few tracks
$m_H > m_{c\bar{c}}$	$b\bar{b},\ c\bar{c},\ \tau^+\tau^-$	acollinear lepton pair; plus two jets
	$q\bar{q}$	Two acollinear jets; large missing E, p_T (for $Z \to \nu\bar{\nu}$)

The background is from the four-fermion decay of Z,

$$Z \to q\bar{q}l\bar{l}.$$

$M_{l\bar{l}}$ peaks at small values for the four-fermion final state and at large values for the $Hl\bar{l}$ final state. Hence a cut on $M_{l\bar{l}}$ can reduce the four-fermion background.

2.6.2 Associated Bjorken process

At centre of mass energies above the Z mass, as for example at LEP2, the Z is produced in a virtual state, and then one should look for its decay into a real Z and H. This is sometimes called the associated Bjorken process. The corresponding differential cross section is given by [33]

$$\frac{d\sigma\,(e^+e^- \to ZH)}{d\cos\theta} \qquad (2.144)$$
$$= \frac{\pi\alpha^2\lambda^{1/2}\left[\lambda\sin^2\theta + 8m_Z^2 s\right]\left[1 + (1 - 4\sin^2\theta_W)^2\right]}{256s^2\sin^4\theta_W\cos^4\theta_W(s - m_Z^2)^2},$$

where
$$\lambda = (s - m_H^2 - m_Z^2)^2 - 4m_H^2 m_Z^2. \qquad (2.145)$$

This integrates to

$$\sigma\,(e^+e^- \to ZH) \qquad (2.146)$$
$$= \frac{\pi\alpha^2\lambda^{1/2}\left[\lambda + 12m_Z^2 s\right]\left[1 + (1 - 4\sin^2\theta_W)^2\right]}{192s^2\sin^4\theta_W\cos^4\theta_W(s - m_Z^2)^2}$$

The cross section peaks at cm energy $\sqrt{s} \approx m_Z + \sqrt{2}m_H$, and it scales with s as $1/\sqrt{s}$. Since the cross section vanishes for asymptotic energies, the process is most useful for searching Higgs bosons in the mass range where the collider energy is of the same order as the Higgs mass, $\sqrt{s} \gtrsim \mathcal{O}(m_H)$.

Since the recoiling Z mass in the two-body reaction $e^+e^- \to ZH$ is monoenergetic, the mass of the Higgs boson can be reconstructed from the energy of the Z, $m_H = s - 2\sqrt{s}E_Z + m_Z^2$, without any need to analyze the decay products of the Higgs boson.

2.6.3 Gauge-boson fusion process

A heavier Higgs, $m_H > 2m_W$, can be produced by WW fusion,

$$e^+e^- \to (W^{+*}\bar{\nu}_e)(W^{-*}\nu_e) \to H\nu_e\bar{\nu}_e. \qquad (2.147)$$

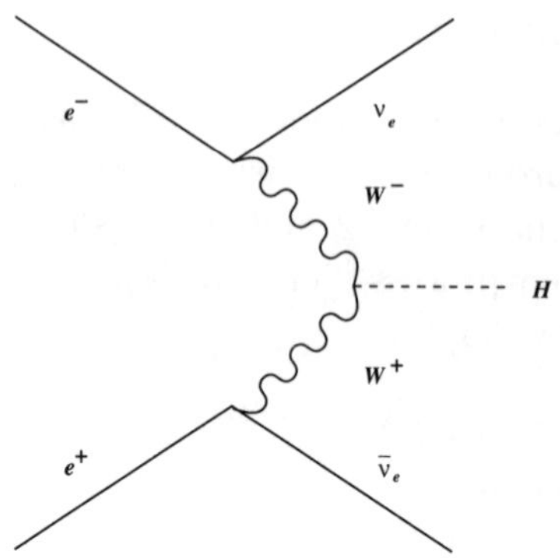

Figure 2.5: The WW fusion process

Two virtual W's fuse to give rise to the Higgs, as shown in Fig. 2.5. If $m_H > 2m_Z$, the process of ZZ fusion can also occur in an analogous fashion:

$$e^+e^- \rightarrow (Z^*e^+)(Z^*e^-) \rightarrow He^+e^-. \qquad (2.148)$$

Because the W and Z propagators peak at low $|q^2|$, the W and Z can be thought of as having $q^2 \approx 0$, and almost real. One can think of a simplified approximate picture, wherein a high-energy electron (or positron) emits a nearly real, nearly collinear W or Z with a certain probability depending on the energy fraction carried by the W or Z. The probability for finding a particle V with a fixed helicity h with a momentum fraction x in an electron is $f_{e/V_h}(x)dx$, where the functions f for the different helicities are given by [34]

$$f_{f/V_\pm}(x) \;=\; \frac{1}{16\pi^2 x}\ln\left(\frac{4E^2}{m_V^2}\right)\left[(v_f \mp a_f)^2 + (1-x)^2(v_f \pm a_f)^2\right],$$

$$f_{f/V_L}(x) \;=\; \frac{1-x}{4\pi^2 x}(v_f^2 + a_f^2).$$

In the above, V is either W or Z, the subscripts on V denote the helicity, with L denoting the "longitudinal" or zero helicity, and E is the energy or the electron beam. Similar expressions may be obtained for positron beams. This is called the equivalent vector boson approximation, and is in the same spirit as the equivalent photon approximation (also known as the Weizäcker-Williams approximation).

Working without approximation, the cross section for the fusion process can be put in the form

$$\sigma(e^+e^- \to \nu_e\bar{\nu}_e H) \tag{2.149}$$
$$= \frac{G_F^3 m_W^4}{4\sqrt{2}\pi^3} \int_{\kappa_H}^1 dx \int_x^1 dy \frac{1}{[1+(y-x)/\kappa_W]^2} f(x,y),$$

where

$$f(x,y) = \left(\frac{2x}{y^3} - \frac{1+3x}{y^2} + \frac{2+x}{y} - 1\right)\left[\frac{z}{1+z} - \ln(1+z)\right]$$
$$+ \frac{x}{y^3}\frac{z^2(1-y)}{1+z}, \tag{2.150}$$

with $\kappa_H = m_H^2/s$, $\kappa_W = m_W^2/s$, and $z = y(x-\kappa_H)/(\kappa_W x)$. This process dominates over the Bjorken process for large s. At asympotic energies, the cross section simplifies to

$$\sigma(e^+e^- \to \nu_e\bar{\nu}_e H) \approx \frac{G_F^3 m_W^4}{4\sqrt{2}\pi^3}\left[\ln\frac{s}{m_H^2} - 2\right]. \tag{2.151}$$

Fig. 2.6, taken from [35], shows the production cross section as a function of the Higgs mass, and of the centre-of-masss energy, in the context of high-energy e^+e^- linear collider. At such a collider, there would be also be higher order proccsses possible, with Higgs produced in association with, for example, $t\bar{t}$ pair.

2.6.4 Photon-photon fusion

The two-photon width is related to the production cross section for polarized γ beams by

$$\sigma(\gamma\gamma \to H) = \frac{16\pi^2\Gamma(H \to \gamma\gamma)}{m_H}\frac{\Gamma_H m_H}{(s_{\gamma\gamma} - m_H^2)^2 + \Gamma_H^2 m_H^2}. \tag{2.152}$$

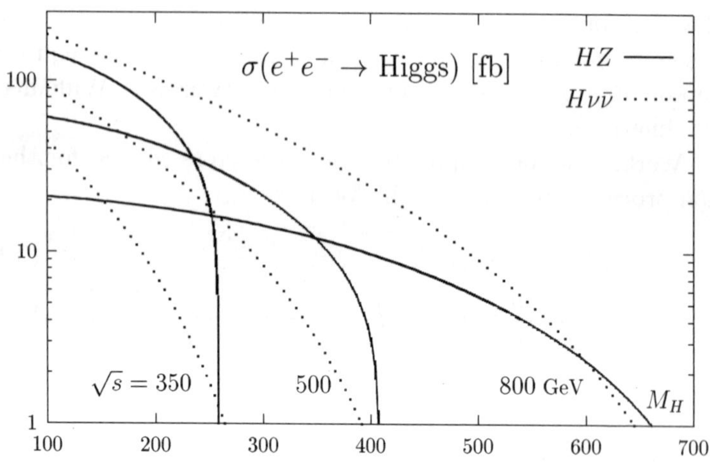

Figure 2.6: Production cross sections of the SM Higgs boson in e^+e^- through the main processes with $\sqrt{s} = 350$, 500 and 800 GeV as a function of M_H.

For a narrow Higgs, the observed cross section is obtained by folding the parton cross section with the invariant $\gamma\gamma$ energy flux $\tau \frac{d\mathcal{L}^{\gamma\gamma}}{d\tau}$ for $J_z^{\gamma\gamma} = 0$ at $\tau = m_H^2/s_{ee}$, where the $\gamma\gamma$ arise from $e^+e^\mp$.

The event rate can be sufficiently large for intense beams of photons obtained by Compton back-scattering of laser beams off high energy electrons.

The large $\gamma\gamma$ continuum background is suppressed in the $J_z^{\gamma\gamma} = 0$ polarization state. For decays into W^+W^-, the continuum background is large; the ZZ background may be under control.

2.6.5   Higgs production at hadron colliders

Possible processes which can contribute to production of Higgs particles at hadron colliders are

- Gluon-gluon fusion: $gg \to H$

- WW, ZZ fusion: $W^+W^- \to H$, $ZZ \to H$

- Bjorken-type process: $q\bar{q} \to W, Z \to W, Z + H$

- Higgs bremsstrahlung off top quark: $q\bar{q}, gg \to t\bar{t} + H$

While gluon fusion plays a dominant role throughout the entire Higgs mass range of the Standard Model, the WW/ZZ fusion process becomes increasingly important with increasing m_H. The last two radiation processes are relevant only for light Higgs masses.

The production cross sections at hadron colliders like the Large Hadron Collider (LHC) are quite large, so that a large sample of Higgs can be produced. However, because of the large background, experimental difficulties are present. These problems may be tackled by triggering on leptonic decays of W, Z and t in the radiation process, or by exploiting the resonance character of Higgs decays $H \to \gamma\gamma$ and $H \to ZZ \to 4l^\pm$. Thus Tevatron is expected to search for Higgs particles in the mass range above that of LEP2, upto about 180 GeV [36]. LHC is expected to cover the entire canonical Higgs mass range $m_H \lesssim 700$ GeV.

2.6.6   Gluon-gluon fusion

At LHC, this provides the dominant production mechanism in the entire m_H range upto about 1 TeV. We will derive an expression for the production cross section for the process.

Define the probability of finding a parton carrying a momentum fraction between x and $x + dx$ of the total proton momentum as $f(x)dx$. Then the cross section in proton-proton collisions is

$$d\sigma(s) = \int dx_1 \int dx_2 f(x_1) f(x_2) d\hat{\sigma}(\hat{s}), \qquad (2.153)$$

where $\hat{s} = x_1 x_2 s$.

For a process corresponding to the production of a particle or a narrow resonance, we will have

$$\hat{\sigma}(\hat{s}) = (2J + 1)\frac{4\pi^2}{M}\Gamma_{\text{partial}}\delta(\hat{s} - M^2), \qquad (2.154)$$

where M is the mass of the resonance, Γ_{partial} its partial decay width into the incoming channel, and J its spin. (This is the Breit-Wigner formulation). In such a case, we have the hadron-level cross section

$$\begin{aligned}
\sigma &= \int_0^1 dx_1 \int_0^1 dx_2 f(x_1) f(x_2)(2J+1) \\
&\quad \times \frac{4\pi^2}{M} \Gamma_{\text{partial}} \delta(sx_1 x_2 - M^2) \\
&= \frac{(2J+1)4\pi^2}{M} \Gamma_{\text{partial}} \frac{1}{s} \int_\tau^1 \frac{dx}{x} f(x) f(\tau/x),
\end{aligned} \qquad (2.155)$$

where $\tau = \frac{M^2}{s}$. The last expression can be rewritten as

$$\sigma = \frac{(2J+1)4\pi^2}{M} \Gamma_{\text{partial}} \tau \frac{d\mathcal{L}}{d\tau}, \qquad (2.156)$$

where

$$\frac{d\mathcal{L}}{d\tau} = \int_\tau^1 f(x) f(\tau/x) \qquad (2.157)$$

is the luminosity spectrum, which is process independent, but depends on the colliding hadrons and the partons.

For the case of $pp \to H + X$, we have the expression:

$$\sigma = \left(\frac{1}{8} \cdot \frac{1}{8} \cdot 2 \right) \frac{4\pi^2}{m_H^2} \frac{\Gamma(H \to gg)}{m_H} \tau \frac{d\mathcal{L}_{gg}}{d\tau}. \qquad (2.158)$$

Here, the first factor of $\frac{1}{8}$ comes from the fact that to make a colour singlet, only one of the eight gluons can annihilate a gluon of a given colour. The second factor of $\frac{1}{8}$ comes because in $\Gamma(H \to gg)$, one sums over all 8 colour channels, but here we are dealing with annihilation of a gluon pair of a single colour. The factor of 2 compensates for the fact that in the calculation of $\Gamma(H \to gg)$ a factor of $\frac{1}{2}$ is included for phase space of identical particles in the final state, whereas here those particles are in the initial state.

The differential luminosity is obtained from the proton structure functions. We can use a simple fit:

$$\tau \frac{d\mathcal{L}_{gg}}{d\tau} = 2 \cdot 10^4 \, e^{-24.8\tau^{0.2}}. \qquad (2.159)$$

The partial width $\Gamma(H \to gg)$ arises at one-loop level from the exchange of quarks in the loop, and is given by

$$\Gamma(H \to gg) = \frac{G_F \alpha_s^2 m_H^2}{36\sqrt{2}\pi^3} \Big| \sum_Q A_Q^H(\tau_Q) \Big|^2, \tag{2.160}$$

where $\tau_Q = 4m_Q^2/m_H^2$, and the sum is over the quarks Q. The form factors A_Q^H are given by

$$A_Q^H(\tau_Q) = \frac{3}{2}\tau_Q[1 + (1 - \tau_Q)f(\tau_Q)] \tag{2.161}$$

where

$$f(\tau_Q) = \begin{cases} \arcsin^2 \frac{1}{\sqrt{\tau_Q}} & (\tau_Q \geq 1) \\ \frac{1}{4}\left[\ln \frac{1+\sqrt{1-\tau_Q}}{1-\sqrt{1-\tau_Q}} - i\pi\right]^2 . & (\tau_Q, 1) \end{cases} \tag{2.162}$$

For small loop masses, the form factor vanishes,

$$A_Q^H(\tau_Q) \approx -\frac{3}{8}\tau_Q \left[\ln(\tau_Q/4) + i\pi\right]^2, \tag{2.163}$$

whereas for large loop masses, it approaches 1.

QCD corrections to the gluon fusion process are very important. They are large and positive, thus increasing the production cross section. Apart from virtual-gluon corrections to $gg \to H$, the real-gluon corrections from $gg \to Hg$, $qg \to Hq$, $q\bar{q} \to Hg$ should also be included. These subprocesses contribute at order α_s^3. The virtual corrections rescale the lowest-order fusion cross section with a coefficient that depends only on the ratios of the Higgs and quark masses. Gluon radiation leads to two-parton final states with invariant energy $\hat{s} > m_H^2$ in the gg, gq and $q\bar{q}$ channels.

The final result for the hadronic cross section can be split into five parts:

$$\sigma(pp \to H + X) = \sigma_0 \left(1 + C\frac{\alpha_s}{\pi}\right)\tau\frac{d\mathcal{L}_{gg}}{\tau} + \Delta\sigma_{gg} + \Delta\sigma_{qg} + \Delta\sigma_{q\bar{q}}. \tag{2.164}$$

σ_0 is the lowest-order result for $gg \to H$. In the heavy quark limit,

$$C = \pi^2 + \frac{11}{2}. \tag{2.165}$$

Also, in this limit, analytic forms are available for $\Delta\sigma_{gg}$, $\Delta\sigma_{qg}$ and $\Delta\sigma_{q\bar{q}}$.

The size of the radiative corrections can be parametrized by defining the K factor

$$K = \frac{\sigma_{\rm NLO}}{\sigma_{\rm LO}}, \tag{2.166}$$

where the subscripts LO and NLO stand respectively "leading order" and "non-leading order". The K factor is around 1.6, with weak dependence on m_H.

Overall, the Higgs production cross section by this mechanism at LHC is around 10 pb, and decreases with increasing m_H because of fall in gg luminosity.

2.6.7 Vector boson fusion

For large Higgs masses, this mechanism becomes competetive with gluon fusion. For intermediate masses, it is smaller by about an order of magnitude.

For large Higgs masses, the two electroweak bosons W and Z are predominantly longitudinally polarized (with helicity 0). The equivalent particle spectra of the longitudinal W and Z bosons in quark beams is given in the equivalent vector boson approximation [34] by

$$f_L^W(x) = \frac{G_F m_W^2}{2\sqrt{2}\pi^2} \frac{1-x}{x}, \tag{2.167}$$

$$f_L^Z(x) = \frac{G_F m_Z^2}{2\sqrt{2}\pi^2} \left[(I_3^q - 2e_q \sin^2\theta_W)^2 + (I_3^q)^2 \right] \frac{1-x}{x}. \tag{2.168}$$

From these particle spectra, the WW and ZZ luminosities can be easily derived.

$$\frac{d\mathcal{L}^{WW}}{d\tau_W} = \frac{G_F^2 m_W^4}{8\pi^4} \left[2 - \frac{2}{\tau_W} - \frac{1+\tau_W}{\tau_W} \ln\tau_W \right],$$

$$\frac{d\mathcal{L}^{ZZ}}{d\tau_Z} = \frac{G_F^2 m_Z^4}{8\pi^4}\left[(I_3^q - 2e_q \sin^2\theta_W)^2 + (I_3^q)^2\right]$$

$$\times \left[(I_3^{q'} - 2e_{q'}\sin^2\theta_W)^2 + (I_3^{q'})^2\right]$$

$$\times \left[2 - \frac{2}{\tau_Z} - \frac{1+\tau_Z}{\tau_Z}\ln\tau_Z\right]. \tag{2.169}$$

Denoting the cross section for $VV \to H$ by $\hat{\sigma}_0$,

$$\hat{\sigma}_0(VV \to H) = \sigma_0 \delta(1 - m_H^2/\hat{s}), \tag{2.170}$$

with

$$\sigma_0 = \sqrt{2}\pi G_F. \tag{2.171}$$

Then

$$\hat{\sigma}(qq \to qqH) = \frac{d\mathcal{L}^{VV}}{d\tau_V}\sigma_0. \tag{2.172}$$

A calculation of this cross section without the equivalent boson approximation was done by Cahn and Dawson [37]

The hadronic cross section is finally obtained by summing the parton cross section over the flux of all possible pairwise combinations of quarks and antiquarks:

$$\sigma(HH' \to VVX \to HX) \tag{2.173}$$

$$= \int_{m_H^2/s}^1 d\tau \sum_{q,q'} \frac{\mathcal{L}^{qq'}}{d\tau}\hat{\sigma}(qq' \to qq'H; \hat{s} = \tau s).$$

Since to lowest order the proton remnants are colour singlets in the WW, ZZ fusion processes, no colour will be exchanged between the two quark lines from which the two vector bosons are radiated. As a result, QCD corrections are already accounted for by the corrections to the quark parton densities.

This process is most important in the upper range of Higgs masses, where the cross section approaches values close to gluon fusion.

2.6.8   Bjorken-like process

The Bjorken-like process $q\bar{q} \rightarrow V^* \rightarrow VH$ $(V = W, Z)$ is an important mechanism for Higgs production. The cross section is smaller than that for gluon fusion. However, leptonic decays of W and Z are extremely useful to filter Higgs signal events out of the huge background. The mechanism is the same as that for e^+e^- colliders, except for the folding with the quark and antiquark densities.

2.6.9   Higgs bremsstrahlung off top quark

This process, $q\bar{q}(gg) \rightarrow t\bar{t}H$, is relevant only for small Higgs masses because of the small phase space available. Analytic expression for the parton level cross section is quite involved. This process is also an interesting process for the measurement of the $t\bar{t}H$ Yukawa coupling.

Figs. 2.7 and 2.8, taken respectively from [38] and [39], show the cross sections for various prodution mechanisms at Tevatron and at the proposed LHC.

2.7   Present status and outlook

The Higgs has not been seen so far. The most important limits come from the e^+e^- experiment LEP at CERN in Geneva. The LEP began in 1989 and started taking data at the Z resonance. Later, the cm energy was pushed up in steps to 209 GeV. There were four experiments, ALEPH, DELPHI, L3 and OPAL. Until shutdown in November 2000, about 1000 pb^{-1} of data were delivered to each of the four experiments. The analysis of the data for SM Higgs are optimized for Higgs bosons decaying in to b quarks, and thus use b identification as a major tool to suppress background from signal. The final states considered can be characterized as $Hq\bar{q}$, $H\nu\bar{\nu}$, He^+e^-, $H\mu^+\mu^-$ and $H\tau^+\tau^-$. The combined limit on the SM Higgs mass at 95% confidence level is set to be 114 GeV.

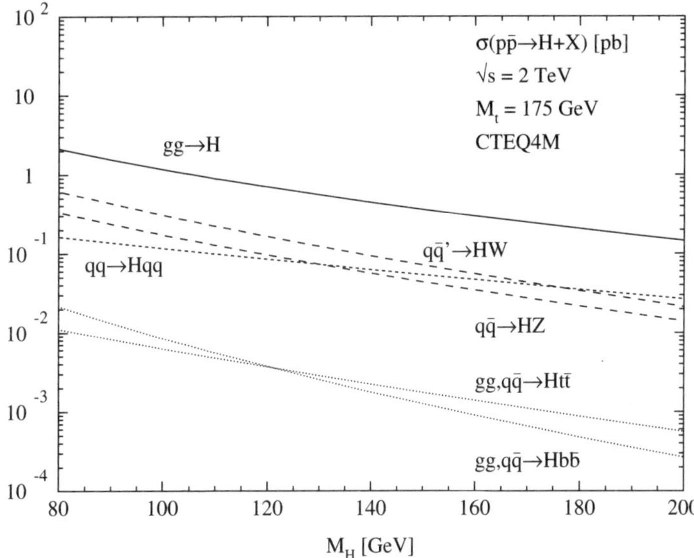

Figure 2.7: Cross sections for contributions of various mechanisms to $p\bar{p} \to HX$ at the Tevatron, taken from [38].

The ongoing Tevatron $p\bar{p}$ experiment at the Fermi National Accelerator Laboratory in the USA is also carrying out a Higgs search. The dominant Higgs production processes at Tevatron are the gluon fusion, $gg \to H$, and associated production with a W or Z boson, $q\bar{q} \to W/ZH$. The gluon fusion process has a higher cross section, but it is very difficult to distinguish from the overwhelming QCD background in the light Higgs region ($m_H \leq$ 130 GeV), where $H \to b\bar{b}$ is the dominant decay channel. In the case of associated production, the vector boson provides a handle to suppress the background, but the cross section $\sigma(p\bar{p} \to W/ZH) \approx 300$ fb for $m_H = 110$ GeV, is approximately 20 times smaller than the $t\bar{t}$ cross section. According to the decay modes of the vector bosons, four different signatures are considered. The search strategy is similar for all decay channels. After identifying the W or Z through its decay, at least two extra jets from the

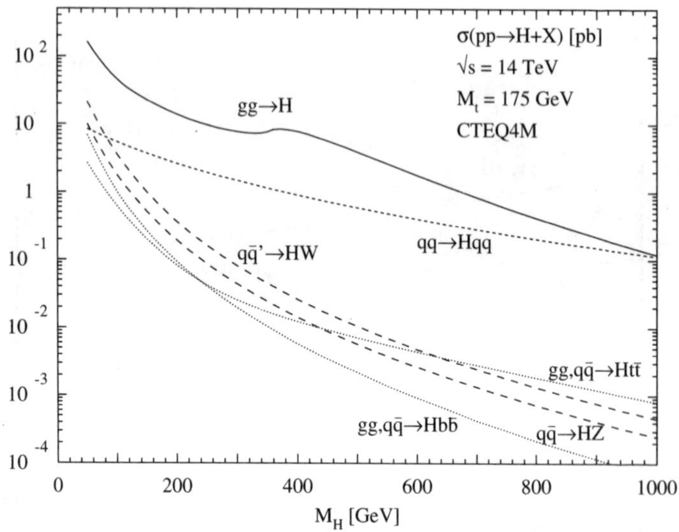

Figure 2.8: Cross sections for contributions of various mechanisms
to $pp \rightarrow HX$ at the LHC, taken from [39].

$H \rightarrow b\bar{b}$ decay are required, and, depending on the analysis, one or
both jets must be tagged as b jets. The Higgs mass is reconstructed
using the two b tagged jets or the two most energetic jets.

No Higgs signal has been seen, and the observations are consistent with the expected background events. It ought to be able
to put a limit of about 180 GeV by the end of its run. If the Higgs
is not found at Tevatron, the Large Hadron Collider (LHC), a pp
collider, should be able to explore the SM Higgs upto about 700
GeV.

If the Higgs is discovered at any of these machines, its properties could be studied in more detail and with greater accuracy at
a future e^+e^- linear collider with cm energy of 500 GeV or more,
which seems likely to be built in the coming years.

If the SM Higgs is not found upto the perturbative limit on its
mass, the implication would be that the Higgs sector is strongly
interacting, and interesting new physics would be waiting to be
discovered.

Bibliography

[1] M.S. Chanowitz, Ann. Rev. Nucl. Part. Sci. **38**, 323 (1988).

[2] R.N. Cahn, Rep. Prog. Phys. **52**, 398 (1989).

[3] R.N. Cahn in *"Où est le Higgs", Lecture Notes of L'École d'été de physique des particules*, 3–7 September 1990, Centre de Recherches Nucléaires de Strasbourg.

[4] J.F. Gunion, H.E. Haber, G. Kane and S. Dawson, *Higgs Hunter's Guide*, Addison-Wesley (1990).

[5] M.B. Einhorn (ed.), *The Standard Model Higgs Boson*, North-Holland (1991).

[6] J. Goldstone, Nuov. Cim. **19**, 154 (1961);
J. Goldstone, A. Salam and S. Weinberg, Phys. Rev. **127**, 965 (1962).

[7] P.W. Higgs, Phys. Lett. **12**, 132 (1964);
Phys. Rev. Lett. **13**, 508 (1964);
Phys. Rev. **145**, 1156 (1966);
F. Englert and R. Brout, Phys. Rev. Lett. **13**, 321 (1964);
G.S. Guralnik, C.R. Hagen and T.W.B. Kibble, Phys. Rev. Lett. **13**, 585 (1964);
T.W.B. Kibble, Phys. Rev. **155**, 1554 (1967).

[8] A.D. Linde, JETP Lett. **23**, 63 (1976);
Phys. Lett. **B62**, 435 (1976).

[9] S. Weinberg, Phys. Rev. Lett. **36**, 294 (1976).

[10] S. Coleman and E. Weinberg, Phys. Rev. **D7**, 1888 (1973).

[11] H. Politzer and S. Wolfram, Phys. Lett. **B82**, 242 (1978);
Phys. Lett. **B83**, 421 (1979) (E).

[12] P.Q. Hung, Phys. Rev. Lett. **42**, 873 (1979).

[13] N. Cabibbo, L. Maiani, G. Parisi, and R. Petronzio, Nucl. Phys. **B158**, 295 (1979).

[14] A.D. Linde, Phys. Lett. **B70**, 306 (1977).

[15] M. Lindner, Zeit. Phys. **C31**, 295 (1986);
M. Sher, Phys. Rep. **C179**, 273 (1989);
M. Lindner, M. Sher, and H.W. Zaglauer, Phys. Lett. **B 228**, 139 (1989);
M. Sher, Phys. Lett. **B317**, 159 (1993);
G. Altarelli and G. Isidori, Phys. Lett. **B337**, 141 (1994);
J. Espinosa and M. Quiros, Phys. Lett. **B353**, 257 (1995);
T. Hambye and K. Riesselmann, Phys. Rev. **D55**, 7255 (1997).

[16] E. Accomando *et al.*, in *Physics at LEP2*, vol. 1, p 351, CERN Yellow Report CERN-96-01 (1996).

[17] K.G. Wilson, Phys. Rev. **B4**, 3184 (1971);
K.G. Wilson and J. Kogut, Phys. Rep. **12**, 75 (1974).

[18] R. Dashen and H. Neuberger, Phys. Rev. Lett. **50**, 1897 (1983);
A. Hasenfratz, K. Jansen, C. Lang, T. Neuhas and H. Yoneyama, Phys. Lett. **B199**, 531 (1987);
J. Kuti, L. Liu and Y. Shen, Phys. Rev. Lett. **61**, 678 (1988);
M. Lüscher and P. Weisz, Nucl. Phys. **B318**, 705 (1989).

[19] D.A. Dicus and V.S. Mathur, Phys. Rev. **D7**, 3111 (1973).

[20] B.W. Lee, C. Quigg and H.B. Thacker, Phys. Rev. **D16**, 1519 (1977);
M. Chanowitz and M.K. Gaillard, Nucl. Phys. **B261**, 379 (1985).

[21] M. Veltman, Acta Phys. Pol. **B8**, 475 (1977);
Phys. Lett. **B70**, 253 (1977).

[22] M.B. Einhorn and J. Wudka, Phys. Rev. **D39**, 2758 (1989).

[23] The LEP and SLC Working Groups, hep-ex/0112021.

[24] J. Ellis, M.K. Gaillard and D.V. Nanopoulose, Nucl. Phys. **B106**, 292 (1976).

[25] E. Braaten and J.P. Leveille, Phys. Rev. **D22**, 715 (1980); N. Sakai, Phys. Rev. **D22**, 2220 (1980); T. Inami and T. Kubota, Nucl. Phys. **B179**, 171 (1981).

[26] B.W. Lee, C. Quigg and H.B. Thacker, Phys. Rev. Lett. **38**, 883 (1977); Phys. Rev. **D16**, 1519.

[27] H. Leutwyler and M.A. Shifman, Phys. Lett. **B221**, 384 (1989).

[28] A. Vainshtein, M.B. Voloshin, V.I. Zakharov and M.A. Shifman, Sov. J. Nucl. Phys. **30**, 711 (1979); A. Vainshtein, V.I. Zakharov and M.A. Shifman, Sov. Phys. Usp. **23**, 429 (1980); M. Voloshin, Sov. J. Nucl. Phys. **44**, 478 (1986).

[29] B.A. Kniehl and M. Spira, Z. Phys. **C69**, 77 (1995).

[30] B.A. Kniehl, Int. J. Mod. Phys. **A17**, 1457 (2002).

[31] J.D. Bjorken, Stanford report, SLAC-198 (1976).

[32] B.L. Ioffe and V.A. Khoze, Sov. J. Part. Nucl. **9**, 50 (1978).

[33] R.L. Kelly and T. Shimada, Phys. Rev. **D23**, 1940 (1981).

[34] S. Dawson, Nucl. Phys. **B249**, 42 (1985); G.L. Kane, W.W. Repko and W.B. Rolnick, Phys. Lett. **B148**, 367 (1984); R.M. Godbole and S.D. Rindani, Phys. Lett. **B190**, 192 (1987); Z. Phys. Lett. **C36**, 395 (1987).

[35] ECFA/DESY LC Physics Working Group (J.A. Aguilar-Saavedra et al.), *TESLA: The Superconducting electron*

positron linear collider with an integrated x-ray laser laboratory. Technical design report. Part 3. Physics at an e+ e-linear collider, hep-ph/0106315.

[36] M. Carena *et al.*, Fermilab-Conf-00/279-T, hep-ph/0010338.

[37] R.N. Cahn and S. Dawson, Phys. Lett. **B136**, 196 (1984).

[38] M. Spira, hep-ph/9810289.

[39] M. Spira, Contribution to the proceedings of the International Europhysics Conference on High-Energy Physics, 19–26 August 1997, Jersusalem, Israel, hep-ph/9711394.

Chapter 3

CP-violation: A Pedagogical Introduction

Biswarup Mukhopadhyaya

3.1 Introduction

The role of symmetries in the study of elementary particles can hardly be over-emphasised. While complex and elaborate calculations are often required for the prediction and interpretation of various phenomena, equally often do we find useful guidelines, either on the underlying structure of the theory or on the mutual relation among various types of phenomena, from considerations of symmetry.

Technically, a symmetry means the existence of some transformation which keeps the action of the theory (defined as $\int \mathcal{L} d^4 x$ where $\mathcal{L}$ is the Lagrangian density) invariant. Using symmetries, we can relate the rates of different processes, since symmetry operations relate states consisting of particles whose corresponding fields from the Lagrangian. A symmetry usually has a conserved charge associated with it, which forbids the transition between certain states.

Two kinds of symmetry can be defined in general: discrete and continuous, depending on the operation under which invariance of the theory is sought. Since space-time enter into the Lagrangian implicitly, it is interesting to investigate how symmetry operations on space-time influence the underlying physics.

Symmetry operations usually form groups. The most obvious example of a continuous space-time symmetry is the Poincare group, which is generated by the Lorenz group generators together with the four-momentum. Discrete space-time symmetry finds its most common examples in space inversion or parity (P), and time reversal (T). Both of these transformations preserve the line-element ds^2 in Minkowski space. Another discrete symmetry operation which is discussed simultaneously with the above two is charge conjugation or particle-antiparticle transformation (C).

Our discussion is centred around the three discrete symmetries mentioned above. We begin by recalling that out of the four fundamental types of interactions known in nature so far, three (namely, electromagnetic, strong and weak interactions) are described by Lorenz invariant, renormalisable, local quantum field theories. The Lagrangians corresponding to all these interactions come under the purview of the CPT theorem which requires an invariance under the combined operations of C, P and T. In addition, electromagnetic and strong interactions conserve each of C, P, T, both in the structures of their respective theories and in experimental observation. Weak interaction, on the other hand, has long been known to violate C and P. Furthermore, parity violation in weak interaction, as we shall subsequently see, is universal in character, and is also 'maximal' *i.e.*, the weak interaction Lagrangian makes the maximum possible distinction between left- and right-handedness, the two archetypes connected by the parity transformation.

A further interesting discovery came in 1964, when CP-violation (*i.e.*, a phenomenon which could only come from non-invariance of the Lagrangian under the combined operations of C and P) was discovered in weak processes involving the neutral kaon (K-meson) system. Since we are still functioning under the framework of an underlying CPT-invariant theory, this was tantamount to the discovery of time reversal violation as well.

One may ask: what was so unusual about the non-invariance under T when the other two discrete symmetries were already violated? To answer this, we note that compared, for example, with parity violation, the phenomenon of CP-violation has the following remarkable differences:

- As opposed to maximal parity violation, CP-violation is extremely small (about 1 part in 1000).

- In contrast with the universal character of parity violation within the domain of weak interaction, CP-violation was initially observed in *only one system*, the neutral kaon. Although it has subsequently been observed in charged kaon decays as well as in the B-meson system, the phenomenon of CP-violation has by and large maintained an exclusive character, requiring considerable experimental finesse to uncover it.

The observation of CP-violation raises some more interesting issues. It has been an unsolved puzzle as to why our universe shows such a large excess of baryons over antibaryons. In other words, why are we, made of matter, do not get instantly annihilated through recombination with an equally abundant supply of antimatter. Such a baryon asymmetry necessitates a few conditions which include the violation of both C and CP. Chances are thus quite high that we owe our existence to the phenomenon of CP-violation.

Concerted efforts to observe CP-violation outside the kaon system have been successful in the B-factories at Stanford, USA and Tsukuba, Japan. The other curious possibility opened up by CP-violation is the existence of a non-vanishing electric dipole moment for the neutron, since the dipole moment operator violates time reversal and thus CP. The present experimental upper limit on such a dipole moment for the neutron is extremely small, but such a limit acts as Occam's razor in constraining new theories which predict CP-violation of larger magnitude than is predicted in the standard electroweak model of Glashow, Salam and Weinberg (henceforth to be called the GSW model or the Standard Model).

Although the Standard Model includes the possibility of CP-violation, a better understanding of the situation is still required. One does not have a final answer yet as to whether the mechanism allowed by the Standard Model is the only one responsible for CP-violation in nature, or whether some physics beyond the Standard Model is playing a role in it. The study of CP-violation in a larger class of phenomena can hopefully resolve this puzzle.

In these lectures, we shall try to understand the basic principles involved in CP-violation, and elucidate how the components built into the theory can lead to the relevant experimental observables. We shall mostly use the neutral kaon system for this, since this has been the fountainhead of the entire framework. In addition, we shall very briefly touch upon CP-violation in the B-meson system. As the starting point, however, we need to remind ourselves of the meaning of the symmetry operations C, P, T and how these operations affect physical observables. Then we shall examine how the GSW model incorporates the possibility of CP-violation, and embark on the implications in neutral meson systems.

I thank Partha Konar and Subhendu Rakshit for their invaluable help in preparing this manuscript.

3.2   Parity, charge conjugation and time reversal

These discrete symmetry transformations are *prima facie* expected to be unitary, since the norm of a state in Hilbert space should be left unaltered by them. There is a further subtlety in the case of time reversal, but we shall come to that later.

3.2.1   Parity P

Parity means a shift from a left-handed system of co-ordinates to a right-handed one. With three spatial dimensions, this corresponds to the reversal of each space coordinate. Thus we define

$$P : (t, \vec{x}) \rightarrow (t, -\vec{x}).$$

For any field ϕ, one can in general expect a transformation to ϕ^P under parity:

$$\phi(t, \vec{x}) \xrightarrow{P} \phi^P(t, -\vec{x}), \tag{3.1}$$

with

$$P^\dagger = P^{-1} = P.$$

The minus sign in the argument is really of no consequence, since the integration over the coordinates in the action makes $\vec{x}$ a dummy variable, and since the arguments of all fields in the parity-tranformed form have $-\vec{x}$ in place of $\vec{x}$, thereby causing at most an overall sign change in the entire Lagrangian. Moreover, if ϕ is an eigenstate of such an operator P, then $\phi^P(t, -\vec{x}) = P_\phi \phi(t, -\vec{x})$, ($P_\phi^2 = 1$). Such a state ϕ is said to be one of *definite parity*, the latter being odd or even, depending on whether P_ϕ is $+$ or $-$.

Parity of a two-particle system $(\phi_1 \phi_2)$ is given by

$$P_{\phi_1 \phi_2} = (-1)^\ell\, p_1 p_2, \tag{3.2}$$

where $(-1)^\ell$ is the orbital parity (coming from the angular part of the two-particle wave function) and p_1, p_2 are the intrinsic parities. We shall explicitly demonstrate this below in the case of spin-zero fields.

Take a scalar (or pseudoscalar) field ϕ and consider its transformation under parity:

$$\begin{aligned}
P\phi_s(t, \vec{x})P &= \eta\phi_s(t, -\vec{x}) \\
P\phi_P(t, \vec{x})P &= -\eta\phi_P(t, -\vec{x}),
\end{aligned} \tag{3.3}$$

where η is a phase factor. The freedom associated with the choice of η can be exploited in (re)-defining the intrinsic parity of a field.

Remember now the expansion:

$$\phi(x) = \sum_{\vec{p}} \left[a(\vec{p})e^{-ip\cdot x} + b^\dagger(\vec{p})e^{ip\cdot x} \right] \tag{3.4}$$

with $a = b$ for neutral fields.

The very definition of parity implies that for a scalar, respectively a pseudoscalar, (ignoring the phase freedom),

$$\text{for}\quad \phi_S : Pa(\vec{p})P = a(-\vec{p})$$
$$Pb(\vec{p})P = b(-\vec{p})$$
$$\text{for}\quad \phi_P : Pa(\vec{p})P = -a(-\vec{p})$$
$$Pb(\vec{p})P = -b(-\vec{p})$$

The same is true for $a^\dagger$ and $b^\dagger$. Note that $\vec{p}$ is reversed but $p \cdot x$ is unaffected by P. Therefore,

$$
\begin{aligned}
P|\phi(\vec{p_1})\phi(\vec{p_2})\rangle &= Pa^\dagger(\vec{p_1})a^\dagger(\vec{p_2})|0\rangle \\
&= Pa^\dagger(\vec{p_1})PPa^\dagger(\vec{p_2})PP|0\rangle \\
&= p_1 p_2 a^\dagger(-\vec{p_1})a^\dagger(-\vec{p_2})|0\rangle \\
&= p_1 p_2 |\phi(-\vec{p_1})\phi(-\vec{p_2})\rangle \\
&= p_1 p_2 (-)^\ell |\phi(\vec{p_1})\phi(\vec{p_2})\rangle,
\end{aligned}
$$

where we have to remember that $P|0\rangle = |0\rangle$. The above calculation shows how the orbital angular momentum affects the parity of a two-particle system.

Similarly, for a spin-1 field, parity can be defined in the following way (neglecting the phase η):

$$\text{vector:}\quad V_\mu(t, \vec{x}) \;\rightarrow\; V^\mu(t, -\vec{x})$$
$$\text{axial vector:}\quad A_\mu(t, \vec{x}) \;\rightarrow\; -A^\mu(t, -\vec{x}),$$

where the chosen metric is $(1, -1, -1, -1)$.

Noting that

$$V_\mu(x) = \sum_{\vec{p},\lambda} \left[a_\lambda(\vec{p})\epsilon_\mu^\lambda(\vec{p})e^{-ip\cdot x} + b_\lambda^\dagger(\vec{p})\epsilon_\mu^{\star\lambda}(\vec{p})e^{ip\cdot x} \right], \qquad (3.5)$$

and remembering that $\epsilon_\mu(\vec{p}) \rightarrow -\epsilon^\mu(\vec{p})$, one can obtain the following transformation properties under parity for the creation and annihilation operators:

$$
\begin{aligned}
Pa_V^\lambda(\vec{p})P &= -a_V^\lambda(-\vec{p}), \\
Pa_A^\lambda(\vec{p})P &= +a_A^\lambda(-\vec{p}).
\end{aligned}
\qquad (3.6)
$$

Next, let us consider the behaviour of Dirac spinors under parity. Considering such a spinor field ψ, and neglecting the phase factor, the rule for transformation is

$$
\begin{aligned}
\psi(t, \vec{x}) &\rightarrow \gamma^0 \psi(t, -\vec{x}), \\
\bar{\psi}(t, \vec{x}) &\rightarrow \bar{\psi}(t, -\vec{x}) \gamma^0.
\end{aligned}
\tag{3.7}
$$

To see the origin of such a transformation, take a look at the expansion of the field operator ψ:

$$
\psi(x) = \sum_{\vec{p},s} \left[a_s(\vec{p}) u_s(p) e^{-ip \cdot x} + b_s^\dagger(\vec{p}) v_s(p) e^{ip \cdot x} \right]
\tag{3.8}
$$

and use

$$
\begin{aligned}
P a_s(\vec{p}) P &= \eta_a a_s(-\vec{p}), \\
P b_s(\vec{p}) P &= \eta_b b_s(-\vec{p}).
\end{aligned}
\tag{3.9}
$$

Here $\eta_a^2, \eta_b^2 = \pm 1$ as two successive P's should correspond to a phase rotation through 2π, which takes ψ to ψ or $-\psi$.

Eq. 3.9 implies that

$$
P\psi(t, \vec{x}) P = \sum_{\vec{p},s} \left[\eta_a a_s(-\vec{p}) u_s(p) e^{-ip \cdot x} + \eta_b^\star b_s^\dagger(-\vec{p}) v_s(p) e^{ip \cdot x} \right].
$$

Let $\tilde{p} = (p^0, -\vec{p})$ such that $p \cdot x = \tilde{p}(t, -\vec{x})$.

$$
u(p) = \begin{pmatrix} \sqrt{p \cdot \sigma}\, \xi \\ \sqrt{p \cdot \bar{\sigma}}\, \xi \end{pmatrix} = \begin{pmatrix} \sqrt{\tilde{p} \cdot \bar{\sigma}}\, \xi \\ \sqrt{\tilde{p} \cdot \sigma}\, \xi \end{pmatrix},
\tag{3.10}
$$

where detailed forms for the spinor functions u and v are obtained in terms of the Dirac matrices

$$
\begin{aligned}
\sigma^\mu &= (\mathbf{1}, \bar{\sigma}), & \bar{\sigma}^\mu &= (\mathbf{1}, -\bar{\sigma}), \\
\gamma^\mu &= \begin{pmatrix} 0 & \sigma^\mu \\ \bar{\sigma}^\mu & 0 \end{pmatrix}, & \gamma^0 &= \begin{pmatrix} 0 & 1 \\ 1 & 0 \end{pmatrix},
\end{aligned}
\tag{3.11}
$$

which have the properties $p \cdot \sigma = \tilde{p} \cdot \bar{\sigma}$ and $p \cdot \bar{\sigma} = \tilde{p} \cdot \sigma$.

In terms of these matrices,

$$
u(p) = \gamma^0 \begin{pmatrix} \sqrt{\tilde{p} \cdot \bar{\sigma}}\, \xi \\ \sqrt{\tilde{p} \cdot \sigma}\, \xi \end{pmatrix} = \gamma^0 u(\tilde{p}).
\tag{3.12}
$$

Similarly,

$$v(p) = \begin{pmatrix} \sqrt{\tilde{p}\cdot\bar{\sigma}}\,\xi \\ -\sqrt{\tilde{p}\cdot\sigma}\,\xi \end{pmatrix} = -\gamma^0 \begin{pmatrix} \sqrt{\tilde{p}\cdot\bar{\sigma}}\,\xi \\ \sqrt{\tilde{p}\cdot\sigma}\,\xi \end{pmatrix} = -\gamma^0 v(\tilde{p}). \qquad (3.13)$$

This gives, $P\psi(t,\vec{x})P$ as:

$$\sum_{\vec{p},s} \left[\eta_a a_s(-\vec{p})\gamma^0 u_s(\tilde{p})e^{-i\tilde{p}\cdot(t,-\vec{x})} - \eta_b^\star b_s^\dagger(-\vec{p})\gamma^0 v_s(\tilde{p})e^{i\tilde{p}\cdot(t,-\vec{x})} \right],$$

which is a constant matrix times ψ if and only if $\eta_a = -\eta_b^\star$. Thus for the spinor field to have a definite parity, one requires $\eta_a\eta_b = -\eta_a\eta_a^\star = -\mathbf{1}$ $(\eta_a\eta_a^\star = 1)$, which implies

$$P\psi(t,\vec{x})P = \eta_a\gamma^0\psi(t,-\vec{x}). \qquad (3.14)$$

The relation $\eta_a\eta_b = -1$ has an interesting meaning. Although the actual value of the intrinsic parity of a spinor cannot be definitely known, the above relation tells us that particles and antiparticles should always have opposite intrinsic parities. This is further significant since any observable quantity related to fermions is expressed in terms of some of the so-called Dirac (fermion) bilinears, formed out of ψ and its conjugate. The properties η_a and η_b, as obtained above, allow us to assign the following definite parities to each of these bilinears:

$$
\begin{array}{rccc}
 & (t,\vec{x}) & \to & (t,-\vec{x}) \\
S: & \bar{\psi}_1\psi_2 & \to & \bar{\psi}_1\psi_2 \\
P: & \bar{\psi}_1\gamma_5\psi_2 & \to & -\bar{\psi}_1\gamma_5\psi_2 \\
V: & \bar{\psi}_1\gamma_\mu\psi_2 & \to & \bar{\psi}_1\gamma^\mu\psi_2 \\
A: & \bar{\psi}_1\gamma_\mu\gamma_5\psi_2 & \to & -\bar{\psi}_1\gamma^\mu\gamma_5\psi_2 \\
T: & \bar{\psi}_1\sigma_{\mu\nu}\psi_2 & \to & -\bar{\psi}_1\sigma^{\mu\nu}\psi_2,
\end{array}
\qquad (3.15)
$$

where $\sigma_{\mu\nu} = \frac{i}{2}[\gamma_\mu,\gamma_\nu]$.

As we have seen above, any term in the Lagrangian that apparently violates parity can be made parity-conserving by applying suitable phase rotations to different fields. What then does parity violation mean? To understand this, we have to remember that the same field can appear in more than one terms in the Lagrangian. It may so happen that the exercise of the phase freedom may restore even parity to one of the infringing terms, but alters another term (or set of terms) when the transformation is applied. Such situations, where $\mathcal{L}$ (or more precisely, the action) *cannot be kept invariant in spite of the phase freedom*, lead to the violation of parity.

As an example, consider the following Lagrangian containing a scalar field ϕ and a spinor ψ:

$$\mathcal{L} = \frac{1}{2}\partial^\mu\phi\partial_\mu\phi + V(\phi^2) + i\bar{\psi}\gamma^\mu\partial_\mu\psi - m\bar{\psi}\psi + \bar{\psi}(a + i\,b\,\gamma_5)\psi\phi.$$

It can be clearly seen that whichever way we choose the phase of the parity-transformed form of ϕ, either of the two terms proportional to a or b changes sign under parity, using the transformation rules discussed above. This is a case of parity violation in the interaction between the fermion and the scalar.

3.2.2 Charge conjugation C

Charge conjugation, denoted usually by the operator C, stands for particle-antiparticle replacement for free fields. In terms of the creation and annihilation operators, this can be looked upon as the transformation $a(\vec{p}) \leftrightarrow b(\vec{p})$ (with momenta etc. unchanged).

For spin-zero fields (scalar or pseudoscalar) charge conjugation implies $\phi(t,\vec{x}) \xrightarrow{C} \phi^\dagger(t,\vec{x})$ (upto a phase). Similarly, for a vector field: $V_\mu \rightarrow -V_\mu^\dagger$. Such a transformation may be justified by considering a photon whose interaction with charged particles is experimentally found to be C-invariant. This is achieved if, under C, $A_\mu \rightarrow -A_\mu$ because all charges reverse sign under conjugation.

For a Dirac spinor ψ undergoing charge conjugation, $\psi \rightarrow \psi^C$ where ψ^C should have the same equation of motion as ψ. It is straightforward to see that this is possible provided $\psi^C = C\bar{\psi}^T$ where $C^{-1}\gamma_\mu C = -\gamma_\mu^T$. A good choice for the operator C, satisfying the above property, is $C = i\gamma^2\gamma^0$.

Alternatively, we may impose the condition $Ca_s(\vec{p})C = -b_s(\vec{p})$ and use the properties

$$u_s(p) = -i\gamma^2 v_s^\star(\vec{p}) \tag{3.16}$$

and

$$v_s(p) = -i\gamma^2 u_s^\star(\vec{p}) \tag{3.17}$$

to obtain

$$\begin{aligned}
C\psi(t,\vec{x})C &= \sum_{\vec{p},s}\left[i\gamma^2\left\{b^s(\vec{p})v_s^\star(p)e^{-ip\cdot x} + a_s^\dagger u_s^\star(p)e^{ip\cdot x}\right\}\right] \\
&= i\gamma^2\psi^\star = i\gamma^2(\psi^\dagger)^T \\
&= i\gamma^2(\bar{\psi}\gamma^0)^T \\
&= i\gamma^2\gamma^0\psi.
\end{aligned}$$

Again, any study of the charge conjugation invariance (or otherwise) of a theory containing fermions involves the transformation properties of the Dirac bilinears. A straightforward application of the result derived above yields the following transformation rules for the bilinears:

$$\begin{aligned}
S: &\quad \bar{\psi}_1\psi_2 &\rightarrow&\quad \bar{\psi}_2\psi_1 \\
P: &\quad \bar{\psi}_1\gamma_5\psi_2 &\rightarrow&\quad \bar{\psi}_2\gamma_5\psi_1 \\
V: &\quad \bar{\psi}_1\gamma_\mu\psi_2 &\rightarrow&\quad -\bar{\psi}_2\gamma_\mu\psi_1 \\
A: &\quad \bar{\psi}_1\gamma_\mu\gamma_5\psi_2 &\rightarrow&\quad \bar{\psi}_2\gamma_\mu\gamma_5\psi_1 \\
T: &\quad \bar{\psi}_1\sigma_{\mu\nu}\psi_2 &\rightarrow&\quad \bar{\psi}_2\sigma_{\mu\nu}\psi_1.
\end{aligned} \tag{3.18}$$

3.2.3   Time reversal T

Time reversal corresponds to the transformation

$$\begin{aligned}
\vec{x} &\rightarrow \vec{x}, \\
t &\rightarrow -t, \tag{3.19}
\end{aligned}$$

hence, $p \cdot x = -p \cdot x$. In analogy with charge conjugation and parity, one has the immediate impulse to classify T as a unitary transformation. However, here the situation is somewhat different.

In order to see this difference, let us first assume T to be a unitary operator. Then, with any field operator $\psi(t, \vec{x}) = e^{iHt}\psi(0, \vec{x})e^{-iHt}$

$$T\psi(t, \vec{x})T = e^{iHt}T\psi(0, \vec{x})Te^{-iHt},$$

H being the Hamiltonian. It follows that, with $|0\rangle$ denoting the vacuum,

$$\begin{aligned}
T\psi(t, \vec{x})T|0\rangle &= e^{iHt}[T\psi(0, \vec{x})T]e^{-iHt}|0\rangle \\
&= e^{iHt}[T\psi(0, \vec{x})T]|0\rangle \\
&= e^{iHt}\psi^T(0, \vec{x})|0\rangle.
\end{aligned}$$

But on the other hand one expects $T\psi(t, \vec{x})T|0\rangle = \psi^T(-t, \vec{x})|0\rangle = e^{-iHt}\psi^T(0, \vec{x})|0\rangle$, where ψ^T is the time-reversed form of the wave function. Thus one ends up with a self-contradictory prediction for the time evolution of the system.

The problem stems from the fact that T interchanges the initial and final states in any process. As a result, a proper prescription is required to reverse the time evolution of a field, which is controlled by a complex exponential function. The solution lies in treating T as a anti-unitary operator, which means $T(\text{c number}) = (\text{c number})^* T$. With such a prescription,

$$T[e^{iHt}\psi(0, \vec{x})e^{-iHt}]T = e^{-iHt}[T\psi(0, \vec{x})T]e^{iHt} \qquad (3.20)$$

and consistency is restored on the whole.

The transformation properties of different fields under time reversal can be obtained following a procedure very similar to that in the case of parity. For a spinor field, for example, the property

$$\psi(t, \vec{x}) \rightarrow -\gamma^1\gamma^3\psi(-t, \vec{x}) \qquad (3.21)$$

follows from the demand

$$\begin{aligned}
Ta_s(\vec{p})T &= +a_{-s}(-\vec{p}) \\
Tb_s(\vec{p})T &= -b_{-s}(-\vec{p}), \qquad (3.22)
\end{aligned}$$

modulo an overall phase. Also, particles and antiparticles can be shown to have opposite T-transformation properties.

Using the above transformation rules, one can easily write down the time reversal rules from Dirac bilinears to be as follows:

$$
\begin{aligned}
(t, \vec{x}) &\rightarrow (-t, \vec{x}) \\
S: \quad \bar{\psi}_1\psi_2 &\rightarrow \bar{\psi}_1\psi_2 \\
P: \quad \bar{\psi}_1\gamma_5\psi_2 &\rightarrow -\bar{\psi}_1\gamma_5\psi_2 \\
V: \quad \bar{\psi}_1\gamma_\mu\psi_2 &\rightarrow \bar{\psi}_1\gamma_\mu\psi_2 \\
A: \quad \bar{\psi}_1\gamma_\mu\gamma_5\psi_2 &\rightarrow \bar{\psi}_1\gamma_\mu\gamma_5\psi_2 \\
T: \quad \bar{\psi}_1\sigma_{\mu\nu}\psi_2 &\rightarrow -\bar{\psi}_1\sigma_{\mu\nu}\psi_2
\end{aligned}
\tag{3.23}
$$

3.2.4　Basic principle (P, C or T)

At this stage it may be helpful to clearly understand the meaning of violation of one of the above symmetries. The phase freedom always exists in defining the transformation properties of various fields. But if the Lagrangian contains more than one terms involving a given field (or a set of fields), which transform differently under a particular symmetry operation, then one term or other must violate that symmetry, no matter how we 'rephase' the fields. That is the situation when the corresponding symmetry is said to have been violated.

As an example, consider the following interaction Lagrangian for a spinor ψ with a spin-one field V^μ:

$$
\mathcal{L} = \bar{\psi}\gamma_\mu \frac{1 - \gamma_5}{2} \psi V^\mu.
\tag{3.24}
$$

It is straightforward to see that this interaction cannot be made invariant under C and P, whichever way you choose the phase in the transformation of V^μ (because both $\bar{\psi}\gamma_\mu\psi$ and $\bar{\psi}\gamma_\mu\gamma_5\psi$ are involved).

Next, let us examine the consequence of 'symmetry breaking' in the above sense. Taking a symmetry operation $\mathcal{O}$ ($\mathcal{O} = C, P$ or CP), consider the transition matrix element for $i \rightarrow f$ (The transition matrix T_s is defined by the relation $S = \mathbf{1} + iT_s$. This matrix element is given by

$$
\mathcal{M} = \langle f | T_s | i \rangle.
\tag{3.25}
$$

Invariance under $\mathcal{O}$ means $\mathcal{O}^\dagger T_s \mathcal{O} = T_s$, and therefore

$$\langle f|T_s|i\rangle = \langle f|\mathcal{O}^\dagger T_s \mathcal{O}|i\rangle = \langle f^0|T_s|i^0\rangle. \tag{3.26}$$

The above equality is not satisfied if the corresponding symmetry is broken. Thus the consequence of symmetry breaking is that the amplitude between two states is not equal to that between the states transformed by the symmetry operation.

But if $\langle f|T_s|i\rangle$ and $\langle f^o|T_s|i^0\rangle$ differ only by a phase, it does not show up in the transition probability, and violation of the symmetry is in practice unobservable. In order to retain observability of symmetry breaking, we therefore require that unless the matrix element between the transformed states is different in its absolute value, *a part of the T_s element should change under $\mathcal{O}$*. Interference between different terms in the amplitudes then shows up in the observable quantities, such as the transition probability.

As an example, consider the decay of a negatively charged pion into a muon and an antineutrino. The observed fact that only a left-handed μ^- is produced in this process can be translated into the statement

$$\mathcal{M}(\pi^- \to \mu_L^- \bar\nu_\mu) \neq \mathcal{M}(\pi^- \to \mu_R^- \bar\nu_\mu) = 0.$$

The origin of this can be traced to the appearance of the pion decay matrix element, namely

$$\mathcal{M} = f_\pi[\bar\mu\gamma_\mu(1 - \gamma_5)\nu_\mu]p_\pi^\mu, \tag{3.27}$$

which in turn follows from the $(V-A)$ character of the interaction lagrangian. Here, one of the two terms proportional to γ_μ and $\gamma_\mu\gamma_5$ must violate parity, in spite our freedom of 'rephasing' the fields.

As a corollary, one may say that if the states $|i\rangle$ and $|f\rangle$ have different eigenvalues $(1, -1)$ for $\mathcal{O} = C, P$ and CP, then $\mathcal{O}$ invariance implies

$$\begin{aligned}
\langle f|T_s|i\rangle &= \langle f|\mathcal{O}^\dagger T_s \mathcal{O}|i\rangle \\
\langle f^0|T_s|i^0\rangle &= -\langle f|T_s|i\rangle \\
i.e., \quad \langle f|T_s|i\rangle &= 0,
\end{aligned} \tag{3.28}$$

which implies that the transition $i \rightarrow f$ is not possible.

A different condition, however, follows for time reversal. Using the antiunitary character of T, it is straightforward to see that the time reversal invariance means

$$\langle f|T_s|i\rangle = \langle i^T|T_s|f^T\rangle \qquad (3.29)$$

In other words, the transition matrix elements between the original and T-transformed states are equal after the initial and final states are exchanged.

3.3   A complex phase in the Lagrangian causes CP-violation

The central idea in studying CP-violation is that it is caused by the existence of a complex phase in the Lagrangian. We first wish to convince the reader of the truth in this statement. One way of looking at it is via the CPT theorem. Without entering into a rigorous discussion of this theorem, it is sufficient here to note that a Lorentz invariant local field theory is also invariant under the combined operations of C, P and T. Since we have already seen that time reversal involves a simultaneous conjugation of all complex phases, a Lagrangian containing phases is clearly not time-reversal invariant. The CPT-theorem then implies that there should be a concomitant violation of CP as well, so that overall CPT invariance can be ensured.

The important point to note in the above argument is that any complex phase occurring as a c-number in the Lagrangian gets reversed under T. But there is no conjugation of the fields which are treated as operators here.

Now, exactly the reverse argument can be applied when one is thinking in terms of the successive operations of C and P. The fields and not the complex term multiplying them are conjugated under C (and CP). When there are complex coefficients multiplying the fields in the Lagrangian, there is a palpable change in the phase relationships of the fields and the coefficients in different terms, which in turn implies CP-violation.

To understand this more clearly, consider the Lagrangian

$$\mathcal{L} = h\phi\bar{\psi}_1\psi_2 + h^*\phi^*\bar{\psi}_2\psi_1 + g\left(V_\mu\bar{\psi}_1\gamma^\mu\psi_2 + V_\mu^\dagger\bar{\psi}_2\gamma^\mu\psi_1\right), \quad (3.30)$$

where ϕ is a complex scalar, V a vector field, ψ a spin-1/2 field and h corresponds to a complex coupling. Under the successive operations of parity and charge conjugation, the first two terms transform as follows:

$$h\phi\bar{\psi}_1\psi_2 \xrightarrow{P} h\phi\bar{\psi}_1\psi_2 \xrightarrow{C} h\phi^*\bar{\psi}_2\psi_1,$$
$$h^*\phi^*\bar{\psi}_2\psi_1 \xrightarrow{P} h^*\phi^*\bar{\psi}_2\psi_1 \xrightarrow{C} h^*\phi\bar{\psi}_1\psi_2.$$

Hence,

$$\mathcal{L}_h \xrightarrow{CP} h\phi^*\bar{\psi}_2\psi_1 + h^*\phi\bar{\psi}_1\psi_2 \qquad (3.31)$$

The Lagrangian can be seen to transform in the same way upon T-transformation. If h were not complex, hermiticity would have restored CP and T invariance. It is furthermore not possible to restore symmetry by 'rephasing' the fields, because then that phase will show up in the interaction with V_μ, again destroying invariance of the Lagrangian.

Thus hermiticity, coupled with the antiunitary character of T and the requirement of complex conjugation of fields under C, lead us to conclude that a complex phase in the coefficient of some term in the Lagrangian is a necessary and sufficient condition for CP- and T-violation.

3.4 CP-violation in the Glashow-Salam-Weinberg model

The standard electroweak model named after Glashow, Salam and Weinberg (GSW), has no C- or P-violation in the gauge and Higgs sectors. It is in the interaction of fermions that the non-invariance under parity and charge-conjugation becomes manifest, the first of these underlining the *left-handed* nature of weak interaction. We, therefore, begin by considering the coupling of fermions with both of the above sectors. As far as the basic interactions are concerned, the forms are identical for both the quark and lepton sectors; however, quarks exhibit the additional phenomenon of mixing for which a conclusive evidence is yet to emerge in the domain of leptons (although recent data on neutrinos strongly suggest similar phenomena there as well). In view of the above, we start by considering just the first family of quarks:

$$Q_L = \begin{pmatrix} u_L \\ d_L \end{pmatrix}, \quad u_R, \quad d_R.$$

After spontaneous symmetry breaking takes place, the gauge and Yukawa (*i.e.*, Higgs) interactions of these quarks, with the $SU(2) \times U(1)$ neutral gauge bosons written in the basis (W_3, B), are given by

$$\mathcal{L}(\text{gauge-Yukawa}) = \bar{Q}_L \gamma^\mu \left(i\partial_\mu + \frac{g_2 \sigma^i}{2} W_\mu^i + \frac{g_1}{6} B_\mu \right) Q_L$$

$$+ \bar{u}_R \gamma^\mu \left(i\partial_\mu + \frac{2g_1}{3} B_\mu \right) u_R$$

$$+ \bar{d}_R \gamma^\mu \left(i\partial_\mu - \frac{g_1}{3} B_\mu \right) d_R \qquad (3.32)$$

$$+ h^u (\bar{u}_L u_R + \bar{u}_R u_L)(v + \phi)$$

$$+ h^d (\bar{d}_L d_R + \bar{d}_R d_L)(v + \phi).$$

Here ϕ is the physical Higgs field and its vacuum expectation value (vev) is denoted by v.

The last two terms of the above equation can be written as

$$h_u (v + \phi)\bar{u}u + h_d (v + \phi)\bar{d}d \qquad (3.33)$$

The fact that gauge bosons belong to real representations of the gauge group restricts the gauge couplings to be real. Also, any phase in h_u or h_d can be absorbed by redefining u_R / d_R. Such redefinition should not show up in any physical observable, since the right-handed quarks participate only in neutral current (weak or electromagnetic) interactions, and any additional phase in them is liable to get canceled. Thus there is no net CP-violating phase in the set of interactions involving the quarks.

After introducing the 'physical' gauge bosons Z_μ and A_μ, the above Lagrangian is rewritten as

$$
\begin{aligned}
\mathcal{L} &\equiv \mathcal{L}_W + \mathcal{L}_Z + \mathcal{L}_\gamma + \mathcal{L}_Y \\
&= \frac{g_2}{\sqrt{2}} \bar{u} \gamma^\mu \frac{1 - \gamma_5}{2} d W_\mu + \text{h.c.} \\
&\quad + \frac{g}{\cos \theta_W} \left[\bar{u} \gamma_\mu (a + b \gamma_5) u Z_\mu + \bar{d} \gamma_\mu (a' + b' \gamma_5) d Z_\mu \right] \\
&\quad + e \left(q_u \bar{u} \gamma_\mu u + q_d \bar{d} \gamma^\mu d \right) A_\mu \\
&\quad + h_u (v + \phi) \bar{u} u + h_d (v + \phi) \bar{d} d,
\end{aligned}
\tag{3.34}
$$

where, $q_u = \frac{2}{3}$, $q_d = -\frac{1}{3}$ and a, b, a', b' are functions of Q and T_3. Clearly, both $\mathcal{L}_\gamma$ and $\mathcal{L}_Y$ are invariant under C and P.

Now, take a look at $\mathcal{L}_W$. With

$$
\mathcal{L}_W \sim \bar{u} \gamma^\mu (1 - \gamma_5) d W_\mu + \bar{d} \gamma^\mu (1 - \gamma_5) u W_\mu^\dagger,
\tag{3.35}
$$

under P it becomes

$$
\mathcal{L}_W^P \sim \bar{u} \gamma_\mu (1 + \gamma_5) d W_\mu + \bar{d} \gamma_\mu (1 + \gamma_5) u W_\mu^\dagger.
\tag{3.36}
$$

A further application of C yields

$$
\mathcal{L}_W^C \sim -\bar{d} \gamma^\mu (1 + \gamma_5) u W_\mu^\dagger - \bar{u} \gamma^\mu (1 + \gamma_5) d W_\mu.
\tag{3.37}
$$

On the other hand, successive operations of C and P lead to

$$
\mathcal{L}_W^{CP} \sim -\bar{u} \gamma_\mu (1 - \gamma_5) d W_\mu - \bar{d} \gamma_\mu (1 - \gamma_5) u W_\mu^\dagger.
\tag{3.38}
$$

Thus the one-generation Standard Model violates C and P but conserves CP. A similar conclusion follows on examining the Z-interaction terms as well.

The *central point here is that V- and A-type bilinears have opposite transformation properties both under C and P, and her-miticity takes care of the change caused by charge conjugation.* That is how CP-conservation is ensured under the combined operations, which can also be seen clearly from the fact that the interaction terms contain no complex phase after all.

We have already learnt that CP-invariance is broken if and only if there are complex factors multiplying the charged current terms in $\mathcal{L}_W$. Such phases will be reversed under hermitian conjugation but not under C, breaking down the invariance under CP, as discussed in the previous paragraph.

The question is, how can such complex phases arise? In the rest of this section, we shall find out the answer to this question, which basically lies in the fact that complexity is introduced through Yukawa couplings when we have not one but several families of quarks, as is indeed the situation in real life.

Let us examine the quark mass terms together with the Yukawa couplings:

$$
\begin{aligned}
\mathcal{L}_{(m+Y)} \quad &\sim \quad h_u(v+\phi)\bar{u}u + h_d(v+\phi)\bar{d}d \\
&= \quad h_u\phi(\bar{u}u + \bar{d}d) + m_u\bar{u}u + m_d\bar{d}d.
\end{aligned}
$$

The full Lagrangian, including the mass and kinetic terms, gauge and Yukawa couplings, will in general involve a sum over all quark families, of which three have been known to exist so far. Taking this sum into account, let $U^0 = \begin{pmatrix} u^0 \\ c^0 \\ t^0 \end{pmatrix}$, $D^0 = \begin{pmatrix} d^0 \\ s^0 \\ b^0 \end{pmatrix}$. U^0, D^0 define the basis in which $\mathcal{L}_W$ is flavour diagonal. This basis is given various names such as the flavour basis, the weak interaction basis or the current eigenstate basis.

In this basis,

$$
\begin{aligned}
\mathcal{L}_{W+\gamma+Z} \quad \sim \quad & \bar{U}^0\gamma_\mu \frac{1-\gamma_5}{2} D^0 W^\mu + \text{h.c.} \\
& + \left[\bar{U}^0\gamma_\mu(a+b\gamma_5)U^0 Z^\mu + \bar{U}^0\gamma_\mu U^0 A^\mu \right] \quad (3.39) \\
& + \left[U^0 \to D^0, a, b \to a', b' \right].
\end{aligned}
$$

Now consider $\mathcal{L}_{(m+Y)}$ in this basis:

$$
\mathcal{L}_{(m+Y)} = h_{ij}^u \bar{U}_i^0 U_j^0 (v+\phi) + h_{ij}^d \bar{D}_i^0 D_j^0 (v+\phi), \qquad (3.40)
$$

where h_{ij}^u, h_{ij}^d are not necessarily diagonal in this basis. Since the mass matrices $M^u = h^u v$ and $M^d = h^d v$ are proportional to the

Yukawa coupling matrices, the mass terms in the flavour basis are also non-diagonal:

$$\mathcal{L}_m = \bar{U}_i^0 M_{ij}^u U_j^0 + \bar{D}_i^0 M_{ij}^d D_j^0. \tag{3.41}$$

In general, M_{ij}^u is not proportional to M_{ij}^d and each is non-zero for $i \neq j$. In this sense, the fields U^0, D^0 are not "physical", since they do not propagate as states with definite masses. In order to obtain the corresponding "physical" quark fields, we need to diagonalise the mass matrices and go to the "mass eigenstate" basis.

$$V_L^u M^u V_R^{u\dagger} = \hat{M}_u, \qquad V_L^d M^d V_R^{d\dagger} = \hat{M}_d,$$

where the V's are unitary matrices.

Now,

$$\begin{aligned} \mathcal{L}_m &= \bar{U}^0 V_L^{u\dagger} \hat{M}^u V_R^u U^0 + \bar{D}^0 V_L^{d\dagger} \hat{M}^d V_R^d D^0 \\ &= \bar{U} \hat{M}^u U + \bar{D} \hat{M}^d D \end{aligned} \tag{3.42}$$

and one has diagonal mass terms. Thus the mass eigenstates — or "physical fields" — are defined as

$$U = V_L^u U^0, \qquad D = V_L^d D^0.$$

The charged current lagrangian can be written in terms of these fields as

$$\begin{aligned} \mathcal{L}_W &= \frac{g}{\sqrt{2}} \bar{U}^0 \gamma_\mu \frac{1 - \gamma_5}{2} D^0 W^\mu + \text{h.c.} \\ &= \frac{g}{\sqrt{2}} \bar{U} V_L^u \gamma_\mu \frac{1 - \gamma_5}{2} V_L^{d\dagger} D W^\mu + \text{h.c.} \\ &= \frac{g}{\sqrt{2}} (V_L^u V_L^{D\dagger})_{ij} \bar{U}_i \gamma_\mu \frac{1 - \gamma_5}{2} D_j W_\mu + \text{h.c.} \end{aligned} \tag{3.43}$$

where the 3×3 unitary matrix $K_{ij} = (V_L^u V_L^{D\dagger})_{ij}$ is known as the Cabibbo-Kobayashi-Maskawa (CKM) matrix. It diagonalises M_u in a basis in which M_d is already diagonal. It is of course not necessary for M_u and M_d to be real; thus K can naturally contain complex phases. We can also see from above that it is possible to have off-diagonal charged current interactions between mass eigenstates, or, in other words, weak interaction allows a transition between quark families if we write them in terms of the mass eigenstates.

It also follows from above that no mixing matrix like K occurs in Z and γ couplings, since the interactions are proportional to $\bar{D}D$ and $\bar{U}U$ in the flavour space and $V_{L(R)}^i{}^\dagger V_{L(R)}^i = \mathbf{1}$, $(i = u, d)$. Therefore, neutral current interaction is diagonal in the mass eigenstate basis as well. Such absence of tree-level flavour changing neutral currents (FCNC) is a notable feature of the Standard Model, the underlying principle, as explained above, being known as the Glashow-Illiopoulos-Maiani (GIM) mechanism.

Using our earlier conclusion, it is now obvious that any complex phase(s) in the CKM matrix is a source of CP-violation in the GSW model. Before we convince ourselves that this is actually expected, let us take note of the following corollaries of the above results, which can easily be verified:

- K is the identity matrix if all U's or all D's are degenerate. This implies that there is no lepton mixing if neutrinos are massless.

- An $SU(2)$ singlet quark added to either the up or the the down sector will cause FCNC in the down sector, since it will have a different coupling with the Z compared to a doublet, and thus a gauge coupling matrix (diagonal, but not scalar) will be introduced in the flavour space, which will not commute with the diagonalising matrices V_L^u or V_L^d.

Now, let us once more shift our attention to the matrix K, and observe that K is an $n \times n$ unitary matrix if there are n families of quarks. Such a matrix has n^2 independent parameters, with $n(n-1)/2$ angles and $n(n+1)/2$ phases. Since three families of quarks have already been discovered, one should expect three angles and six phases in K. However, K always occurs in the interaction Lagrangian where it is placed between the up U- and down d-type quark fields whose phases can be redefined to absorb some of the CKM phases. Since there are six quarks altogether, they carry five physically relevant phases (an overall phase being immaterial). On 'rephasing', the CKM matrix is thus ultimately left with $(6 - 5) = 1$ complex phase, which is the sole CP-violating phase of the electroweak sector. It is straightforward to check at this point that a two-family scenario allows one to absorb all the phases of the mixing matrix via field redefinition, and that is why a single 'Cabibbo' angle suffices for the description of such a scenario, admitting of no CP-violation.

Complex Yukawa interactions, together with the existence of (at least) three families of fermions, thus emerge as the sources of CP-violation in the GSW theory. K can in general be parameterized in different ways which are physically equivalent to each other. The phase can occur is different elements of the matrix; the fact that these choices are physically equivalent can be understood by realising that all observable manifestations of CP-violation require interference of CP-conserving and CP-violating amplitudes, and that such amplitudes for any process will include more than one terms involving various CKM elements, whereby the phase, even though located differently in different conventions, will show up identically through interference terms.

A standard parameterisation consists in K expressed in terms of three angles, θ_{12}, θ_{23} and θ_{13}, together with the phase δ, in the following way:

$$K = \begin{pmatrix} c_{12}c_{13} & s_{12}c_{13} & s_{13}e^{-i\delta} \\ -s_{12}c_{23} - c_{12}s_{23}s_{13} & c_{12}c_{23} - s_{12}s_{23}s_{13}e^{i\delta} & s_{23}c_{13} \\ s_{12}s_{23} - c_{12}c_{23}s_{13}e^{i\delta} & s_{23}c_{12} - s_{12}c_{23}s_{13}e^{i\delta} & c_{23}c_{13} \end{pmatrix}$$

with

$$K_{us} \approx \sin\theta_c \approx 0.22 \quad \text{[Measured from } K\text{-decay]}$$
$$K_{ud} \approx 0.97 \quad \text{[Measured from } \beta\text{-decay]},$$

where $\sin\theta_c$ is the Cabibbo angle.

An alternative parameterisation which owes itself to Wolfenstein shows clearly the relative magnitudes of the different elements:

$$K = \begin{pmatrix} 1 - \frac{\lambda^2}{2} & \lambda & A\lambda^3(\rho - i\eta) \\ -\lambda & 1 - \frac{\lambda^2}{2} & A\lambda^2 \\ A\lambda^3(1 - \rho - i\eta) & -A\lambda^2 & 1 \end{pmatrix} + \mathcal{O}(\lambda^4), \quad (3.44)$$

where $\lambda \approx \sin\theta_c$. This shows that the mixing between the first two families is approximately given by the Cabibbo angle, whereas that between the second and third, and the first and third families is of the order of the square and cube of the angle respectively.

3.5   The unitarity triangles

The unitarity of K imposes orthonormality conditions on the various pairs of rows and columns. These lead to a set of equations constraining the elements of K, which correspond to triangular relationships in the complex plane. The resulting triangles (altogether six of them) are called unitarity triangles.

Consider, for example, the relation

$$K_{ud}K_{ub}^* + K_{cd}K_{cb}^* + K_{td}K_{tb}^* = 0. \qquad (3.45)$$

Thus we have a triangle defined by the three terms above in the complex plane. Such conditions, essentially arising from complex numbers or planar vectors adding to zero, enable us to draw some important conclusions concerning CP-violation, as described below.

- The area of the triangle is zero if all CKM elements are real (no CP-violation).

- The area of the triangle also acts as a measure CP-violation.

- Sides of a unitary triangle are 'rephasing' invariant.

- The angles of a unitarity triangle, named α, β and γ in the example given above, are expressible in terms of the relative phases of the products of CKM elements. These can be measured from the rates of various weak processes. Such

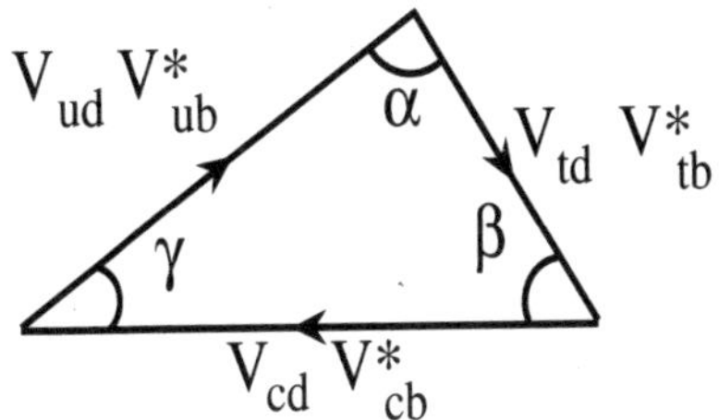

Figure 3.1: The unitarity triangle.

measurements not only give us crucial information on the magnitude of the CP-violating phase, but also allow us to determine whether the three-generation CKM phase is ultimately responsible for CP-violation in nature. If there are more families or additional CP-violating phases, the sides of the unitarity triangle will not close in the manner suggested above.

3.6 The neutral kaon system: where CP-violation was first observed

The neutral kaon states with definite strangeness are

$$|K^0\rangle = |\bar{s}d\rangle, \qquad |\bar{K}^0\rangle = |s\bar{d}\rangle, \tag{3.46}$$

with $s = 1$ and -1 respectively. Their CP-transformation properties are phase convention dependent but mutually related.

K^0 and $\bar{K}^0$ are not CP-eigenstates, which is obvious from the fact that quarks become antiquarks on charge conjugation. Let us use the convention

$$CP|K^0\rangle = -|\bar{K}^0\rangle, \qquad CP|\bar{K}^0\rangle = -|K^0\rangle.$$

Now, it is well-known that strangeness is not conserved in weak interaction. Supposing that CP is conserved (in weak interaction), we can try to construct CP-eigenstates in the following way, as linear combinations of K^0 and $\bar{K}^0$:

$$|K_1\rangle = -\frac{1}{\sqrt{2}}(|K^0\rangle + |\bar{K}^0\rangle), \qquad CP|K_1\rangle = -|K_1\rangle,$$

$$|K_2\rangle = -\frac{1}{\sqrt{2}}(|K^0\rangle - |\bar{K}^0\rangle), \qquad CP|K_2\rangle = +|K_1\rangle. \qquad (3.47)$$

If these were the 'physical' neutral K-states that decay weakly, the decay products would have to be CP-eigenstates. Experimentally, the physical states having definite decay modes were identified as K_L and K_S (where L and S stand for long and short), with mean lifetimes

$$\tau(K_L) \approx 5.0 \times 10^{-8}\,\text{sec},$$
$$\tau(K_S) \approx 0.9 \times 10^{-10}\,\text{sec}.$$

Their principal decay channels were initially identified as

$$\begin{aligned}
K_L &\rightarrow \pi^0\pi^0\pi^0, \pi^0\pi^+\pi^- \quad (CP = -1), \\
K_S &\rightarrow \pi^0\pi^0, \pi^+\pi^- \quad\;\; (CP = +1).
\end{aligned}$$

Thus everything was fitting in, with $|K_L\rangle = |K_1\rangle$, $|K_S\rangle = |K_2\rangle$ and CP known as a symmetry of nature, until in 1964, Christenson, Cronin, Fitch and Turley observed the decay $K_L \rightarrow \pi\pi$. In fact, if one defines the following quantities involving the amplitudes of the errant modes:

$$\eta_{+-} = \frac{\langle \pi^+\pi^- |T_{wk}|K_L\rangle}{\langle \pi^+\pi^- |T_{wk}|K_S\rangle}$$

$$\eta_{00} = \frac{\langle \pi^0\pi^0 |T_{wk}|K_L\rangle}{\langle \pi^0\pi^0 |T_{wk}|K_S\rangle},$$

then it is experimentally found that,

$$|\eta_{+-}| \approx (2.285 \pm 0.019) \times 10^{-3},$$
$$|\eta_{00}| \approx (2.275 \pm 0.019) \times 10^{-3}.$$

It may be remembered that if $CP|i\rangle = -|i\rangle$, $CP|f\rangle = +|f\rangle$, where $|i\rangle = |K_L\rangle$, and $|f\rangle = |\pi\pi\rangle$, then CP-conservation would imply

$$\langle f|T_{wk}|i\rangle = \langle f|(CP)^{-1}T_{wk}(CP)|i\rangle = -\langle f|T_{wk}|i\rangle,$$

giving $\langle f|T_{wk}|i\rangle = 0$.

Thus physicists were led to the unavoidable conclusion that *weak decays of neutral kaons violate CP*. There could be two possible ways in which this could happen:

- K_L is a mixture of $CP = +1$ and $CP = -1$ states (K_1, K_2) (indirect CP-violation).

- CP-violation takes place in the *decay process itself* (direct CP-violation).

We shall see in the following section that a careful measurement of η_{+-} and η_{00} can in principle reveal both these effects.

3.7   CP-violation in neutral kaons: observables

Let us start by considering the time evolution of a $K^0\bar{K}^0$-system:

$$i\frac{d}{dt}\begin{pmatrix} K^0(t) \\ \bar{K}^0(t) \end{pmatrix} = \begin{pmatrix} H_{11} & H_{12} \\ H_{21} & H_{22} \end{pmatrix}\begin{pmatrix} K^0(t) \\ \bar{K}^0(t) \end{pmatrix}, \qquad (3.48)$$

where H is the effective hamiltonian of the two-level system. The off-diagonal elements of H originate from weak interactions, since strangeness is not conserved there.

Thus, for example,

$$H_{11} = \langle K^0|H|K^0\rangle, \qquad H_{12} = \langle K^0|H|\bar{K}^0\rangle. \qquad (3.49)$$

CPT-invariance of the hamiltonian further implies that

$$\langle K^0|H|K^0\rangle = \langle K^0|(CPT)^{-1}H(CPT)|K^0\rangle = \langle \bar{K}^0|H|\bar{K}^0\rangle.$$

The hamiltonian H, representing an unstable system, in general contains a hermitian and an anti-hermitian part:

$$H = M - \frac{i}{2}\Gamma,$$

where M and Γ are the mass and decay matrices respectively. Thus if $\psi = \begin{pmatrix} K^0 \\ \bar{K}^0 \end{pmatrix}$,

$$\begin{aligned} \psi(t) &= e^{-iHt}\psi(0) = e^{-\frac{\Gamma}{2}}e^{-iMt}\psi(0) \\ |\psi(t)|^2 &= e^{-\Gamma}|\psi(0)|^2. \end{aligned} \qquad (3.50)$$

The hermiticity of M and Γ, together with the condition $H_{11} = H_{22}$, means

$$M_{11} = M_{22}, \qquad \Gamma_{11} = \Gamma_{22},$$
$$M_{21} = M_{12}^*, \qquad \Gamma_{21} = \Gamma_{12}^*.$$

A non-hermitian H in this case admits non-orthogonal eigenstates. Such eigenstates can be found on diagonalising H, the "physical" states thus obtained being

$$|K_{L(S)}\rangle = \frac{p|K^0\rangle \mp q|\bar{K}^0\rangle}{\sqrt{|p|^2 + |q|^2}}, \tag{3.51}$$

where,

$$\frac{q}{p} = \sqrt{\frac{M_{12}^* - i\Gamma_{12}^*/2}{M_{12} - i\Gamma_{12}/2}} = \sqrt{\frac{H_{21}}{H_{12}}}. \tag{3.52}$$

The corresponding eigenvalues are

$$M_{L(S)} = M \pm \mathrm{Re}\, Q, \qquad \Gamma_{L(S)} = \Gamma \pm 2\,\mathrm{Im}\, Q,$$

with

$$Q = \sqrt{\left(M_{12} - \frac{i\Gamma_{12}}{2}\right)\left(M_{12}^* - \frac{i\Gamma_{12}^*}{2}\right)} = \sqrt{H_{21} H_{12}}$$
$$\Delta M = 2\,\mathrm{Re}\, Q, \qquad\qquad \Delta\Gamma = 4\,\mathrm{Im}\, Q.$$

The following observations are in order:

- $K_S = K_1$, $K_L = K_2$ if $\frac{p}{q} = 1$.

- The quantities p and q, complex in general, depend on the phase convention. For example, a phase transformation $|K^0\rangle \to e^{i\phi}|K^0\rangle$, $|\bar{K}^0\rangle \to e^{-i\phi}|\bar{K}^0\rangle$ will change p and q.

- As measures of CP-violation, we need to find quantities that are independent of phase conventions and are directly related to observable quantities.

One such quantity satisfying the second criterion above is $\eta = |q/p|$. CP-conservation means that there is no phase in M_{12} or Γ_{12}, which in turn gives $\eta = 1$.

It follows from above that

$$
\begin{aligned}
\langle K_L | K_S \rangle &= \frac{\left(\langle K^0 | p^* - \langle \bar{K}^0 | q^* \right) \left(p | K^0 \rangle + q | \bar{K}^0 \rangle \right)}{p^2 + q^2} \\
&= \frac{|p|^2 - |q|^2}{|p|^2 + |q|^2} = \frac{1 - \eta^2}{1 + \eta^2}, \\
\eta &= 1 + \frac{2|\Gamma_{12}|^2}{4|M_{12}|^2 + |\Gamma_{12}|^2} \, \mathrm{Im} \left(\frac{M_{12}}{\Gamma_{12}} \right).
\end{aligned}
\tag{3.53}
$$

One can draw the following conclusions now:

- $\eta = |q/p| = 1$ means that there is no CP-violation through mixing.

- $(1 - \eta)$ is a measure of CP-violation (though we still need to relate it to η_{+-} and η_{00}).

- $\eta \neq 1$ implies that K_L and K_S are non-orthogonal and hence there is indirect CP-violation.

It is also straightforward to see that the time evolution of the system in terms of the mass eigenstates is given by

$$
K_{L(S)}(t) = e^{-\Gamma_{L(S)}t} e^{iM_{L(S)}t} K_{L(S)}(0).
\tag{3.54}
$$

Let us now go back to

$$
\begin{aligned}
\eta_{+-} &= \frac{\langle \pi^+ \pi^- | T_{wk} | K_L \rangle}{\langle \pi^+ \pi^- | T_{wk} | K_S \rangle} \\
\eta_{00} &= \frac{\langle \pi^0 \pi^0 | T_{wk} | K_L \rangle}{\langle \pi^0 \pi^0 | T_{wk} | K_S \rangle}
\end{aligned}
$$

and try to express them in terms of the elements of $(M - \frac{i\Gamma}{2})$. This will enable us to ultimately link the measures of CP-violation to the basic parameters (masses, angles and phase) of our underlying theory.

For such a purpose, we need to write the amplitudes for different processes in terms of K^0 and $\bar{K}^0$. We should also remember that the formation of the final state pions involves not only weak interactions but also strong interaction effects. The latter is in play when, say, two pions that are produced within a very short distance separate out. Such interactions are called final state interaction and they in general introduce additional phases (arising as phase shifts in the elastic scattering amplitudes). These are the so-called strong phases and are *not* reversed, unlike the CKM phase, on charge conjugation, a fact whose origin can be traced to the charge conjugation invariance of strong interactions. The contributions to the amplitudes (and the strong phase shifts) from such interactions depends on the total isospin (I) of the final state ($\pi\pi$ in our case). Therefore, we need to decompose the amplitudes into those for transitions to different states in the isospin basis.

A two-pion final state can have $I = 0, 2$, since one must have a symmetric state. With this in mind, we define

$$\langle(\pi\pi)_{I=0}|T_{wk}|K^0\rangle = a_0 e^{i\delta_0},$$
$$\langle(\pi\pi)_{I=2}|T_{wk}|K^0\rangle = a_2 e^{i\delta_2},$$

where δ_0, δ_2 are the strong phases for transition into the two allowed isospin states. Any weak phases are assumed to be contained within a_0 and a_2, *i.e.*,

$$a_0 = |a_0|e^{i\theta_0} \qquad a_2 = |a_2|e^{i\theta_2}. \tag{3.55}$$

Using $CP|K^0\rangle = -|\bar{K}^0\rangle$ and $(CPT)^{-1}H(CPT) = H$.

$$\langle(\pi\pi)_{I=0}|T_{wk}|\bar{K}^0\rangle = -a_0^* e^{i\delta_0},$$
$$\langle(\pi\pi)_{I=2}|T_{wk}|\bar{K}^0\rangle = -a_2^* e^{i\delta_2}.$$

CP-conservation implies $\theta_1, \theta_2 = 0$ or π. A relative complex (non-trivial phase between a_2 and a_0 can thus be linked CP-violation in the decay (direct CP-violation).

Now use Eq. 3.51, to obtain

$$\langle(\pi\pi)_{I=0}|T_{wk}|K^{L(S)}\rangle = \frac{e^{i\delta_0}}{\sqrt{|p|^2 + |q|^2}}(pa_0 \mp qa_0^*) = a_{0L(S)},$$

$$\langle(\pi\pi)_{I=2}|T_{wk}|K^{L(S)}\rangle \;=\; \frac{e^{i\delta_2}}{\sqrt{|p|^2+|q|^2}}(pa_2\mp qa_2^*) \;=\; a_{2L(S)}.$$

One can at the same time use the standard Clebsch-Gordon coefficients:

$$|\pi^0\pi^0\rangle \;=\; \frac{1}{\sqrt{2}}|\pi\pi\rangle_{I=0} - \sqrt{\frac{2}{3}}|\pi\pi\rangle_{I=2}$$

$$|\pi^+\pi^-\rangle \;=\; \frac{1}{2}\left(|\pi^+\pi^-\rangle+|\pi^-\pi^+\rangle\right) = \sqrt{\frac{2}{3}}|\pi\pi\rangle_{I=0} + \frac{1}{\sqrt{2}}|\pi\pi\rangle_{I=2},$$

thus obtaining

$$\eta_{+-} \;=\; \frac{\langle\pi^+\pi^-|T_{wk}|K_L\rangle}{\langle\pi^+\pi^-|T_{wk}|K_S\rangle} \;=\; \frac{\sqrt{2}a_{0L}+a_{2L}}{\sqrt{2}a_{0S}+a_{2S}},$$

$$\eta_{00} \;=\; \frac{\langle\pi^0\pi^0|T_{wk}|K_L\rangle}{\langle\pi^0\pi^0|T_{wk}|K_S\rangle} \;=\; \frac{a_{0L}-\sqrt{2}a_{2L}}{a_{0S}-\sqrt{2}a_{2S}}.$$

Expressing a_{0L}, a_{2L}, a_{0S} and a_{2S} in terms of a_2, a_0, θ_2, θ_0, p and q, we finally get:

$$\eta_{+-} \;=\; \frac{(1-y)[1+w\cos(\theta_2-\theta_0)]+i(1+y)w\sin(\theta_2-\theta_0)}{(1+y)[1+w\cos(\theta_2-\theta_0)]+i(1-y)w\sin(\theta_2-\theta_0)},$$

$$\eta_{00} \;=\; \frac{(1-y)[1-2w\cos(\theta_2-\theta_0)]-2i(1+y)w\sin(\theta_2-\theta_0)}{(1+y)[1-2w\cos(\theta_2-\theta_0)]-2i(1-y)w\sin(\theta_2-\theta_0)},$$

where $y = \frac{q}{p}e^{-2i\theta_0}$, $w = \frac{1}{\sqrt{2}}\left|\frac{a_2}{a_0}\right|e^{i(\delta_2-\delta_0)}$, ($|y|=\eta$). It should be noted here that CP can be violated indirectly (through mixing) even if $\theta_2 = \theta_0$,

Next we make use of some empirical inputs:

- Since $(1-y)$ is a measure of CP-violation, and $y=1$ corresponds to no (indirect) CP-violation, $|1-y| \ll 1$.

- K^0 has $I = \frac{1}{2}$, so that a_2 corresponds to $\Delta I = 3/2$ and a_0 corresponds to $\Delta I = 1/2$.

In nonleptonic kaon decays in general, $\Delta I = \frac{3}{2}$ is suppressed compared to $\Delta I = \frac{1}{2}$. This is the so-called $\Delta I = \frac{1}{2}$ rule.

To be more specific, it is observed that the decay rate for $K^+ \to \pi^+\pi^0$ is much smaller than that for $K_S^0 \to \pi^+\pi^-$. The former decay involve $I_{\text{initial}} = 0$, $I_{\text{final}} = 0$ or 2. Since $I_3 = 0$ for the final state, the only allowed possibility is $I_{\text{final}} = 2$, whence we know that the process corresponds to $\Delta I = 3/2$. In case of the neutral kaon decay, on the other hand, $I_{\text{initial}} = 0$ whereas I_{final} can be both 0 or 2. The immediate conclusion, therefore, is that the $\Delta I = \frac{1}{2}$ transition is responsible for the enhancement of the rate for $(K_S^0 \to \pi^+\pi^-)$, thereby vindicating the empirical rule, for which a likely explanation is larger QCD corrections for the corresponding final states.

General guidelines based on the $\Delta I = \frac{1}{2}$ rule tell us that $|w| \leq \frac{1}{25}$. Thus, in η_{+-} and η_{00}, the terms proportional to $(1-y)w$ can be dropped to a good approximation, and one can write

$$\eta_{+-} \approx \frac{1-y}{1+y} + \frac{iw\sin(\theta_2 - \theta_0)}{1 + w\cos(\theta_2 - \theta_0)} = \epsilon + \frac{\epsilon'}{1 + \frac{w}{\sqrt{2}}},$$

$$\eta_{00} \approx \frac{1-y}{1+y} - \frac{2iw\sin(\theta_2 - \theta_0)}{1 - 2w\cos(\theta_2 - \theta_0)} = \epsilon - \frac{2\epsilon'}{1 + \frac{w}{\sqrt{2}}},$$

where,

$$\epsilon = \frac{1-y}{1+y} = \frac{a_{0L}}{a_{0S}},$$

$$\epsilon' = \frac{i}{\sqrt{2}} e^{i(\delta_2 - \delta_0)} \frac{a_2}{a_0} \sin(\theta_2 - \theta_0).$$

With $w \ll 1$,

$$\eta_{+-} = \epsilon + \epsilon', \qquad\qquad \eta_{00} = \epsilon - 2\epsilon'$$

gives:

$$\epsilon = \frac{1}{3}(2\eta_{+-} + \eta_{00}), \qquad \epsilon' = \frac{1}{3}(\eta_{+-} - \eta_{00}). \tag{3.56}$$

ϵ and ϵ' are phase convention independent measures of CP-violation. ϵ gives the extent of CP-violation through mixing, contributing to both η_{+-} and η_{00}, while ϵ' is a measure of direct CP-violation, showing up in the splitting between η_{+-} and η_{00}.

Let us elaborate on ϵ. The expression

$$\epsilon \approx \frac{1-y}{1+y} = \frac{p - qe^{-2i\theta_0}}{p + qe^{-2i\theta_0}},$$

where p and q are expressed in terms of M_{12} and Γ_{12}. If one uses further empirical inputs, namely, $\Delta\Gamma \approx \Gamma_s \approx 2\Delta m$, $\mathrm{Im}\,\Gamma_{12} \ll \mathrm{Im}\,M_{12}$ and $\tan 2\theta_0 \ll \mathrm{Im}\,M_{12}/\mathrm{Im}\,M_{12}$, one obtains, after some algebra,

$$\epsilon \approx \frac{1}{\sqrt{2}} \frac{\mathrm{Im}\,M_{12}}{\Delta M} e^{i\pi/4}. \tag{3.57}$$

Using the additional information $\delta_2 - \delta_0 \approx \frac{\pi}{4}$ (obtained from pion scattering data)

$$\epsilon' \approx \frac{i}{\sqrt{2}} \left|\frac{a_2}{a_0}\right| \sin(\theta_2 - \theta_0) e^{i\pi/4}. \tag{3.58}$$

In the next section, we shall indicate how ϵ and ϵ' can be estimated in the Weinberg-Salam model.

3.8 CP-violating parameters from the Standard Model

3.8.1 The calculation of ϵ

Using elementary perturbation theory, we can write

$$\left(M - \frac{i\Gamma}{2}\right)_{12} = \frac{1}{2m_k}\left\langle K^0|H_w|\bar{K}^0\right\rangle$$
$$+ \sum_n \frac{1}{2m_k} \frac{\langle k^0|H_w|n\rangle\langle n|H_w|\bar{K}^0\rangle}{m_k - E_n + i\epsilon},$$

plus higher order terms. In the above, $|n\rangle$ represents all possible intermediate states with energy eigenvalues E_n and $\frac{1}{m_k - E_n + i\epsilon} = P\left(\frac{1}{m_k - E_n}\right) - i\pi\delta(E_n - m_k)$. Therefore,

$$\left(M - \frac{i\Gamma}{2}\right)_{12} = \frac{1}{2m_k}\left\langle K^0|H_w|\bar{K}^0\right\rangle$$

$$+\frac{1}{2m_k}P\left\{\sum_n \frac{\langle k^0|H_w|n\rangle\langle n|H_w|\bar{K}^0\rangle}{m_k-E_n}\right\}$$

$$-i\pi\sum_n \delta(E_n-m_k)\langle k^0|H_w|n\rangle\langle n|H_w|\bar{K}^0\rangle.$$

Equating the Hermitian and anti-Hermitian parts, we get

$$M_{12} = \frac{1}{2m_k}\Big[\langle K^0|H_w|\bar{K}^0\rangle$$

$$+ \sum_n \langle K^0|H_w|n\rangle\langle n|H_w|\bar{K}^0\rangle\, P\frac{1}{m_k-E_n}\Big],$$

$$\Gamma_{12} = 2\pi\sum_n \delta(E_n-m_k)\langle k^0|H_w|n\rangle\langle n|H_w|\bar{K}^0\rangle,$$

where Γ_{12} is the absorptive part of the $K^0\bar{K}^0$-transition matrix element and the second term in M_{12} is the 'dispersive part' (the long distance contribution, coming from intermediate hadronic states). The dispersive contribution to ϵ can be shown to be small.

Thus, Im $M_{12} \approx \langle K^0|H_w|\bar{K}^0\rangle$, where H_w is calculable from SM by obtaining the quark-level effective Hamiltonian. Re M_{12}, on the other hand, has large long-distance contributions. Consequently, $2\mathrm{Re}\, M_{12}$ is replaced by the measured value of the K_L-K_S mass difference in the expression for ϵ. For $\langle K^0|H_w|\bar{K}^0\rangle$, we obtain the lowest order quark-level effective interaction for $s\bar{d}\to\bar{s}d$, and take the matrix element by sandwiching the effective interaction between the K^0 and $\bar{K}^0$ states.

It has been shown in the literature that the dominant contributions come form the 'box diagrams' shown above. Thus, to a high degree of accuracy,

$$\epsilon = \frac{e^{1\pi/4}}{\sqrt{2}\Delta m}\frac{1}{2m_k}\,\mathrm{Im}\,\langle K^0|H_{\mathrm{eff}}(\mathrm{box})|\bar{K}^0\rangle. \tag{3.59}$$

The procedure to obtain the relevant matrix element is as follows:

- Calculate the amplitude from the Standard Model box diagrams.

- Replace the spinors (u,v) by the fields (s,d).

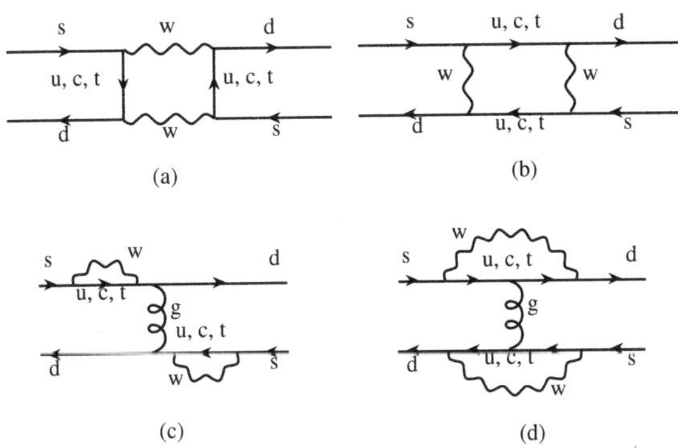

Figure 3.2: Feynman diagrams contributing to $K^0 \bar{K}^0$-mixing in the Standard Model.

- Take the dimension-six effective interaction term thus obtained, and extract its hadronic matrix element, whose imaginary part (induced by the phase in the CKM matrix) is relevant for our purpose.

On explicit evaluation (to be outlined presently), one has

$$H_{\text{eff}}(\text{box}) = I \bar{s}^\alpha \gamma_\mu (1 - \gamma_5) d^\alpha \bar{s}^\beta \gamma^\mu (1 - \gamma_5) d^\beta, \qquad (3.60)$$

with α, β are colour indices and I is the loop integral. Therefore,

$$\begin{aligned}
\left\langle K^0 | H_{eff} | \bar{K}^0 \right\rangle &= I \left\langle K^0 | \bar{s}^\alpha \gamma_\mu (1 - \gamma_5) d^\alpha \bar{s}^\beta \gamma^\mu (1 - \gamma_5) d^\beta | \bar{K}^0 \right\rangle \\
&= I \left\langle K^0 | O_{12} O_{34} | \bar{K}^0 \right\rangle \\
&= I \left\langle K^0 | O_{12} | i \right\rangle \left\langle i | O_{34} | \bar{K}^0 \right\rangle. \qquad (3.61)
\end{aligned}$$

The evaluation of the hadronic matrix element is, of course, model-dependent. Here we demonstrate the results in the so-called vacuum insertion approximation, which basically consists in the postulate

$$\Sigma |i\rangle\langle i| \approx |0\rangle\langle 0|, \qquad (3.62)$$

which essentially means that the dominant contribution comes from vacuum as the intermediate state.

Under this approximation,

$$M_{12} = \frac{I}{2m_K} \langle K^0 | \bar{s}^\alpha \gamma_\mu (1 - \gamma_5) d^\alpha | 0 \rangle \langle 0 | \bar{s}^\beta \gamma_\mu (1 - \gamma_5) d^\beta | \bar{K}^0 \rangle.$$

In addition, one needs to take into account the possibility of inserting the vacuum in all possible ways between the above current-current product, with every allowed colour combination. Thus, writing

$$\bar{s}^\alpha \gamma_\mu (1 - \gamma_5) d^\alpha | 0 \rangle \langle 0 | \bar{s}^\beta \gamma_\mu (1 - \gamma_5) d^\beta = O_{12} O_{34}$$

and remembering its Fierz transformation, namely

$$\bar{s}^\alpha \gamma_\mu (1 - \gamma_5) d^\beta | 0 \rangle \langle 0 | \bar{s}^\beta \gamma_\mu (1 - \gamma_5) d^\alpha = O_{14} O_{32},$$

we get:

$$\begin{aligned}
M_{12} &= \frac{I}{2m_K} (\langle K^0 | O_{12} | 0 \rangle \langle 0 | O_{34} | \bar{K}^0 \rangle \\
&\quad + \langle K^0 | O_{34} | 0 \rangle \langle 0 | O_{12} | \bar{K}^0 \rangle \\
&\quad + \langle K^0 | O_{34} | 0 \rangle \langle 0 | O_{12} | \bar{K}^0 \rangle \\
&\quad + \langle K^0 | O_{14} | 0 \rangle \langle 0 | O_{32} | \bar{K}^0 \rangle \\
&= \langle K^0 | O_{34} | 0 \rangle \langle 0 | O_{12} | \bar{K}^0 \rangle.
\end{aligned}$$

Since the only vector available to characterise the neutral kaon is its four-momentum p^μ (and since no axial vector can be constructed out of it, in the absence of any polarisation tensor for the scalar particle), we can write

$$\langle K^0 | \bar{s}^\alpha \gamma_\mu (1 - \gamma_5) d^\alpha | 0 \rangle = f_k p_\mu, \tag{3.63}$$

where f_k is the kaon decay constant. Thus

$$\langle K^0 | O_{12} | 0 \rangle \langle 0 | O_{34} | \bar{K}^0 \rangle = f_k^2 m_K^2, \tag{3.64}$$

and similarly for the term $O_{34} O_{12}$. For each of the Fierz transformed terms, however, there will be a colour suppression factor of

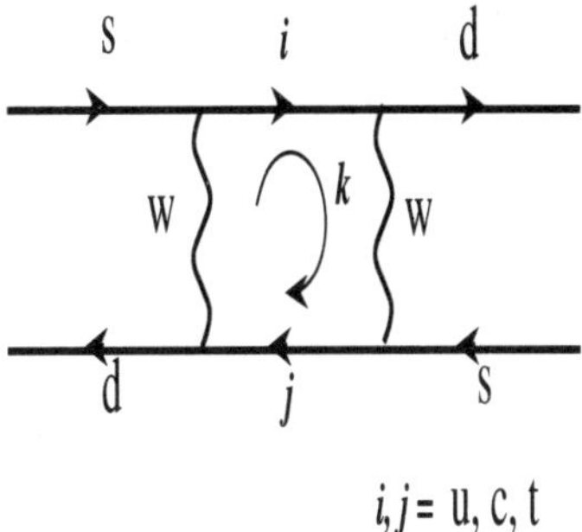

Figure 3.3: The notations relevant for the computation of a typical box diagram.

1/3 since colour is a conserved quantum number and the vacuum has to be colourless. Therefore, one finally obtains:

$$M_{12} = \frac{I}{2m_K} \frac{2.4}{3} f_k^2 m_K^2 = \frac{4}{3} f_K^2 m_K I \qquad (3.65)$$

in the vacuum insertion approximation.

The model dependence of this result can be encapsulated in the slightly generalised formula

$$M_{12} = \frac{4}{3} f_k^2 B_k m_k I, \qquad (3.66)$$

where B_k incorporates the uncertainties in the calculation of the hadronic matrix element. In most estimates, the value of B_k is predicted to lie between 0.3 and 1.2.

3.8.2   Evaluation of I

We demonstrate a few salient features of the calculation by taking a sample box diagram, as shown in the figure. All external momenta and masses are neglected in this calculation.

Simple power counting shows that the box diagram is finite. We shall evaluate the quark-level transition amplitude in the unitarity gauge. In that gauge, with the CKM matrix denoted by V, the box diagram amplitude is given by

$$
\begin{aligned}
A &= \sum_{i,j} \int \frac{d^4k}{(2\pi)^4} \left(\frac{ig}{\sqrt{2}}\right)^4 (V_{is}^* V_{id} V_{js}^* V_{jd})\, \bar{s}\gamma_\mu \frac{(1-\gamma_5)}{2} \frac{i}{\not{k}-m_i} \\
&\quad \times \gamma_\nu \frac{(1-\gamma_5)}{2} d\, \bar{s}\gamma_\alpha \frac{(1-\gamma_5)}{2} \frac{i}{\not{k}-m_j} \gamma_\beta \frac{(1-\gamma_5)}{2} d \\
&\quad \times \frac{i\left[-g^{\nu\alpha}+k^\nu k^\alpha/m_w^2\right]}{k^2-m_w^2} \frac{i\left[-g^{\mu\beta}+k^\mu k^\beta/m_w^2\right]}{k^2-m_w^2} \\
&= (\text{constant}) \times \sum_{i,j} \lambda_i \lambda_j \int d^4k \left[\bar{s}\gamma_\mu\, \not{k}\gamma_\nu \frac{(1-\gamma_5)}{2} d\right] \\
&\quad \times \left[\bar{s}\gamma_\alpha\, \not{k}\gamma_\beta \frac{(1-\gamma_5)}{2} d\right] \\
&\quad \times \frac{(-g^{\nu\alpha}+k^\nu k^\alpha/m_w^2)\left(-g^{\mu\beta}+k^\mu k^\beta/m_w^2\right)}{(k^2-m_i^2)\left(k^2-m_j^2\right)(k^2-m_w^2)^2},
\end{aligned}
$$

where $\lambda_i = V_{is}^* V_{id}$. The condition $V^\dagger V = 1$ implies $\sum_i \lambda_i = 0$. Therefore, any piece in A that does not depend on both i and j must vanish, just as a manifestation of the GIM mechanism mentioned earlier.

Next, replace $1/(k^2 - m_i^2)$ by

$$
\frac{1}{k^2-m_i^2} - \frac{1}{k^2-m_j^2} \equiv \frac{m_j^2-m_i^2}{(k^2-m_i^2)(k^2-m_j^2)}.
$$

It is obvious that the extra piece has no i-dependence and does not contribute since $\sum_i \lambda_i = 0$. A similar consideration can be applied to $1/(k^2 - m_j^2)$, leading to an equivalent description of the loop integral, with $\frac{1}{(k^2-m_j^2)(k^2-m_j^2)}$ replaced by

$$
\frac{1}{2}\left[\frac{1}{k^2-m_i^2} - \frac{1}{k^2-m_j^2}\right]\left[\frac{1}{k^2-m_j^2} - \frac{1}{k^2-m_i^2}\right],
$$

so that

$$
A = (\text{constant}) \times \sum_{i,j} \lambda_i \lambda_j \int d^4k \left(\bar{s}\gamma_\mu\, \not{k}\, \gamma_\nu \frac{1-\gamma_5}{2} d\right)
$$

$$\times \left(\bar{s}\gamma_\alpha \not{k} \gamma_\beta \frac{1-\gamma_5}{2} d \right)$$

$$\times \frac{\left(m_j^2 - m_i^2\right)^2 \left(-g^{\nu\alpha} + k^\nu k^\alpha/m_w^2\right) \left(-g^{\mu\beta} + k^\mu k^\beta/m_w^2\right)}{(k^2 - m_i^2)^2 \left(k^2 - m_j^2\right)^2 (k^2 - m_w^2)^2} ,$$

which is manifestly divergence-free.

The following points round up the calculation:

- $I = 0$, if $m_u = m_c = m_t$, *i.e.*, a complete degeneracy of the quarks running in the loops causes the amplitude to vanish. This is again a consequence of the GIM mechanism.

- Larger mass splitting among the quarks means larger contribution to I, which also means that heavier quarks contribute more. However, we also have the $\lambda_i \lambda_j$ as overall multiplicative factors. Thus, diagrams with heavier quarks can still give less contributions due to more CKM suppression. In practice, Re M_{12} indeed turns out to be charm-dominated, but Im M_{12} has the largest contribution from the top quark loop.

- The u-contribution is in any case negligible. Thus one finally obtains

$$I \simeq \left[\lambda_c^2 \eta_1 F(x_c, x_c) + \lambda_t^2 \eta_2 F(x_t, x_t) + 2\lambda_c \lambda_t \eta_3 F(x_c, x_t) \right],$$

where $x_i = (m_i^2/m_W^2)$ and $F(x_i, x_j)$ is the function arising out of the integral in the loop with the ith and jth quarks in the internal lines. Furthermore, the above contributions are subject to QCD corrections, thus acquiring the multiplicative factors η_i.

- Since $\epsilon \sim$ Im I and $\lambda_u + \lambda_c + \lambda_t = 0$, one can write Im $\lambda_c = -$Im λ_t since $\lambda_u = V_{us}^* V_{ud}^*$ can always be made real in the CKM matrix. Thus the value of ϵ is found to be proportional to Im λ_t. As has been mentioned above, this can be contrasted with the real part of the amplitude where the charm contribution dominated over the top-induced one since $\lambda_c \gg \lambda_t$.

- The above discussion also demonstrates that ϵ can be non-zero only because there are three families of quarks.

The present experimental value of ϵ is

$$\epsilon = (2.28 \pm 0.013) \times 10^{-3}. \qquad (3.67)$$

ϵ turns out to be a function of s_{12}, s_{23}, s_{13}, f_K, B_K, m_t and δ. It is the still persisting uncertainty in the values of the different parameters, particularly the 'bag factor' B_K, which prevents us from obtaining an estimate of the CP-violating phase δ in the CKM matrix. It is also to be noted that δ suffers from a 'quadratic ambiguity', *i.e.*, two values for it emerge by solving the equation in terms of the remaining parameters.

3.8.3 ϵ'/ϵ in the Standard Model

The other relevant quantity, characterising 'direct' CP-violation, is parameterised by ϵ'/ϵ, where ϵ' has been defined earlier. A non-zero value of this parameter means that η_{00} and η_{+-} are not equal. Its current experimental value is $\epsilon'/\epsilon = (28.0 \pm 4.1) \times 10^{-4}$. From equation (134), $\epsilon'/\epsilon \neq 0$ means:

(a) $a_2 \neq 0$, or in other words, the $\Delta I = 1/2$ rule operates in some way.

(b) $\sin(\theta_2 - \theta_0) \neq 0$, that is to say, a_2 and a_0 must not be relatively real.

Due to lack of space, we refrain from entering into a detailed discussion on the actual calculation of ϵ'/ϵ in the Glashow-Salam-Weinberg model. However, the different diagrams which contribute to the process $K_L \rightarrow \pi\pi$ are shown in Fig. 3.4.

It should be noted that both tree and loop diagrams are operative; however, the loop diagrams, popularly know as 'penguin diagrams', are required to ensure the relative complexity of a_2

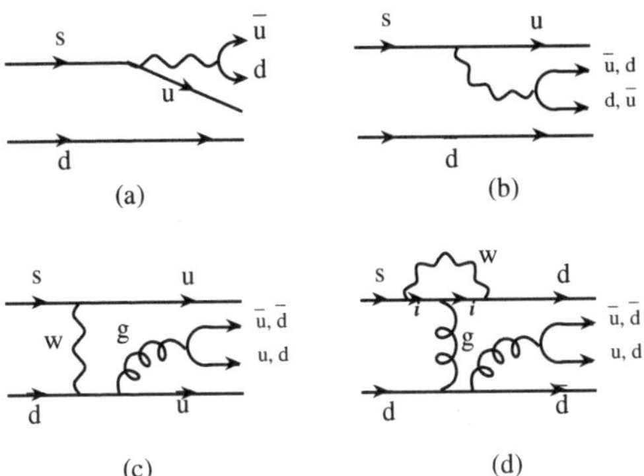

Figure 3.4: Contributions to ϵ'/ϵ in the Standard Model.

and a_0. This is because diagrams (a) and (b) (between which (b) is colour-suppressed) are proportional to λ_u as defined earlier, and are therefore are completely real. A similar statement can be made about diagram (c). Therefore, the contributions from (d) plays a crucial role in the Standard Model contribution. In addition, one has to take into account the 'electroweak penguins' where the gluons in (d) are replaced by the Z or the photon. Uncertainties related to hadronic matrix elements are of course unavoidable here also. And finally, the QCD corrections to ϵ'/ϵ are extremely important in relating its theoretical predictions to precision measurements, and in any search for the signature of physics beyond the Standard Model in direct CP-violation. A technical discussion of such corrections is beyond the scope of these lectures.

3.9 CP-violation in B-meson decays

Before we end this discussion, a few remarks on the possibility of CP-violation in the B-meson system are warranted, since it is on this system that the current experimental efforts are focused. While we shall shall give a brief outline which motivated such efforts, we consciously avoid entering into a detailed discussion, for which a separate course of lectures is certainly required.

The lowest-lying spin-0 mesons containing a b-quark are $B_d^0 = \bar{b}d$, $B_s^0 = \bar{b}s$, $B^+ = \bar{b}u$. Physicists are interested in studying the B-system to find out more about CP-violation for the following reasons:

- CP-violation requires a specific role of the *three-family character* of the quark mixing matrix. The K-system, studied earlier, contains the first two families only. To a first approximation, the FCNC processes involving them may not know about the third family which is connected with the sd-system only via small CKM elements. The B-meson system (B_d^0, for example) can give rise to processes involving all three quark families, where the role of the full 3×3 CKM matrix is much more central. Thus one may expect to see larger CP-violating effects in the B-system. Furthermore, if it happens to be the case that the CKM phase is actually large but the small effects seen in the K-system is due to the secondary role of the third family, then, too, one can expect to see enhanced effects by studying B-mesons.

- A study of the B-system yields additional information on CKM elements. Such information may complete the construction of the unitarity triangle. One may therefore say with a higher degree of confidence whether the CKM phase alone is responsible for all observed CP-violation, or whether new physics is responsible for it.

It is with such goals that the the so-called B-factories have started operation, the two most recent enterprises being the ones in Stanford, USA, and Tsukuba, Japan. Together with the Cornell Electron Synchrotron Ring (CESR), USA and B-related experiments at hadron machines (BTeV, LHC-B), these are expected to answer the questions asked above to a fair degree of precision.

3.9.1 Charged B-decays

Consider the decay

$$B^+ \to f \tag{3.68}$$

and the conjugate process

$$B^- \to \bar{f}. \tag{3.69}$$

If a CP-violating phase shows up in the decay matrix element, the amplitudes of the above processes are not the same. However, this does not necessarily mean an asymmetry in the corresponding rates. That is to say, if $\mathcal{M}_+ = A + Be^{i\theta}$, and $\mathcal{M}_- = A + Be^{-i\theta}$, we shall still have $|\mathcal{M}_+|^2 = |\mathcal{M}_-|^2$.

As we have already discussed, for an asymmetry to show up, it is necessary to have (a) interference of more than one diagrams, and (b) a phase which does not get reversed on CP-conjugation. This is possible only through the strong interaction phase shifts, or final-state interactions.

To understand the point, let us assume

$$\mathcal{A}(B^+ \to f) = a + a', \tag{3.70}$$

with

$$a = Me^{i\delta_w}e^{i\delta_s}, \qquad a' = M'e^{i\delta'_w}e^{i\delta'_s},$$

where δ_W (δ'_W) and δ_s (δ'_s) are respectively the weak phases and the strong interaction phase shifts for the two amplitudes.

Then

$$
\begin{aligned}
|\mathcal{A}(B^+ \to f)|^2 &= |a|^2 + |a'|^2 + 2\mathrm{Re}\,(a^*a') \\
&= M^2 + M'^2 + 2MM' \cos\left[(\delta_w - \delta'_w) + (\delta_s - \delta'_s)\right].
\end{aligned}
$$

Similarly, with

$$\mathcal{A}(B^- \to \bar{f}) = Me^{-i\delta_w}e^{i\delta_s} + M'e^{-i\delta'_w}e^{i\delta'_s},$$

one obtains:

$$|\mathcal{A}(B^- \to \bar{f})|^2 = M^2 + M'^2 + 2MM' \cos\left[(\delta_w - \delta'_w) - (\delta_s - \delta'_s)\right].$$

Thus, $\Gamma(B^+ \to f) - \Gamma(B^- \to \bar{f}) \equiv MM' \sin(\delta_w - \delta'_w)\sin(\delta_s - \delta'_s)$ and a rate asymmetry shows up.

The problem here is in the experimental extraction of all the quantities of relevance, before any definitive statements can be made on the CKM phase. From the above expressions, one basically needs to find the values of M, M', $(\delta_w - \delta'_w)$ and $(\alpha_s - \alpha'_s)$. This is not an easy proposition, given the different types of uncertainties underlying any process.

3.9.2   Neutral B-decay: cleaner grounds

The neutral B-system offers a relatively easier path to the study of CP-violation. Consider the $B^0_d - \bar{B}^0_d$ system, as these are the lowest-lying neutral B-mesons. Given the composition of B^0_d, $B^0_d \bar{B}^0_d$-mixing also can take place through box diagrams, much in the same way as the $K^0 - \bar{K}^0$ system. The only difference here is that the top quark contributions (which are by far the largest ones to the loop integrals) are favoured by a diagonal CKM element in one vertex each.

Analogous to the $K^0 \bar{K}^0$-system, the time evolution of the $B^0_d - \bar{B}^0_d$ system is governed by,

$$i\frac{d}{dt}\begin{pmatrix} B^0(t) \\ \bar{B}^0(t) \end{pmatrix} = \left(M - \frac{i\Gamma}{2}\right)\begin{pmatrix} B^0(t) \\ \bar{B}^0(t) \end{pmatrix}. \tag{3.71}$$

Diagonalising the Hamiltonian, the physical states are found to be

$$\begin{aligned} |B_L\rangle &= p\left|B^0\right\rangle + q\left|\bar{B}^0\right\rangle, \\ |B_H\rangle &= p\left|B^0\right\rangle - q\left|\bar{B}^0\right\rangle. \end{aligned} \tag{3.72}$$

B_L and B_H have similar decay widths ($\Gamma_H \sim \Gamma_L$) so that $\Delta\Gamma$ can be neglected to a good approximation. Here we note the difference with the neutral kaon system where K_L and K_S have widely disparate lifetimes. This difference makes the separation of the mass eigenstates via their decay channels difficult, and we need to look for other kinds of observables bearing the signature of CP-violation.

After solving the coupled differential equations one finds the
the flavour eigenstates B^0 and $\bar{B}^0$ to evolve in time as follows:

$$|B^0(t)\rangle = g_+(t)|B^0(0)\rangle + \frac{q}{p}g_-(t)|\bar{B}^0(0)\rangle,$$

$$|\bar{B}^0(t)\rangle = \frac{p}{q}g_-(t)|B^0(0)\rangle + g_+(t)|\bar{B}^0(0)\rangle, \qquad (3.73)$$

where,

$$g_+(t) = e^{-\Gamma t/2}e^{iMt/2}\cos\left(\frac{\Delta M t}{2}\right)$$

$$g_-(t) = i\,e^{-\Gamma t/2}e^{iMt/2}\sin\left(\frac{\Delta M t}{2}\right)$$

$$M = \frac{M_H + M_L}{2} \quad \text{and} \quad \Delta M = M_H - M_L.$$

As the flavour composition changes in the above manner, let us
consider the production of a $B^0 - \bar{B}^0$ pair in electron-positron
collision at the $\upsilon(4s)$ resonance (where $\upsilon = b\bar{b}$). First, we need to
select a final state f into which both B^0 and $\bar{B}^0$ can decay, f being
a CP-eigenstate. Define $A = \langle f|T_W|B^0(t)\rangle$, $\bar{A} = \langle f|T_W|\bar{B}^0(t)\rangle$,
where T_W is the weak transition matrix.

A measure of CP violation, which is time-dependent, is the
asymmetry

$$a_{fCP} = \frac{\Gamma(B^0(t) \to f) - \Gamma(\bar{B}^0(t) \to f)}{\Gamma(B^0(t) \to f) + \Gamma(\bar{B}^0(t) \to f)}.$$

a_{fCP} can be the result of

- Direct CP-violation (inequality of A and $\bar{A}$);

- CP-violation through mixing ($\frac{q}{p} \neq 1$).

The relevant parameter characterising the above two effects is $\lambda = q\bar{A}/pA$, in terms of which one can write

$$\langle f|T_W|B^0(t)\rangle = A\left(g_+(t) + \lambda g_-(t)\right)\langle f|T_W|\bar{B}^0(t)\rangle$$

$$= \frac{pA}{q}\left(g_-(t) + \lambda g_+(t)\right). \qquad (3.74)$$

Now,

$$
\begin{aligned}
\Gamma(B^0(t) \to f) \;&\sim\; |\langle f|T_W|B^0(t)\rangle|^2 \\
&= |A|^2 e^{-\Gamma t}\Big[\frac{(1+|\lambda|^2)}{2} + \frac{(1-|\lambda|^2)}{2}\cos(\Delta M t) \\
&\qquad - \operatorname{Im}\,\lambda\sin(\Delta M t)\Big],
\end{aligned}
\tag{3.75}
$$

and similarly,

$$
\begin{aligned}
\Gamma(\bar{B}^0(t) \to f) \;&\sim\; |A|^2 e^{-\Gamma t}\Big[\frac{(1+|\lambda|^2)}{2} + \frac{(1-|\lambda|^2)}{2}\cos(\Delta M t) \\
&\qquad + \operatorname{Im}\,\lambda\sin(\Delta M t)\Big],
\end{aligned}
\tag{3.76}
$$

so that

$$
a_{fcp} = \frac{(1-|\lambda|^2)\cos(\Delta M t) - 2\operatorname{Im}\,\lambda\sin(\Delta M t)}{1+|\lambda|^2}.
\tag{3.77}
$$

Now, $|\frac{q}{p}| = |\frac{M_{12}}{M_{12}^*}| = 1$ if $\Gamma_{12} \ll 1$, in which case $|A| = |\bar{A}| = 1$ and $|\lambda| = 1$. This happens when all amplitudes contributing to the decay carry the same phase factor, and one has

$$
a_{fCP} = -\operatorname{Im}\,\lambda\sin(\Delta M t).
\tag{3.78}
$$

In general, once $|\lambda|$ and $\operatorname{Im}\,\lambda$ can be calculated for a given f, it is possible to predict the CP-asymmetry in terms of the CKM parameters. The applicability of this method can be checked in the decay $B^0(\bar{B}^0) \to \Psi k_s$, rather distinct final state observable at B-factories.

The question, however, is, how does one ascertain whether, in an experimental situation, the final state f has come from $B^0(t) \to f$ or $\bar{B}^0(t) \to f$?

For that purpose, consider the semileptonic decays of $B^0(t)$ and $\bar{B}^0(t)$. While the allowed decay is $b \to \bar{c}l^+\nu_l$ for the former, the latter decays in the channel $b \to cl^-\bar{\nu}_l$. Since they are both spinless, one cannot, out of a $B^0 - \bar{B}^0$ pair produced at a B-factory, have both B^0 or both B^0 in course of oscillation at a given time, otherwise their combined wavefunction will be antisymmetric and Bose statistics will not hold for the system of identical B-mesons.

Therefore, the identity of the meson decaying into the final state f (say, $J/\psi K_s$) can be established from whether one has an l^+ or l^- on the other side. Thus

$$a_{fCP} = \frac{(\text{no. of events with } l^+) - (\text{no. of events with } l^-)}{(\text{no. of events with } l^+) + (\text{no. of events with } l^-)}.$$

One has to still remember, however, that the decays $\bar{B} \to l^-$ and $B \to l^+$ at the two ends may occur at different times, say, t_- and t_+. Now, if $t = 0$ corresponds to the moment when the B_d-pair is formed, The probabilities of one meson decaying to l^+ at t_+ is $e^{-\Gamma t_l}$ and of the other meson surviving till then are each given by $e^{-\Gamma t_-}$. If thus an l^+ is detected at t_+, the other meson at that point of time is $\bar{B}^0$. Taking t_- as the origin of time, the probability of this meson decaying to f at t_+ is given by

$$P_{t_f} = e^{-\Gamma(t_+ - t_-)} \left[\frac{(1 + |\lambda|^2)}{2} + \text{Im } \lambda \sin \Delta M(t_+ - t_-) \right].$$

Thus, the probability of a $B^0 \bar{B}^0$ pair formed at $t = 0$, and one going to l^+ at t_- and the other going to f as $\bar{B}^0$ at t_+ is

$$\begin{aligned}
P_{t_- t_+} &= e^{-\Gamma t_-} e^{-\Gamma t_-} e^{-\Gamma(t_+ - t_-)} \left[\frac{(1 + |\lambda|^2)}{2} \right.\\
&\qquad \left. + \text{Im } \lambda \sin \Delta M(t_+ - t_-) \right]\\
&\sim \frac{d^2\Gamma(\bar{B}^0 \to f)}{dt_+ dt_-}.
\end{aligned} \qquad (3.79)$$

In order to see CP-violation, we have to obtain the asymmetry between $d^2\Gamma(\bar{B}^0 \to f)/dt_f dt_{l+}$ and $d^2\Gamma(\bar{B}^0 \to f)/dt_+ dt_-$. This asymmetry turns out to be proportional to $\text{Im } \lambda \sin \Delta M(t_+ - t_-)$.

It is obvious from above that the above kind of asymmetry vanishes when integrated over time. The need of the day is therefore to have a situation where the decays at different times can be recorded and thus the double differentials mentioned above can be measured to a reliable degree, given the finite resolution of the experiments. This becomes extremely difficult at the centre-of-mass frame of the colliding e^+e^- system, where the υ must be produced at rest. In this frame, the average length each B_d travels before decaying is on the order of 30 μm, which is too small a distance to achieve the requisite resolution in time.

The solution lies in obtaining the υ boosted in the laboratory frame. One, therefore, aims at an *asymmetric beam B*-factory where the electron and the positron are colliding with unequal momenta. In this manner, it is possible to have the B-mesons travel over larger distances, as measured in the laboratory frame, before they decay, whereby it becomes feasible to obtain the the time-dependence of the final state asymmetry.

Bibliography

[1] *CP-violation*, C. Jarlskog (ed), World Scientific (1989).

[2] A. Buras, Lectures given at Lake Louise Winter Institute: Electroweak Physics, Lake Louise, Alberta, Canada (1999). In *Lake Louise 1999, Electroweak physics* 1–83 (hep-ph/9905437).

[3] B. Kayser, Talk given at ICTP Summer School in High-energy Physics and Cosmology, Trieste, Italy (1995). Published in Trieste HEP Cosmology 1995: 432–478 (hep-ph/9702264).

[4] B. Winstein and L. Wolfenstein, *The search for direct CP violation*, Rev. Mod. Phys. **65**, 1113–1148 (1993).

[5] J. Rosner, Invited talk at 2nd Tropical Workshop on Particle Physics and Cosmology: Neutrino and Flavor Physics, San Juan, Puerto Rico (2000). Published in *San Juan 2000, Particle physics and cosmology* 283–304 (hep-ph/0005258).

[6] *CP violation*, I. Bigi and A. Sanda, Cambridge Monograph on Particle Physics, Nuclear Physics and Cosmology, 9: 1–382, 2000.

[7] R. Fleischer, *Theoretical review of CP violation*, hep-ph/0310313.

Apart from these review articles, some important papers are:

[8] J.H. Christenson, J.W. Cronin, V.L. Fitch, R. Turlay, *Phys. Rev. Lett.* **13**, 138 (1964).

[9] S.L. Glashow, J. Iliopoulos and L. Maiani, *Phys. Rev.* **D2**, 1285 (1970).

[10] M. Kobayashi and T. Maskawa, *Prog. Thoer. Phys.* **49**, 652 (1973).

[11] I. Bigi and A. Sanda, *Nucl. Phys.* **B193**, 85 (1981).

[12] P. Ginsparg and M. Wise, *Phys. Lett.* **B127**, 265 (1983).

[13] P. Ginsparg, S. Glashow and M. Wise, *Phys. Rev. Lett.* **50**, 1415 (1983).

[14] J.F. Donoghue and B. Holstein, *Phys. Rev.* **D29**, 2088 (1984).

[15] J.F. Donoghue, B. Holstein and G. Valencia, *Int. J. Mod. Phys.* **A2**, 319 (1987).

Chapter 4

Lectures on Perturbative Quantum Chromodynamics

Compiled by Debashis Ghoshal & V. Ravindran

The following notes, meant as a pedagogical introduction to the subject of perturbative quantum chromodynamics (pQCD), are based on the lectures by Ashoke Sen in the XVI SERC Main School on Theoretical High Energy Physics at the HRI, Allahabad. The notes were taken by Anindya Dutta, Debashis Ghoshal, Dileep Jatkar, Swapan Majhi, Partha Konar, Subhendu Rakshit and V. Ravindran; and compiled in their final form by Debashis Ghoshal and V. Ravindran.

4.1 Lecture I

Quantum Chromodynamics (QCD) is the theory of strong interactions. It is to some extent similar to Quantum Electrodynamics (QED), the theory of electrons and photons. We will concentrate on the perturbative aspects of QCD. Perturbative QCD (pQCD for short) involves a Taylor series expansion in terms of the strong coupling constant, a parameter which enters in the QCD Lagrangian and characterise strong interaction. The perturbative approach works only when this coupling constant is small, which is true for QED. The relevant coupling constant in QED is α_{EM}

which is of the order of $1/137$. This is not true for QCD where the coupling constant is not too small. In addition, QCD being the theory for quarks and gluons which cannot seen in nature, it is not only that the application of perturbation theory is non-trivial, but also it is not obvious why perturbation theory is valid.

The strongly interacting particles, collectively called hadrons are made up of quarks, anti-quarks, and gluons. There are six quark flavours (u, c, t, d, s, b). The first three have electromagnetic charge $+2/3$ and the last three have $-1/3$. We assign a spin-1/2 Dirac field, $\psi_s^{\alpha i}$ for each of these particles. Here s is the flavour index which runs from 1 to 6, α is the spinor index and i is the colour index, which runs from 1 to 3. In terms of these quark fields, the proton wave function can be written as: $p \sim \epsilon_{ijk} u^i u^j d^k$. Since QCD interaction in the proton changes the colour indices of the quarks, the actual proton wave-function is a linear combination of all such terms.

Before going into the details of the QCD interaction (which is a non-abelian gauge interaction), let us recapitulate QED. The two basic fields which enter here are ψ, a Dirac field describing an electron and A_μ the vector potential which, on quantisation, gives the photon. The Lagrangian density for QED can be written as

$$\mathcal{L} = -\frac{1}{4} F_{\mu\nu} F^{\mu\nu} + \bar{\psi} \left[i\gamma^\mu (\partial_\mu - ieA_\mu) - m \right] \psi, \qquad (4.1)$$

where, $\partial_\mu \equiv \partial/\partial x^\mu$ and $F_{\mu\nu} = \partial_\mu A_\nu - \partial_\nu A_\mu$. Also $F^{\mu\nu} = \eta^{\mu\mu'} \eta^{\nu\nu'} F_{\mu'\nu'}$ where, $\eta^{\mu\nu} = \mathrm{diag}(1, -1, -1, -1)$ in our conventions.

The action is given by $\mathcal{S} = \int d^4x \mathcal{L}$, which is Lorentz invariant. The action is also invariant under a global phase transformation on ψ:

$$
\begin{aligned}
\psi(x) &\rightarrow e^{ie\lambda} \psi(x), \\
\psi^\dagger(x) &\rightarrow e^{-ie\lambda} \psi^\dagger(x), \\
\bar{\psi}(x) = \psi^\dagger(x)\gamma^0 &\rightarrow e^{-ie\lambda} \bar{\psi}(x), \\
A_\mu &\rightarrow A_\mu.
\end{aligned}
\qquad (4.2)
$$

In fact, action has a larger symmetry called *gauge symmetry* under which the fields transform as follows,

$$\begin{aligned}
\psi(x) &\rightarrow e^{ie\lambda(x)}\psi(x), \\
\bar\psi(x) &\rightarrow e^{-ie\lambda(x)}\bar\psi(x), \\
A_\mu &\rightarrow A_\mu + \partial_\mu\lambda(x).
\end{aligned} \tag{4.3}$$

It can easily be checked that under gauge transformation the action is invariant. So gauge transformation is a special kind of phase transformation in which the phase parameter is a function of space-time. If λ is independent of space-time, the free Dirac Lagrangian itself is invariant under the phase transformation. But when we try to extend this global symmetry to a local one it becomes necessary to introduce a vector field A_μ which modifies the derivative term. This vector field has a certain transformation property defined above.

In a similar vein, we start with the free Dirac Lagrangian for the quarks:

$$\mathcal{L} = \sum_{i,s} \bar\psi_s^i(i\gamma_\mu\partial^\mu - m_s)\psi_s^i \tag{4.4}$$

In the following, we will drop the flavour index s for simplicity and work with one quark only. Consider, the (colour) transformation

$$\psi^k \rightarrow U_{kl}\psi^l, \qquad \bar\psi^k \rightarrow U_{kl}^*\bar\psi^l.$$

If this is to be a symmetry of the action, we must have

$$U_{mi}U_{ni}^* = \delta_{mn},$$

i.e., $U^\dagger U = \mathbf{1}_{3\times3}$, in other words U is a unitary matrix. Since we want to preserve the norm of the quark fields, we will focus on a subset of U, for which $\det U = 1$. Therefore, U is an $SU(3)$ transformation matrix. It should be emphasised that the choice of this particular symmetry is solely guided by experiments.

If we choose U to be $\mathbf{1}$, the transformation of ψ is trivial. So let us choose U to be close to the unity so that,

$$\psi_k \rightarrow \psi_k + \delta\psi_k,$$

where $\delta\psi_k$ is infinitesimal. This is obtained by choosing U such that,

$$U = \mathbf{1} - i\epsilon\,\Omega,$$

where ϵ is some infinitesimal real parameter and Ω is a 3×3 matrix. Now, the condition $U^\dagger U = \mathbf{1}_{3\times3}$ implies that

$$\Omega^\dagger = \Omega + \mathcal{O}(\epsilon^2).$$

Moreover, $\det U = 1$ implies that, to order ϵ, $\mathrm{Tr}\,\Omega = 0$. In other words, Ω is a traceless hermitian matrix.

Let $T_1, T_2, \cdots, T_8$ be eight linearly independent 3×3 traceless hermitian matrices. Clearly they form a basis for the SU(3) algebra. A particularly convenient choice are the so called Gell-Mann matrices. Let us choose the normalisation of these matrices such that,

$$\mathrm{Tr}\left(T^a T^b\right) = \frac{1}{2}\delta^{ab}.$$

The commutators of the matrices T^a are

$$\left[T^a, T^b\right] = i f^{abc}\, T_c. \tag{4.5}$$

From the definition, f^{abc} is anti-symmetric in its first two indices. However, due to our choice of normalisation f is totally anti-symmetric under the exchange of any two of its indices.

Using the basis, $\epsilon\Omega = \sum \epsilon_a T^a$ and U can be written as,

$$U = \mathbf{1} - i\sum \epsilon_a T^a.$$

To the lowest order in ϵ

$$\delta\psi^k = -i\epsilon_a(T^a)_{kl}\psi^l, \qquad \delta\bar{\psi}^k = i\epsilon_a(T^a)_{lk}\bar{\psi}^l. \tag{4.6}$$

It is easy to check that the free Dirac action is invariant under the global, *i.e.*, when ϵ_a's are independent of space time, SU(3) transformation.

Now we want to make this global symmetry of the action into a local symmetry:

$$\delta\psi^k = -i\epsilon_a(x)(T^a)_{kl}\,\psi^l. \tag{4.7}$$

Under the above transformation $\mathcal{L}_{\text{free}} \to \mathcal{L}_{\text{free}} + \delta\mathcal{L}_{\text{free}}$, where

$$\delta\mathcal{L}_{free} = (T^a)_{kl} \, \bar{\psi}^k \gamma^\mu \psi^l \, \partial_\mu \epsilon_a(x). \tag{4.8}$$

So we have to add something to the Lagrangian to cancel this extra term. Adding a new interaction term

$$\mathcal{L}_{\text{int}} = g \, \bar{\psi}^k \gamma^\mu \psi^l \, A_{a\mu} \, (T^a)_{kl}$$

and demanding that $A_{a\mu}$ transforms as

$$\delta A_\mu^a = -\frac{1}{g}\partial_\mu \epsilon^a(x) + f^{abc}\epsilon_b(x)A_{c\mu},$$

it is easy to show the full action is gauge invariant.

Now we want the action of the gauge fields in the non-abelian case. If we define non-abelien $F^{\mu\nu}$ as in QED, it would not be gauge invariant as δA_μ^a contains a term proportional to f^{abc}. Let us define the field strength $F^{\mu\nu}$ in case of non-abelian gauge theory as

$$F_{\mu\nu}^a = \partial_\mu A_\nu^a - \partial_\nu A_\mu^a + g f^{abc} A_{b\nu} A_{c\nu}.$$

Under the gauge transformation

$$\delta F_{\mu\nu}^a = g f^{abc}\theta_b F_{c\mu\nu},$$

where, $\theta_a = \epsilon^a(x)/g$. The field strength so defined is gauge covariant. In order to check that it is so, one has to use the following identity (Bianchi identity)

$$f^{abc} f^{ckl} + f^{acl} f^{ckb} - f^{ack} f^{bcl} = 0.$$

Let us use a convenient notation to summarise these. Define $\underline{A}_\mu = A_{a\mu}T^a$, $\underline{F}_{\mu\nu} = F_{a\mu\nu}T^a$ and $\underline{\theta} = \theta_a T^a$. Also, $\underline{\psi} = \begin{pmatrix} \psi_1 \\ \psi_2 \\ \psi_3 \end{pmatrix}$ and $\underline{\bar{\psi}} = (\,\psi_1, \psi_2, \psi_3\,)$ In term of these matrices, the QCD Lagrangian is written as

$$\mathcal{L}_{QCD} = i\underline{\bar{\psi}}\gamma^\mu \left(\partial_\mu - ig\underline{A}_\mu\right)\underline{\psi} - m\underline{\bar{\psi}}\underline{\psi} - \frac{1}{2}Tr\left(\underline{F}_{\mu\nu}\underline{F}^{\mu\nu}\right), \tag{4.9}$$

where, $\underline{\gamma}^\mu = \text{diag}\,(\gamma^\mu, \gamma^\mu, \gamma^\mu)$. In the new notation,

$$
\begin{aligned}
\underline{F}_{\mu\nu} &= \partial_\mu \underline{A}_\nu - \partial_\nu \underline{A}_\mu - ig\left[\underline{A}_\mu, \underline{A}_\mu\right], \\
\delta\underline{\theta} &= -\partial_\mu\underline{\theta} - ig\left[\underline{\theta}, \underline{A}_\mu\right], \\
\delta\underline{F}_{\mu\nu} &= -ig\left[\underline{\theta}, \underline{F}_{\mu\nu}\right], \\
\delta\underline{\psi} &= -ig\,\underline{\theta}\,\underline{\psi}, \\
\delta\underline{\bar{\psi}} &= ig\,\underline{\theta}\,\underline{\bar{\psi}}.
\end{aligned}
\tag{4.10}
$$

4.2   Lecture II

In this lecture, we shall recapitulate the quantization of field theories.

Consider a classical field theory with a set of fields. These can be a set of scalars or (the components of) vectors or spinor fields. We shall denote these collectively by $\{\phi_1, \phi_2 ... \phi_N\}$. The action

$$
\mathcal{S} = \int d^4x \, \mathcal{L}(\phi_1, \cdots, \phi_N; \partial_\mu\phi_1, \cdots, \partial_\mu\phi_N)
\tag{4.11}
$$

is a function of the fields and their first derivatives. The canonical quantization proceeds as follows. Treat each field ϕ_i as an operator $\hat{\phi}_i$. The quantities we are interested in are the Green's functions (also called correlation functions) defined by

$$
\left\langle 0\left|T\left(\hat{\phi}_{i_1}(x_{i_1}) \cdots \hat{\phi}_{i_n}(x_{i_n})\right)\right|0\right\rangle = \left\langle \phi_{i_1}(x_{i_1}) \cdots \phi_{i_n}(x_{i_n})\right\rangle.
\tag{4.12}
$$

Recall that one needs to compute the Green's function to obtain the S-matrix elements.

The path integral prescription to compute the Green's functions is given by

$$
\left\langle \phi_{i_1}(x_1) \cdots \phi_{i_n}(x_n)\right\rangle = \frac{\displaystyle\int \prod_{i=1}^{N} \mathcal{D}\phi_i(x_i) \, e^{i\mathcal{S}} \, \phi_{i_1}(x_1) \cdots \phi_{i_n}(x_n)}{\displaystyle\int \prod_{i=1}^{N} \mathcal{D}\phi_i(x_i) \, e^{i\mathcal{S}}},
$$

where, $\mathcal{D}\phi_i(x_i)$ means the sum over all possible field configurations.

The simplest way to interpret the path integral is to discretize spacetime. We write $(x^0, x^1, x^2, x^3) \equiv (an^0, an^1, an^2, an^2)$, where a is a small number and $n^0, \cdots$ are arbitrary integers. So $\{\phi_i(x)\}$ is completely specified by $\{\phi_i(an^0, an^1, an^2, an^3)\}$ for every integers n^0, n^1, n^2, n^3. We also have to give a meaning to the derivatives after discertization:

$$\partial_0 \phi_i = \frac{1}{a} \left[\phi_i(a(n^0+1), an^1, an^2, an^3) - \phi_i(an^0, an^1, an^2, an^3) \right],$$

and similarly for the others. The integral is replaced by a sum:

$$\int d^4x \to a^4 \sum_{n^0, n^1, n^2, n^3}, \tag{4.13}$$

while the delta-function can be written as

$$\delta^{(4)}(x - x') = \frac{1}{a^4} \, \delta_{n_0 n_0'} \, \delta_{n_1 n_1'} \, \delta_{n_2 n_2'} \, \delta_{n_3 n_3'}. \tag{4.14}$$

The integral over all field configuration now becomes an infinite number of ordinary integrals:

$$\int \prod_{i=1}^{N} [\mathcal{D}\phi_i(x_i)] \to \int \prod_{i, n^0, n^1, n^2, n^3} d\phi_i(an^0, an^1, an^2, an^3) \tag{4.15}$$

This interpretation of path integral defines the Lattice Field Theory (see the lectures by Rajiv V. Gavai in this volume).

Perturbation theory becomes simpler if one works in momentum space. To this end, we define the Fourier transformation

$$\phi_i(x) = \frac{1}{(2\pi)^4} \int d^4k \, e^{-ikx} \tilde{\phi}_i(k). \tag{4.16}$$

We will now write the path integral in momentum space using the (Fourier) conjugate variables. For this we discretize momentum space:

$$(k^0, k^1, k^2, k^3) \equiv (bn^0, bn^1, bn^2, bn^2), \tag{4.17}$$

where b is a small number and n's are integers. Then

$$\langle \phi_{i_1}(x_1) \cdots \phi_{i_n}(x_n) \rangle = \int \frac{d^4 k_1}{(2\pi)^4} e^{-ik_1 x_1} \cdots \int \frac{d^4 k_n}{(2\pi)^4} e^{-ik_n x_n}$$
$$\times \left\langle \tilde{\phi}_{i_1}(k_1) \cdots \tilde{\phi}_{i_n}(k_n) \right\rangle, \quad (4.18)$$

where,

$$\left\langle \tilde{\phi}_{i_1}(k_1) \cdots \tilde{\phi}_{i_n}(k_n) \right\rangle = \frac{\int \prod_{i=1}^{N} \mathcal{D}\tilde{\phi}_i(k_i)\, e^{iS}\, \tilde{\phi}_{i_1}(k_1) \cdots \tilde{\phi}_{i_n}(k_n)}{\int \prod_{i=1}^{N} \mathcal{D}\tilde{\phi}_i(k_i)\, e^{iS}}.$$

Let us introduce source functions and define the generating functional of the path integral,

$$\mathcal{Z}(J_1, \cdots, J_n) = \int \prod_{i=1}^{N} \mathcal{D}\tilde{\phi}_i(k_i) \, \exp\left(iS + \int d^4 x J_i(x)\phi_i(x) \right)$$
$$= \int \prod_{i=1}^{N} \mathcal{D}\tilde{\phi}_i(k_i) \, \exp\left(iS + \int \frac{d^4 k}{(2\pi)^4} \tilde{J}_i(-k)\tilde{\phi}_i(k) \right),$$

where $\tilde{J}_i(k)$'s are the Fourier transforms of the sources $J_i(x)$s. The functional derivatives with respect to these sources are defined such that

$$\frac{\delta}{\delta \tilde{J}_i(k)} \tilde{J}_j(k') = \delta^4(k - k')\, \delta_{ij}.$$

We shall need the analogue of this in terms of the discretized variables. Note that

$$\delta^4(k - k')\, \delta_{ij} = \frac{1}{b^4} \delta_{ij} \, \delta_{n_0 n_0'} \delta_{n_1 n_1'} \delta_{n_2 n_2'} \delta_{n_3 n_3'} \equiv \frac{1}{b^4} \delta_{ij} \, \delta_{\vec{n}\vec{n}'}, \quad (4.19)$$

and $\frac{\partial}{\partial \tilde{J}_i(k)} \tilde{J}_j(k') = \delta_{ij}\, \delta_{\vec{n}\vec{n}'}$ with $k \equiv (bn^0, bn^1, bn^2, bn^2)$. From the above equations, it follows that $\frac{\delta}{\delta \tilde{J}_i(k)} \equiv \frac{1}{b^4} \frac{\partial}{\partial \tilde{J}_i(k)}$. Hence

$$\frac{(2\pi)^4 \delta}{\delta \tilde{J}_{i_1}(-k_1)} \cdots \frac{(2\pi)^4 \delta}{\delta \tilde{J}_{i_n}(-k_n)} \, \mathcal{Z}(J) \quad (4.20)$$

gives

$$\int \prod_{i=1}^{N} \mathcal{D}\tilde{\phi}_i(k_i) \, e^{i\mathcal{S} + \int \frac{d^4 k}{(2\pi)^4} \tilde{J}_i(-k)\tilde{\phi}_i(k)} \, \tilde{\phi}_{i_1}(k_1) \cdots \tilde{\phi}_{i_n}(k_n) \qquad (4.21)$$

Finally, the Green's function in momentum space can be written as

$$\left\langle \tilde{\phi}_{i_1}(k_1) \cdots \tilde{\phi}_{i_n}(k_n) \right\rangle = \frac{1}{\mathcal{Z}(J)} \frac{(2\pi)^4 \delta}{\delta \tilde{J}_{i_1}(-k_1)} \cdots \frac{(2\pi)^4 \delta}{\delta \tilde{J}_{i_n}(-k_n)} \, \mathcal{Z}(J) \Bigg|_{J=0}$$

We shall now compute $\mathcal{Z}(J)$ in perturbation theory. The full action can be split as

$$\mathcal{S} = \mathcal{S}_0 + \mathcal{S}_I, \qquad (4.22)$$

where $\mathcal{S}_0$ contains only the quadratic (free) part and $\mathcal{S}_I$ involves the cubic and higher order (interaction) terms. As an illustrative example consider a specific field theory, say the ϕ^4 theory defined by the action

$$\mathcal{S} = \int d^4 x \mathcal{L} = \int d^4 x \left[\frac{1}{2}\partial_\mu \phi \, \partial^\mu \phi - \frac{1}{2}m^2 \phi^2 - \frac{\lambda}{4!}\phi^4 \right]. \qquad (4.23)$$

In momentum space

$$\begin{aligned}
\mathcal{S} &= \frac{1}{2}\int \frac{d^4 k}{(2\pi)^4} \, \tilde{\phi}(-k) \, (k^2 - m^2) \, \tilde{\phi}(k) \\
&\quad - \frac{\lambda}{4!}\int \frac{d^4 k_1}{(2\pi)^4} \cdots \frac{d^4 k_4}{(2\pi)^4} \tilde{\phi}(k_1) \cdots \tilde{\phi}(k_4) \, (2\pi)^4 \delta^4(\textstyle\sum k),
\end{aligned}$$

the first line of the action is the free part and the second line is the interacting part. The generating function

$$\mathcal{Z}(J) = \int \prod_{i=1}^{N} \mathcal{D}\tilde{\phi}_i(k_i) \, \exp\left(i\mathcal{S}_0 + \int \frac{d^4 k}{(2\pi)^4} \, \tilde{J}_i(-k)\tilde{\phi}_i(k) \right) e^{i\mathcal{S}_I},$$

where, the exponential of the interaction part

$$e^{i\mathcal{S}_I} = \sum_{m=0}^{\infty} \frac{1}{m!} \left(i\mathcal{S}_I(\tilde{\phi}_i) \right)^m \qquad (4.24)$$

is defined in terms of its Taylor series. To a given order in perturbation theory (say n), one truncates at the nth term, and $\mathcal{S}_I$ is a polynomial in the fields. Hence,

$$\mathcal{Z}(J) = \sum_{m=0}^{\infty} \frac{1}{m!} \left[i\mathcal{S}_I \left(\tilde{\phi}_i(k) \rightarrow \frac{(2\pi)^4 \delta}{\delta \tilde{J}_{i_1}(-k_1)} \right) \right]^m \mathcal{Z}_0(J). \quad (4.25)$$

The computation of $\mathcal{Z}_0(J)$ requires the knowledge of $\mathcal{S}_0$. The most general form of $\mathcal{S}_0$ is

$$\mathcal{S}_0 = \frac{1}{2} \int \frac{d^4 k}{(2\pi)^4} \, \tilde{\phi}_i(-k) \mathcal{M}_{ij} \tilde{\phi}_j(k), \quad (4.26)$$

where $\mathcal{M}_{ij}$ is a matrix. *E.g.*, in case of the ϕ^4 theory

$$\mathcal{M}_{ij} = (k^2 - m^2)\, \delta_{ij}. \quad (4.27)$$

It is easy to check, using symmetry arguements, that

$$\mathcal{M}_{ij}(-k) = \mathcal{M}_{ji}(k). \quad (4.28)$$

Hence,

$$\mathcal{Z}_0(J) \;=\; \int \prod_{i=1}^{N} \mathcal{D}\tilde{\phi}_i(k_i) \, \exp \left[\frac{i}{2} \int \frac{d^4 k}{(2\pi)^4} \left(\tilde{\phi}_i(-k)\,\mathcal{M}_{ij}\,\tilde{\phi}_j(k) \right. \right.$$
$$\left. \left. -2i\tilde{J}_i(-k)\tilde{\phi}_i(k) \right) \right]. \quad (4.29)$$

We can simplify the above. First, let us rewrite the term in the parenthesis

$$\left(\tilde{\phi}_i(-k) - i\tilde{J}_l(-k)(\mathcal{M}^{-1})_{li} \right) \mathcal{M}_{ij} \left(\tilde{\phi}_j(k) - i(\mathcal{M}^{-1})_{jm}\tilde{J}_m(k) \right)$$
$$+ \tilde{J}_l(-k)(\mathcal{M}^{-1})_{lm}\tilde{J}_m(k) \quad (4.30)$$

so as to complete the square. Next, we use the shifted variables

$$\hat{\phi}_i(k) = \tilde{\phi}_i(k) - i(\mathcal{M}^{-1})_{im}\tilde{J}_m(k)$$

and note that $\mathcal{D}\hat{\phi}_i(k) = \mathcal{D}\tilde{\phi}_i(k)$ to arrive at

$$
\begin{aligned}
\mathcal{Z}_0(J) &= \int \prod_{i=1}^{N} \mathcal{D}\hat{\phi}_i(k) \, \exp\left[\frac{i}{2} \int \frac{d^4k}{(2\pi)^4}\left(\hat{\phi}_i(-k)\,\mathcal{M}_{ij}\,\hat{\phi}_j(k)\right.\right.\\
&\qquad\qquad\qquad\left.\left.+\tilde{J}_i(-k)\mathcal{M}_{ij}^{-1}\tilde{J}_j(k)\right)\right]\\
&= \mathcal{N} \, \exp\left[\frac{i}{2} \int \frac{d^4k}{(2\pi)^4}\tilde{J}_i(-k)\,\mathcal{M}_{ij}^{-1}\,\tilde{J}_j(k)\right].
\end{aligned}
\tag{4.31}
$$

Notice that the normalization constant $\mathcal{N}$, arising out of the $\hat{\phi}$ integral independent of the J's, drops out of the Green's functions. In the following, we shall ignore this normalization.

The most general form of $\mathcal{S}_I$ is

$$
\begin{aligned}
\mathcal{S}_I &= \sum_{m\geq 3}\int \frac{d^4k_1}{(2\pi)^4}\cdots\int \frac{d^4k_m}{(2\pi)^4}\,(2\pi)^4\,\delta^4\Big(\sum k_i\Big)\\
&\qquad \times \tilde{\phi}_{i_1}(k_1)\cdots\tilde{\phi}_{i_m}(k_m)\,f^{(m)}_{i_1\cdots i_m}(k_1,k_2,\cdots,k_m).
\end{aligned}
\tag{4.32}
$$

Finally, we obtain

$$
\begin{aligned}
\Big\langle \tilde{\phi}_{i_1}(k_1)\cdots\tilde{\phi}_{i_n}(k_n)\Big\rangle &= \frac{1}{\mathcal{Z}(0)}\frac{(2\pi)^4\delta}{\delta\tilde{J}_{i_1}(-k_1)}\cdots\frac{(2\pi)^4\delta}{\delta\tilde{J}_{i_n}(-k_n)}\\
&\quad \times \left[\sum_{s=0}^{\infty}\frac{1}{s!}\Big(i\sum_{m\geq 3}\int\frac{d^4q_1}{(2\pi)^4}\cdots\int\frac{d^4q_m}{(2\pi)^4}\right.\\
&\qquad\quad \times (2\pi)^4\delta\Big(\sum_i q_i\Big)\,f^{(m)}(q_1,q_2,\cdots,q_m)\\
&\qquad\quad \left.\times \frac{(2\pi)^4\delta}{\delta\tilde{J}_{i_1}(-q_1)}\cdots\frac{(2\pi)^4\delta}{\delta\tilde{J}_{i_m}(-q_m)}\Big)\right]^s \mathcal{Z}_0(J)\Bigg|_{J=0}.
\end{aligned}
$$

4.3 Lecture III

In this lecture we continue our review of quantization of field theories and derive the Feynman rules.

An example of a typical term that we need to calculate is the following.

$$(2\pi)^4 \frac{\delta}{\delta \tilde{J}_{j_1}(-k_1)} \cdots (2\pi)^4 \frac{\delta}{\delta \tilde{J}_{j_m}(-k_m)}$$

$$\exp\left(\frac{i}{2}\int \frac{d^4k'}{(2\pi)^4}\tilde{J}_i(-k')M_{ij}^{-1}(k')\tilde{J}_j(k')\right)\Bigg|_{J=0}.$$

This vanishes if m is an odd integer (due to the symmetry of the theory). We need, therefore, to compute it only for even integers $m = 2p$. Let $\Delta_{ij} = M_{ij}^{-1}$. Then the above is:

$$\left(i\Delta_{j_1 j_2}(k_1)\cdot(2\pi)^4\delta^{(4)}(k_1+k_2)\right)$$

$$\times\left(i\Delta_{j_3 j_4}(k_3)\cdot(2\pi)^4\delta^{(4)}(k_3+k_4)\right)$$

$$\times\cdots\times\left(i\Delta_{j_{2p-1}j_{2p}}(k_{2p-1})\cdot(2\pi)^4\delta^{(4)}(k_{2p-1}+k_{2p})\right)$$

$$+ \text{ all inequivalent pairings of } j_1\cdots j_{2p}.$$

The above can be understood in the simpler context of ordinary differentiation of matrices:

$$\frac{\partial}{\partial x_{i_1}}\cdots\frac{\partial}{\partial x_{i_{2p}}}\exp\left(\frac{1}{2}x_i K_{ij}x_j\right)\Bigg|_{x=0} = K_{i_1 i_2}K_{i_3 i_4}\cdots K_{i_{2p-1}i_{2p}}$$

$$+\text{inequivalent pairings.}$$

A very effective way to keep track of these terms is through a set a set of diagrams (Feynman diagrams) and the associated Feynman rules.

First, each propagator is denoted by a line, sometimes with the momenta and gauge indices displayed explicitly:

$$\underset{i_1 \qquad\qquad i_2}{\overset{k_1 \qquad\qquad k_2}{\longleftarrow\!\!\!\longrightarrow}} \;=\; i\Delta_{i_1 i_2}\cdot(2\pi)^4\delta(k_1+k_2).$$

Secondly, the interaction vertices:

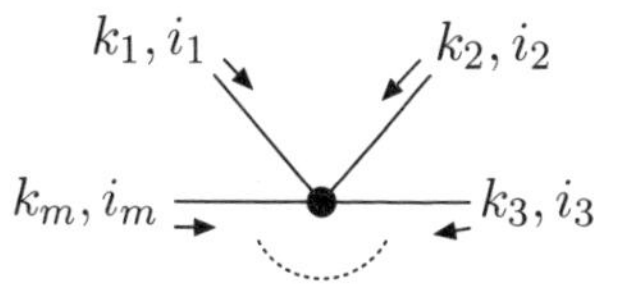

$$= \quad i f^{(m)}_{i_1 i_2 \cdots i_m}(k_1, k_2, \cdots, k_m).$$

Finally, for every $\tilde{\Phi}(k) = \left((2\pi)^4 \frac{\delta}{\delta \tilde{J}_i(k)}\right)$ in the correlation function, we draw an external line:

$$= \; 1.$$

We impose the rule that all lines ending on vertices must be paired by propagators. We also need to examine the constraints on momenta from momentum conservation.

Let us illustrate these in the example of ϕ^4-theory. The action is given by

$$\begin{aligned} S &= \frac{1}{2}\int \frac{d^4k}{(2\pi)^4}\, \tilde{\phi}_i(-k)\,(k^2 - m^2)\,\tilde{\phi}_j(k) \\ &\quad -\frac{\lambda}{4!}\int \frac{d^4k_1}{(\pi)^4}\cdots\frac{d^4k_4}{(2\pi)^4}\,(2\pi)^4\delta(\sum_i k_i)\,\tilde{\phi}(k_1)\cdots\tilde{\phi}(k_4) \end{aligned} \tag{4.33}$$

In this case, $M_{ij}(k) = k^2 - m^2$, since there is only one field. The Feynman rules are the following.

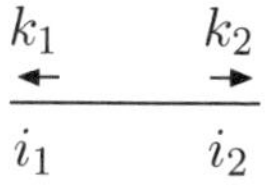

$$= \quad (2\pi)^4\,\delta(k_1 + k_2).$$

The interaction vertex is

$$k_1 \longrightarrow \bullet \longrightarrow k_3 \quad = \quad -\frac{i\lambda}{4!}\,(2\pi)^4\delta(k_1 + k_2 + k_3 + k_4).$$

Let us consider the computation of the four-point function

$$\left\langle \tilde{\phi}(k_1)\tilde{\phi}(k_2)\tilde{\phi}(k_3)\tilde{\phi}(k_4)\right\rangle.$$

We begin by drawing four external legs with appropriate momenta. In the lowest order (terms independent of λ), we simply join these pairwise by propagators. There are three possibilities:

The contribution, to $O(\lambda^0)$, is, therefore

$$(2\pi)^4\delta(k_1 + k_2)\,\frac{i}{k_1^2 - m^2}\,(2\pi)^4\delta(k_3 + k_4)\,\frac{i}{k_1^2 - m^2}$$

plus two other terms from the pairings $(13)(24)$ and $(14)(23)$. In the next order, it is possible to use one vertex, which is to be connected to the external legs by four propagators. This gives the following diagram.

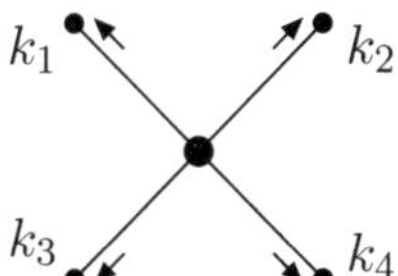

There are actually 4! possible pairings, all of which give the same result. Therefore, to $O(\lambda)$, we get

$$4! \times \left(\prod_{j=1}^{4} \frac{i}{k_j^2 - m^2} \right) \frac{(-i\lambda)^4}{4!} (2\pi)^4 \delta(\sum_j k_j).$$

Some of the momentum integrals in the above are in fact trivial. There are also cancellations between factors of $(2\pi)^4$ appearing in the integrals and δ-functions. It is possible to simplify the calculations by incorporating these from the beginning by the new Feynman rules.

Feynman rules:

1. The propagator:

$$\xrightarrow{\quad k \quad} \; = \; i\Delta_{i_1 i_2}(-k)\cdot.$$

2. The Vertices:

$$k_m, i_m \xrightarrow{\quad} \quad = \quad i f^{(m)}_{i_1 i_2 \cdots i_m}(k_1, k_2, \cdots, k_m).$$

3. The external line:

$$\bullet\!\!-\!\!\!\underset{k}{\longleftarrow} = 1.$$

4. At the end we examine the constraints on momenta as usual. For every constraint of the form $\sum_j k_j = 0$, we put a factor of $(2\pi)^4\delta(\sum_j k_j)$.

5. Every independent momenta ℓ, which is not determined, is to be integrated over. Hence, $\int \frac{d^4\ell}{(2\pi)^4}$ for every independent momenta.

The second example we shall consider is that of the six-point function:

$$\Big\langle \tilde{\phi}(k_1)\tilde{\phi}(k_2)\cdots\tilde{\phi}(k_6)\Big\rangle$$

at $O(\lambda^2)$. We start by drawing the six external legs and two vertices. Next we join the six external legs with two vertices in all possible ways. This produces diagrams of the following type:

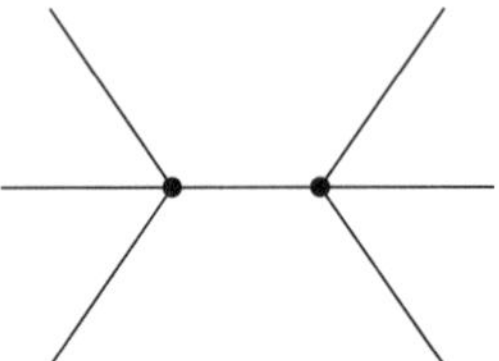

There are various ways to get at this diagram and we need to put appropriate combinatory factors. One can easily convince oneself that the correct combinatorial factor is $8 \times 3 \times 2 \times 4 \times 3 \times 2 = 2 \times (4!)^2$. Now recall the $(\lambda/4!)^2$ term is accompanied by the $1/2!$ from the expansion of the integral. Thus, we get a miraculous cancellation of the factors for this case. This, of course, will not happen all the times. Using usual Feynman rules, the amplitude for this diagram is:

$$(-i\lambda)^2 \, (2\pi)^4 \, \delta \left(\sum_{j=1}^{6} k_j \right) \left[\prod_{j=1}^{6} \frac{i}{k_j^2 - m^2} \right] \frac{i}{(k_1 + k_2 + k_3)^2 - m^2}.$$

As a final example, let us consider a correlator which has an undetermined momentum. This flows in a loop in the corresponding Feynman diagrams. We consider an $O(\lambda^2)$ contribution to the four-point function $\left\langle \tilde{\phi}(k_1) \cdots \tilde{\phi}(k_4) \right\rangle$. A Feynman diagram in this order is the following.

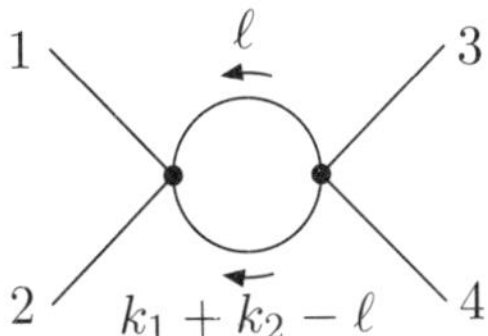

We have the constraint $k_1 + k_2 + k_3 + k_4 = 0$ from momentum consevation, which implies a factor of $(2\pi)^4 \, \delta(k_1 + k_2 + k_3 + k_4)$. However, this still leaves the momentum ℓ flowing in the loop undetermined. Consequently, we should integrate over this momentum. The combinatorial factor for this diagram can be seen to be $8 \times 3 \times 4 \times 3 \times 2 = (4!)^2$. In addition, we have $1/2(4!)^2$ from the perturbative expansion of the exponential. Combining all these, together with the external leg factors, we get the amplitude for this diagram:

$$\frac{(-2\pi i\lambda)^2}{2} \int \frac{d^4\ell}{(2\pi)^4} \frac{i}{\ell^2 - m^2} \frac{i}{(k_1 + k_2 - \ell)^2 - m^2} \left[\prod_{j=1}^{4} \frac{i}{k_j^2 - m^2} \right]$$

In the above, we have suppressed the momentum conservation condition $\delta \left(\sum k \right)$. (Notice that, unlike the previous example, there is no accidental cancellation of combinatorial factors here.)

Let us now calculate $Z[0]$ using Feynman rules. In the perturbative expansion the following Feynman diagrams contribute.

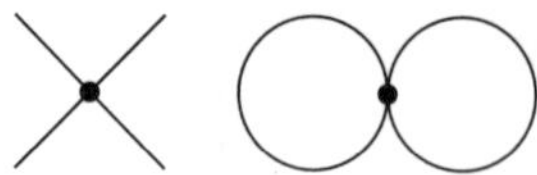

These diagrams are known as the vacuum bubbles.

Dividing a correlation function by $Z[0]$ (the generating function with all sources J set to zero), is equivalent to ignoring all vacuum bubbles from the Feynman graphs of $\langle \tilde{\phi}(k_1)\tilde{\phi}(k_2)\cdots\tilde{\phi}(k_n)\rangle$. For example, at $O(\lambda^2)$, one comes across the graph:

This, and similar graphs, are cancelled when we divide by $Z[0]$.

The method of canonical quantization is equivalent to the quantization via path integrals. However, the latter is better suited for non-abelian gauge theories. Indeed, the problem of canonical methods is already apparent in QED, where one has to impose constraints. This is because all the fields are not independent. The constraints in QED are linear in and hence can be dealt with simply in either method. Those that appear is QCD, on the other hand, are nonlinear and are very difficult to impose in the canonical method.

Exercise: Consider the part of the QCD action dependent on the gauge fields alone:

$$S = \frac{-1}{4}\int d^4x \; F^a_{\mu\nu}F^{\mu\nu}_a.$$

Show that the quadratic terms in A_μ is of the following form.

$$\frac{1}{2}\int \frac{d^4k}{} \; \tilde{A}^a_\mu(-k)\,\delta^{ab}\underbrace{\left(-k^2\eta^{\mu\nu} + k^\mu k^\nu\right)}_{M^{a\mu,b\nu}}\tilde{A}^b_\nu(k).$$

Show that $M^{a\mu,b\nu}$ has zero eigenvalues. (Therefore, the operator M is not invertible.)

Exercise: Let us separate the QCD action in free and interacting parts: $S = S_0 + S_I$). Write S_I in the Fourier basis to show that it takes the following form.

$$
\begin{aligned}
S_I &= \int \frac{d^4k_1}{(2\pi)^4} \cdots \int \frac{d^4k_3}{(2\pi)^4} (2\pi)^4 (2\pi)^4 \delta(k_1 + \dots + k_3) \\
&\quad \times\ V^{abc}_{\lambda\mu\nu}(k_1, k_2, k_3)\tilde{A}^\lambda_a(k_1)\tilde{A}^\mu_b(k_2)\tilde{A}^\nu_c(k_3) \\
&\quad + \int \frac{d^4k_1}{(2\pi)^4} \cdots \int \frac{d^4k_4}{(2\pi)^4} (2\pi)^4 \delta(k_1 + \dots + k_4) \\
&\quad \times\ V^{abcd}_{\lambda\mu\nu\rho}(k_1, k_2, k_3, k_4)\tilde{A}^\lambda_a(k_1)\tilde{A}^\mu_b(k_2)\tilde{A}^\nu_c(k_3)\tilde{A}^\rho_d(k_4)
\end{aligned}
$$

where, $V^{abc}_{\lambda\mu\nu}(k_1, k_2, k_3)$ is symmetric in (λ, μ), (a, b) and (k_1, k_2). Find the expressions for $V^{abc}_{\lambda\mu\nu}(k_1, k_2, k_3)$ and $V^{abcd}_{\lambda\mu\nu\rho}(k_1, k_2, k_3, k_4)$.

4.4 Lecture IV

Let us recall the path integral prescription to compute Green's functions:

$$
\langle \phi_{i_1}(x_1) \cdots \phi_{i_n}(x_n)\rangle = \frac{\displaystyle\int \prod_{i=1}^{N} \mathcal{D}\phi^i\, e^{i\,S[\phi]} \phi_{i_1}(x_1) \cdots \phi_{i_n}(x_n)}{\displaystyle\int \prod_{i=1}^{N} \mathcal{D}\phi^i\, e^{i\,S[\phi]}}
$$

where $\{\phi_i\}$ are the set of fields in the theory. Splitting the action $S = S_0 + S_I$ one develops a diagramatic procedure for finding the correlation functions $\langle \phi_{i_1}(x_1) \cdots \phi_{i_n}(x_n)\rangle$ upto any given order in perturbation theory. But this procedure fails for gauge theories. In order to see this, consider the action of a gauge theory

$$
S_{gauge} = -\frac{1}{2}\int \frac{d^4k}{(2\pi)^4}\, \tilde{A}^a_\mu(-k)\delta_{ab}M_{a\mu,b\nu}A^b_\nu(k)\ +\ S_I, \qquad (4.34)
$$

where, the matrix M is

$$M_{a\mu,b\nu} = \delta_{ab}(k^2\eta^{\mu\nu} - k^\mu k^\nu).\tag{4.35}$$

The propagator is the matrix inverse of M. However, since

$$(k^2\eta^{\mu\nu} - k^\mu k^\nu)k_\nu = 0,\tag{4.36}$$

M has a zero eigenvalue, hence its inverse does not exist. In other words, the gauge field propagator cannot be calculated .

Now consider a more general set of objects:

$$\left\langle 0|T(\hat{\mathcal{O}}_1\cdots\hat{\mathcal{O}}_n)|0\right\rangle = \left\langle \prod_{\alpha=1}^{n}\mathcal{O}_\alpha\right\rangle,\tag{4.37}$$

where, $\mathcal{O}_\alpha$ is some combination of fields. For example, in QCD, we may consider the elementary fields $A_\mu^a(x)$, $\psi_s^k(x)$, $F_{\mu\nu}^a(x)$ or the composites $\bar{\psi}_s^k(x)\gamma^\mu\psi_t^k(x)$, $F_{\mu\nu}^a(x)F_{\rho\sigma}^a(x)$, etc[1]. The path integral prescription to compute the expectation value of such operators is given by

$$\left\langle \prod_\alpha \mathcal{O}_\alpha\right\rangle = \frac{\int \prod_{i=1}^{N}\mathcal{D}\phi_i\, e^{i\,S[\phi]}\prod_\alpha\mathcal{O}_\alpha}{\int \prod_{i=1}^{N}\mathcal{D}\phi_i\, e^{i\,S[\phi]}}.\tag{4.38}$$

The above is also ill-defined for reasons given above.

If two configuration $\{\tilde{\phi}_i(k)\}$ and $\{\tilde{\phi}_i'(k)\}$ are related by a gauge transform defined by

$$U(x) = \exp\left(-ig\theta^a(x)\tau^a\right),$$

then $\{\tilde{\phi}_i\}$ and $\{\tilde{\phi}_i{}'\}$ are said to be on the same gauge orbit (see Fig.4.1).

---

[1]One may ask as to why we are considering these more general operators when we have not solved the problem for the elementary fields. The point is that the composite operators are sometime simpler and often physically more meaningful.

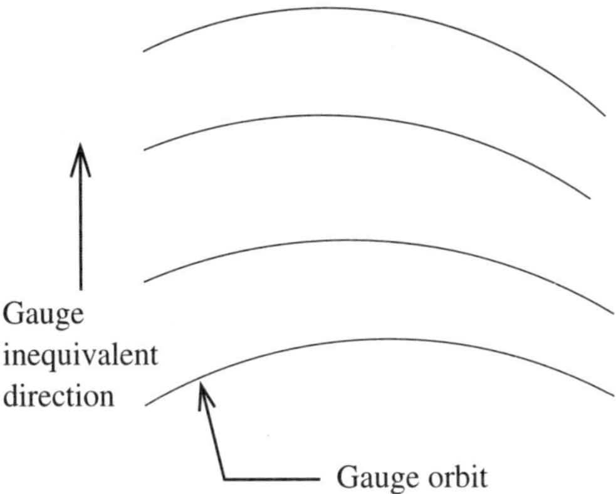

Figure 4.1: The space of gauge field configurations.

Naively on would think that we can decompose the measure as

$$\prod_{i=1}^{N} \mathcal{D}\tilde{\phi}^i = \prod_{i=1}^{N} \mathcal{D}\phi^i\Big|_{ineq} \prod_{i=1}^{N} \mathcal{D}\phi^i\Big|_{orbit} \tag{4.39}$$

so as to integrate along the gauge orbits and along the directions perpendicular to these. Using the fact that $S[\tilde{\phi}]$ is gauge invariant, we arrive at

$$\left\langle \prod_{\alpha} \mathcal{O}_\alpha \right\rangle = \frac{\int \prod_{i}^{N} \mathcal{D}\tilde{\phi}^i\big|_{ineq} e^{i\,S[\tilde{\phi}]} \int \prod_{i}^{N} \mathcal{D}\tilde{\phi}^i\big|_{orbit} \prod_{\alpha} \mathcal{O}_\alpha}{\int \prod_{i}^{N} \mathcal{D}\tilde{\phi}^i\big|_{ineq} e^{i\,S[\tilde{\phi}]} \int \prod_{i}^{N} \mathcal{D}\tilde{\phi}^i\big|_{orbit}}. \tag{4.40}$$

We can integrate along the gauge orbits first, but this fails because there is no convergence factor from an action, indeed the action for this part of the integration is zero. This is the old problem of the non-invertible matrix in the quadratic term in a gauge theory action. However, the infinities $\int \mathcal{D}\tilde{\phi}_i\big|_{orbit}$ from the numerator and the denominator will cancel if we could pull out $\mathcal{O}_\alpha$ out of the integral. This happens if all the $\mathcal{O}_\alpha$'s are gauge invariant. From now on we shall consider gauge invariant operators only (although we shall need to refer to gauge non-invariant ones later).

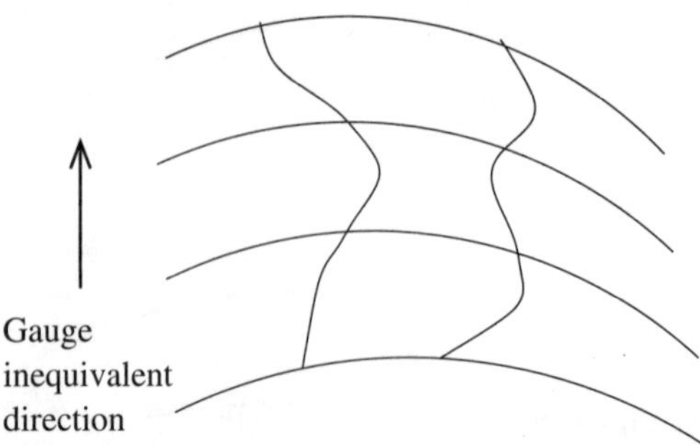

Figure 4.2: Two inequivalent ways of picking representatives from the gauge orbits.

Of the examples above, only $\bar{\psi}_s^k(x)\gamma^\mu\psi_t^k(x)$ and $F_{\mu\nu}^a(x)F_{\rho\sigma}^a(x)$ are allowed.

In order to define the path integral corresponding to the correlators of gauge invariant operators, we do the following:

1. The first step involves gauge fixing, which essentially means picking one point from each gauge orbit (see Fig.4.2).

   Since there are eight functions $\theta^\alpha(x)$ parametrising the gauge orbits of $SU(3)$, we need eight constraints to specify a point. Let us choose eight (gauge non-invariant) functions of fields $G^a(x)$ with constraints $G^a = B^a(x)$, where $B^a(x)$ are some fixed functions. For example we may choose

   $$\begin{aligned} G^a &= n^\mu A_\mu^a(x), \\ \text{or, } G^a &= \partial^\mu A_\mu^a(x), \end{aligned}$$

   where n^μ is an arbitrary vector. After Fourier transform to momentum space, $\tilde{G}^a = \tilde{B}^a(k)$. Roughly speaking, we want

   $$\prod_i \mathcal{D}\tilde{\phi}_i\Big|_{ineq} \sim \prod_i \mathcal{D}\phi_i \prod_{k,a}\left(\tilde{G}^a - \tilde{B}^a(k)\right),$$

   where the δ-function is at each point of the momentum space. However, this is not quite correct as we shall also have to make sure that any other way of gauge fixing will also lead to the same result. In the following we shall learn to manipulate the integral so as to get something like the above from the original expression for the correlators.

2. Define $\phi_{i\theta}$ as the tranformation of $\phi_i(x)$ by the gauge transform $U = e^{-ig\theta^a(x)T^a}$. Let $\phi_{i\theta} + \delta\phi_{i\theta}$ be the tranformation of $\phi_{i\theta}$ by an infinitesimal gauge transformation $(1 - ig\delta\theta^a(x)T^a)$. We denote G^a evaluated for $\phi_{i\theta}$ by G^a_θ and similarly $G^a_\theta + \delta G^a_\theta$ and $\tilde{G}^a_\theta$ have obvious meaning. In general,

$$\delta\widetilde{G^a_\theta} = \int \frac{d^4k'}{(2\pi)^4} \, K_{ab,\theta}(k, k') \, \delta\tilde{\theta}^b(k') \qquad (4.41)$$

For example, for the choice $G^a_\theta(x) = \partial^\mu A^a_{\mu\theta}(x)$, we have $\tilde{G}^a_\theta(k) = -ik^\mu \tilde{A}^a_{\mu\theta}(k)$ and $\delta\tilde{G}^a_\theta(k) = -ik^\mu \delta\tilde{A}^a_{\mu\theta}(k)$. From the gauge transformation of A^a_μ

$$\delta A^a_{\mu\theta}(x) = -\partial_\mu\delta\theta^a(x) + gf^{abc}\delta\theta^b(x)A^c_{\mu\theta}(x)$$

$$\delta\tilde{A}^a_{\mu\theta}(k) = ik_\mu\delta\tilde{\theta}^a(k) + gf^{abc}\int \frac{d^4k'}{(2\pi)^4}\delta\tilde{\theta}^b(k')\tilde{A}^c_{\mu\theta}(k - k'),$$

hence,

$$\begin{aligned} K_{ab,\theta}(k, k') &= k^2\,(2\pi)^4\delta_{ab}\,\delta^4(k - k') \\ &\quad -igf^{abc}k^\mu\tilde{A}^c_{\mu\theta}(k - k'). \end{aligned} \qquad (4.42)$$

At this point, we claim that

$$\begin{aligned} \hat{\mathcal{N}} &\equiv \int \mathcal{D}\tilde{\theta}^a(k) \prod_{k,a} \delta\left(\tilde{G}^a_\theta(k) - \tilde{B}^a(k)\right) \det K_{ab\theta}(k, k') \\ &= \frac{(2\pi)^4}{b^4}, \end{aligned} \qquad (4.43)$$

(recall that b is the lattice size of the discrete momentum space). The above follows from the analogous finite dimensional identity

$$\int \prod_{i=1}^N du_i \prod_{i=1}^N \delta\left(\lambda F_i(\vec{u})\right) \det\left(\frac{\partial F_i(\vec{u})}{\partial u_j}\right) = \lambda^{-N}, \qquad (4.44)$$

where we have assumed that $F_i(\vec{u})$, for all i, vanishes only at one point $\vec{u}_0$. The translation from the finite dimensional case to the functional identity (4.43) is provided by the dictionary:

$$
\begin{aligned}
i &\rightarrow (a, k), \\
u_i &\rightarrow \tilde{\theta}^a(k), \\
F_i(\vec{u}) &\rightarrow \tilde{G}^a_\theta(k) - \tilde{B}^a(k), \\
\lambda &\rightarrow b^4/(2\pi)^4, \\
\frac{\partial F_i}{\partial u_j} &\rightarrow K_{ab\theta}(k, k').
\end{aligned}
$$

The numerator of the path integral, by inserting an '1', can be rewritten as

$$
\int \prod_i^N \mathcal{D}\tilde{\phi}_i \, e^{i\,S[\tilde{\phi}]} \prod_\alpha \mathcal{O}_\alpha(\tilde{\phi}) \times \hat{\mathcal{N}}^{-1}
$$

$$
\times \int \mathcal{D}\tilde{\theta}^a(k) \prod_{k,a} \delta\left(\tilde{G}^a_\theta(k, \tilde{\phi}_\theta) - \tilde{B}^a(k)\right)
$$

$$
\times \det K_{ab\theta}(k, k', \tilde{\phi}_\theta).
$$

The first (original) part of the above integrand is independent of the gauge parameters $\tilde{\theta}^a$. We can therefore change the order of integration to write

$$
\hat{\mathcal{N}}^{-1} \int \prod_a \mathcal{D}\tilde{\theta}^a \int \prod_i \mathcal{D}\tilde{\phi}^i \, e^{i\,S[\tilde{\phi}]} \prod_\alpha \mathcal{O}_\alpha(\tilde{\phi}_\theta)
$$

$$
\times \prod_{k,a} \delta\left(\tilde{G}^a(k, \tilde{\phi}_\theta) - \tilde{B}^a(k)\right) \det K_{ab}(k, k', \tilde{\phi}_\theta).
$$

Now, using the fact that the action $S[\tilde{\phi}]$, the operators $\mathcal{O}_\alpha(\tilde{\phi})$ and the measure $\mathcal{D}\tilde{\phi}$ are all gauge invariant[2], we can replace these by $S[\tilde{\phi}_\theta]$, $\mathcal{O}_\alpha(\tilde{\phi}_\theta)$ and $\mathcal{D}\tilde{\phi}_\theta$ respectively. Now all dependence on θ is through the dummy variable of integration

---

[2]The functional measure is gauge invariant in QCD, but this is not necessarily the case for all gauge theories.

$\tilde{\phi}_\theta$, which can be redefined to be $\tilde{\phi}$. So we make a change of variable from $\tilde{\phi}_\theta$ to $\tilde{\phi}$ and get

$$\hat{\mathcal{N}}^{-1} \int \prod_a \mathcal{D}\tilde{\theta}^a \int \prod_i \mathcal{D}\tilde{\phi}^i \, e^{i \, S[\tilde{\phi}]} \prod_\alpha \mathcal{O}_\alpha(\tilde{\phi})$$

$$\times \prod_{k,a} \delta\left(\tilde{G}^a(k,\tilde{\phi}) - \tilde{B}^a(k)\right) \det K_{ab}(k,k',\tilde{\phi}).$$

as the final expression for the numerator.

4.5   Lecture V

In the previous lecture we have obtained an expression for the correlators of gauge invariant operators in a gauge theory. Actually we manipulated only the numerator. However, a similar manipulation can be done for the denominator as well. It is now easy to see that the integral

$$\int \prod_a \mathcal{D}\tilde{\theta}^a$$

in the numerator and the denominator cancel each other. Parenthetically let us notice that only the absolute value of $\det K_{ab}(k,k')$ matters as its sign can be absorbed in $\hat{\mathcal{N}}^{-1}$ which cancels out in the end. We shall now express $\prod_{k,a} \delta\left(\tilde{G}^a(k) - \tilde{B}^a(k)\right)$ and $\det K_{ab}(k,k')$ in a way so that we can use standard perturbation theory. In order to do this we use

$$\int \prod_a \mathcal{D}\tilde{B}^a(k) \, \exp\left(-\frac{i}{2\alpha} \int \frac{d^4 k}{(2\pi)^4} \tilde{B}^a(-k)\tilde{B}^a(k)\right) = \tilde{\mathcal{N}}(\alpha),$$

$$(4.45)$$

where $\tilde{\mathcal{N}}(\alpha)$ is independent of $\tilde{B}^a(k)$. Inserting $\tilde{\mathcal{N}}^{-1}(\alpha)$ times the LHS above in the correlator we get:

$$\left\langle \prod_\alpha \mathcal{O}_\alpha(\tilde{\phi}) \right\rangle = \frac{1}{\tilde{\mathcal{N}}(\alpha)} \int \prod_a \mathcal{D}\tilde{B}^a(k) e^{-\frac{i}{2\alpha} \int \frac{d^4 k}{(2\pi)^4} \tilde{B}^a(-k)\tilde{B}^a(k)}$$

$$\times \frac{1}{\hat{\mathcal{N}}} \int \prod_a \mathcal{D}\tilde{\theta}^a \int \prod_i \mathcal{D}\tilde{\phi}^i e^{iS[\tilde{\phi}]}$$

$$\times \prod_\alpha \mathcal{O}_\alpha(\tilde\phi) \prod_{k,a} \delta\left(\tilde G^a(k) - \tilde B^a(k)\right) \det K_{ab}(k,k')$$

$$= \mathcal{N}_0 \int \prod_i \mathcal{D}\tilde\phi^i \exp\left(-\frac{i}{2\alpha}\int \frac{d^4k}{(2\pi)^4}\tilde G^a(-k)\tilde G^a(k)\right)$$

$$\times e^{iS[\tilde\phi]} \prod_\alpha \mathcal{O}_\alpha(\tilde\phi) \det K_{ab}(k,k'), \qquad (4.46)$$

where $\mathcal{N}_0$ is a new constant.

We shall now rewrite $\det K_{ab}(k,k')$. To this end, let us recall the results of gaussian integration

$$\int \prod_{i=1}^N dx^i \prod_{j=1}^N dy^j \exp\left(\frac{i}{2}x^i A_{ij}y^j\right) = \frac{1}{\det A}, \qquad (4.47)$$

for commuting variables x, y and

$$\int \prod_{i=1}^N d\xi^i \prod_{j=1}^N d\eta^j \exp\left(\frac{i}{2}\xi^i A_{ij}\eta^j\right), = \det A \qquad (4.48)$$

for anti-commuting (Grassmann) variables ξ, η. We may write

$$\det K_{ab}(k,k') = \int \mathcal{D}\tilde{\bar\chi}^a \mathcal{D}\tilde\chi^a \exp\left[i\int \frac{d^4k}{(2\pi)^4}\frac{d^4k'}{(2\pi)^4}\right.$$

$$\left.\times \tilde{\bar\chi}^a(-k)K_{ab}(k,k')\tilde\chi^b(k')\right],$$

where $\tilde{\bar\chi}^a$ and $\tilde\chi^a$ are two independent spin-zero anti-commutating fields. They do not carry any spinor index (but only gauge index), yet they anti-commute and, hence they violate the spin-statistics theorem. Due to this peculiar property, these fields, which we have introduced for our convenience, are known as *ghost* fields.

Finally we have the desired expression for the correlator:

$$\left\langle \prod_\alpha \mathcal{O}_\alpha(\tilde\phi) \right\rangle = \int \prod_i \mathcal{D}\tilde\phi^i \prod_a \mathcal{D}\tilde{\bar\chi}^a \prod_a \mathcal{D}\tilde\chi^a \, e^{iS_{tot}} \prod_\alpha \mathcal{O}_\alpha \qquad (4.49)$$

where,

$$S_{tot} = S[\tilde\phi] - \frac{1}{2\alpha}\int \frac{d^4k}{(2\pi)^4}\tilde G^a(-k)\tilde G^a(k)$$

$$+ \int \frac{d^4k}{(2\pi)^4} \frac{d^4k'}{(2\pi)^4} \overline{\tilde{\chi}}^a(-k) K_{ab}(k,k') \tilde{\chi}^b(k'). \quad (4.50)$$

Notice that S_{tot} depends on α as well as on $\tilde{G}^a$ although the original action we started with was independent of these. It turns out that the Green's functions of gauge invariant operators actually are independent of α and $\tilde{G}^a$. However, this is not true of Green's functions of operators that are not gauge invariant but appear in the intermediate steps of calculations. This independence provides a nice check on the calculations.

Let us now make an explicit gauge choice, called the Lorenz gauge

$$\tilde{G}^a(k) = -ik^\mu \tilde{A}_\mu(k), \quad (4.51)$$

(which in coordinate space is $G^a(x) = \partial_\mu A^{\mu a}(x)$). For this choice, we have

$$-\frac{1}{2\alpha} \int \frac{d^4k}{(2\pi)^4} \tilde{G}^a(-k)\tilde{G}^a(k) = -\frac{1}{2\alpha} \int \frac{d^4k}{(2\pi)^4} \tilde{A}_\mu^a(-k) k^\mu k^\nu \tilde{A}_\nu^b(k) \delta^{ab},$$
$$(4.52)$$

which is quadratic in gauge fields. The complete quadratic term for A in S_{tot} is

$$\frac{1}{2} \int \frac{d^4k}{(2\pi)^4} \tilde{A}_\mu^a(-k)\delta^{ab} \left(-\eta^{\mu\nu}k^2 + k^\mu k^\nu - \frac{1}{\alpha} k^\mu k^\nu \right) \tilde{A}_\nu^b(k), \quad (4.53)$$

where the first two terms (independent of α) come from the original kinetic term from $S[\tilde{\phi}]$ and the last term is from gauge fixing. It should be noted that the modified kernel

$$\left(-\eta^{\mu\nu}k^2 + k^\mu k^\nu - \frac{1}{\alpha}k^\mu k^\nu \right) \quad (4.54)$$

no longer has a zero eigenvalue. Hence it is is invertible and one can define the propagator of the gauge fields without any difficulty. Indeed, that was the purpose of this exercise. Since the parameter α is arbitrary, we can choose a specific value to simplify computation. A particularly simple choice is $\alpha = 1$, for which the quadratic term in the action is

$$S^{(2)} = \frac{1}{2} \int \frac{d^4k}{(2\pi)^4} \tilde{A}_\mu^a(-k) \left(-\delta^{ab} k^2 \eta^{\mu\nu} \right) \tilde{A}_\nu^b(k). \quad (4.55)$$

Therefore, the gauge field propagator is

$$-i\,\delta^{ab}\,\frac{1}{k^2}\,\eta_{\mu\nu}. \tag{4.56}$$

The gauge fixing condition $G^a(x) = \partial^\mu A_\mu^a(k)$ results in

$$K_{ab}(k, k') = \delta_{ab}(2\pi)^4\delta(k - k')\,k^2 - igk^\mu f^{abc}\tilde{A}_\mu^c(k - k'). \tag{4.57}$$

Hence, the third term in the S_{tot}, which we shall call S_{ghost}, becomes

$$
\begin{aligned}
S_{ghost} \;=\; & \int \frac{d^4k}{(2\pi)^4}\tilde{\bar{\chi}}^a(-k)k^2\,\delta^{ab}\tilde{\chi}^b(k) \\
& -ig \int \frac{d^4k}{(2\pi)^4}\frac{d^4k'}{(2\pi)^4} f^{abc}k^\mu\tilde{\bar{\chi}}^a(-k)\tilde{\chi}^b(k')\tilde{A}_\mu^c(k - k').
\end{aligned} \tag{4.58}
$$

The Feynman rules with the modified action involves two additional diagrams:

The ghost propagator:

$$\bar{\chi} \quad\text{---}\!\!\blacktriangleleft\text{---}\!\!\blacktriangleleft\text{---}\quad \chi \qquad\qquad = \frac{i}{k^2}\delta^{ab}$$

$$a \quad k \qquad\qquad k \quad b$$

Ghost-gauge boson coupling:

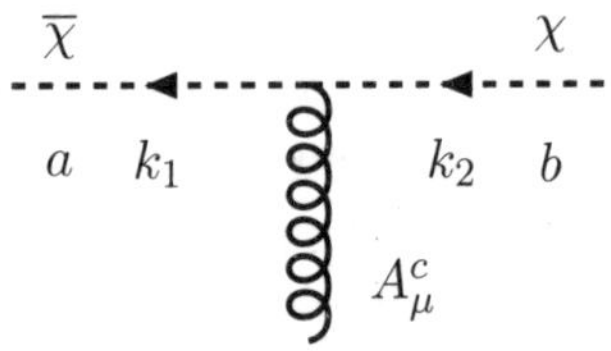

$$\bar{\chi} \quad\text{---}\!\!\blacktriangleleft\text{---}\!\!\blacktriangleleft\text{---}\quad \chi \qquad = -g f^{abc}k_{1\mu}$$

$$a \quad k_1 \qquad\qquad k_2 \quad b$$

$$A_\mu^c$$

The contribution from these diagrams will also have to be taken into account in the computation of amplitudes.

4.6 Lecture VI

We have obtained the Feynman rules of QCD in the path integral formalism. One can now proceed to calculate amplitude for physical processes exactly as in the canonical method. We do, however, encounter divergences at the loop level. These require regularization and renormalization before we can extract physically meaningful expressions. An example of a divergent graph is the following.

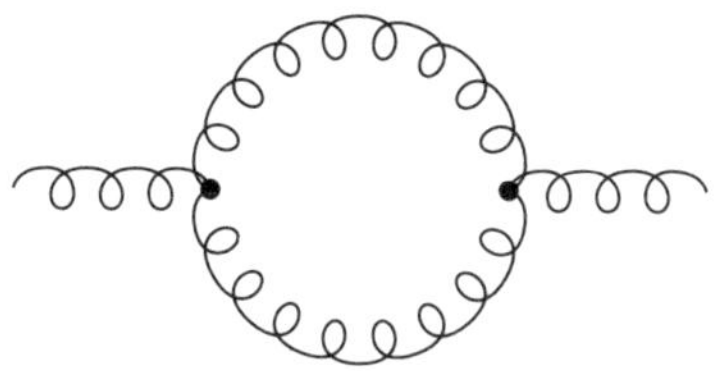

This graph contributes

$$\int \frac{d^4\ell}{(2\pi)^4} \frac{1}{p_1^2\, p_2^2\, \ell^2\, (\ell + p_1)^2}\, \mathcal{V}_{(3)}(\ell)\mathcal{V}_{(3)}(\ell + p_1), \qquad (4.59)$$

where the vertices contain one momentum factor each and the momentum conservation condition $(2\pi)^4 \delta^{(4)}(p_1 + p_2)$ has been suppressed. For large values of internal momentum ℓ, the integral goes as

$$\sim \int \frac{d^4\ell}{(2\pi)^4} \frac{1}{\ell^2} \cdot \frac{1}{\ell^2} \cdot \ell \cdot \ell, \qquad (4.60)$$

which diverges quadratically.

Regularization and renormalization are two steps in the process of renormalization which makes sense of these divergent integrals. In the first step of regularization, the integrals are changed in such a way that they become finite. (It should be pointed out that these are not the same integrals anymore.) There are various ways to do this, for example, with momentum cut-off one puts an upper limit on the value of the internal momenta $|\ell_\mu| < \Lambda$, where

Λ is some fixed mass parameter. One takes $\Lambda \to \infty$ at the end of the renormalization procedure. However, momentum cut-off is not very convenient in QCD. We shall use a different regularization called *dimensional regularization*.

In dimensional regularization, one pretends to work in dimension $D < 4$, where D is chosen such that the integrals are not divergent any longer. Only at the end of renormalization, we take the limit $D \to 4$. We face a problem here since taking the limit is a continuous process. The solution is to write the formulas as functions of D and treat the dimension D (formally) as a continuous variable. In other words, we derive the relevant expressions for general D, without committing to a specific dimension. These are expressions in which D appears as a variable. We pretend that D is continuous and take the $D \to 4$ limit. In anticipation of the fact that we want to take the $D \to 4$ at the end of our calculations, we define

$$\epsilon = 4 - D.$$

This leads to the following modifications

$$\int \frac{d^4 k}{(2\pi)^4} \quad \to \quad \frac{d^{4-\epsilon} k}{(2\pi)^{4-\epsilon}},$$
$$(2\pi)^4 \, \delta^{(4)}(k) \quad \to \quad (2\pi)^{4-\epsilon} \, \delta^{(4-\epsilon)}(k),$$
$$\eta^{\mu\nu} \eta_{\mu\nu} \quad = \quad 4 - \epsilon.$$

We shall also need to deal with the gamma matrices, which satisfy

$$\{\gamma^\mu, \gamma^\nu\} = 2\eta^{\mu\nu} \tag{4.61}$$

in any dimensions. This leads to

$$\gamma^\mu \gamma_\mu = 4 - \epsilon,$$
$$\gamma^\mu \gamma^\nu \gamma_\mu = (2 - D)\gamma^\nu = (\epsilon - 2)\gamma^\nu.$$

There is a subtlety here. What is the dimension of the gamma matrices? While the irreducible representations of Clifford algebra (4.61) is four dimensional in $D = 4$, it is smaller in lower dimensions. However, we can, if we like, use the four dimensional gamma matrices in lower dimensions. (They form a reducible representation.) We shall therefore fix the dimension of the gamma matrices to be four, irrespective of the value of D. This means

$$\text{tr}\,(\gamma^\mu \gamma^\nu) = 4\eta^{\mu\nu} \tag{4.62}$$

will always hold.

The following are useful integration rules in $4 - \epsilon$ dimensions.

$$\int \frac{d^{4-\epsilon}k}{(2\pi)^{4-\epsilon}} \frac{1}{(-k^2 + L - i\varepsilon)^2} = i\,\frac{L^{-\epsilon/2}\Gamma\left(\frac{\epsilon}{2}\right)}{(4\pi)^{2-\frac{\epsilon}{2}}}, \tag{4.63}$$

$$\int \frac{d^{4-\epsilon}k}{(2\pi)^{4-\epsilon}} \frac{1}{-k^2 + L - i\varepsilon} = -i\,\frac{L^{1-\frac{\epsilon}{2}}\Gamma\left(\frac{\epsilon}{2}\right)}{(4\pi)^{2-\frac{\epsilon}{2}}\left(1 - \frac{\epsilon}{2}\right)}, \tag{4.64}$$

where, the first is true for $\epsilon = 1, 2, 3$ and the second only for $\epsilon = 3$.

Exercise: Verify the eqns. (4.63) and (4.64) above.

As yet we have not solved our problem of infinities since $\Gamma\left(\frac{\epsilon}{2}\right)$ is divergent at $\epsilon = 0$.

Now we come to the second step of renormalization. Notice that after the regularization process, the integrals that appear in the calculation of physical amplitudes are still divergent. However, we have some arbitrary parameters in the theory, *e.g.* (m and λ in case of the ϕ^4 theory). Suppose we say that these parameters also depend on ϵ and indeed are divergent as $\epsilon \to 0$ is exactly such a way that they cancel the divergence of the integrals. The reason that it makes sense is due to the fact that the (divergent) parameters in the Lagrangain are not directly observable, but they render physical quantities that are measured in experiments finite in this process.

To be more specific, consider a field theory with fields $\{\phi_i\}$, $i = 1, 2, \cdots, N$; and parameters $\{g_s\}$, $s = 1, 2, \cdots, M$ (these are masses, coupling constants, gauge fixing parameters, etc.). Introduce new parameters and new fields as follows

$$
\begin{aligned}
g_s &= Z_s(\{g_R\}, \epsilon)\, g_{sR} \\
\tilde{\phi}_i &= \tilde{Z}_i^{1/2}(\{g_R\}, \epsilon)\, \tilde{\phi}_{iR},
\end{aligned}
$$

where $Z, \tilde{Z}$ are arbitrary fixed functions of the new parameters and ϵ. (The tilde in $\tilde{Z}$ in the second equation is due to the fact that we consider the fields $\{\tilde{\phi}\}$ in momentum space.) For example, in QED, $e = \sin e_R \left(1 + \frac{1}{\epsilon}\right)$.

Therefore, for a correlation function of renomalized fields, we have

$$\left\langle \prod_{k=1}^{n} \phi_{i_k R}(x_k) \right\rangle = \frac{1}{\displaystyle\prod_{k=1}^{n} \sqrt{Z_{i_k}(\{g_R\}, \epsilon)}} \left\langle \prod_{k=1}^{n} \phi_{i_k}(x_k) \right\rangle$$

$$= F_{i_1 \cdots i_n}(x_1, \cdots, x_n; \{g_R\}, \epsilon) \qquad (4.65)$$

We would like to know if it is possible to choose $Z_s(\{g_R\}, \epsilon)$ and $Z_i(\{g_R\}, \epsilon)$ such that $F_{i_1 \cdots i_n}(x_1, \cdots, x_n; \{g_R\}, \epsilon)$ are finite as $\epsilon \to 0$ at fixed g_R. If so, all Green's functions are finite in terms of the new fields and the theory is called a *renormalizable theory* and g_R's *renormalized* coupling constants. Actually this done in two steps. Firstly one shows that some Green's functions can be made finite by appropriate choice of Z_s and Z_i's. Then one shows that with these choices *all* other Green's functions are finite. This proves the *renormalizability* of a theory. For example, ϕ^4 theory and QCD are both *renormalizable*.

Notice that to lowest order in perturbation theory, there is no need to renormalize. This is because, to this order, there are only tree graphs, *i.e.* there are no loops with undetermined momenta, hence there are no divergences. This implies that

$$Z_s = 1 + \text{powers of renormalized coupling constants,}$$
$$\tilde{Z}_i = 1 + \text{powers of renormalized coupling constants.}$$

In the above, we have taken the leading term to be 1 for simplicity. This could have been chosen to be any other constants. Using above

$$Z_s g_{sR} = g_{sR} + (Z_s - 1)g_{sR},$$
$$\tilde{Z}_i^{1/2} \tilde{\phi}_{iR} = \tilde{\phi}_{iR} + \left(\tilde{Z}_i^{1/2} - 1\right)\tilde{\phi}_{iR}, \qquad (4.66)$$

which lets us rewrite the Lagrangian density as follows.

$$\mathcal{L}\left(\phi_i, g_s\right) = \mathcal{L}\left(Z_i^{1/2}\phi_{iR}, Z_s g_s\right) = \mathcal{L}\left(\phi_{iR}, g_{sR}\right) + \mathcal{L}_{ct}, \qquad (4.67)$$

where, $\mathcal{L}_{ct}$, called the *counterterm* Lagrangian contains all the other terms in the Taylor series expansion in renormalized coupling constants. The first term in the RHS is obviously dominant in the limit in which the coupling constants go to zero. The action can be expressed as

$$S = S_0 + S_I, \qquad (4.68)$$

where, S_0 contains only terms quadratic in fields from the Lagrangian $\mathcal{L}\left(\{\phi_{iR}\}, \{g_{sR}\}\right)$, (the first in the RHS of (4.67), and S_I has all the rest of the terms, *i.e*, the interaction terms as well as new terms from the counterterms. Notice that the new *interaction term S_I* may now contain terms quadtratic in fields, but these already have at least one power of coupling constants.

Renomalization of ϕ^4 theory

The ϕ^4 theory is an interacting theory of one real scalar field ϕ in four dimensions, defined by the action

$$S = \int d^4x \left(\frac{1}{2}\partial_\mu\phi\,\partial^\mu\phi - \frac{1}{2}m^2\phi^2 - \frac{1}{4!}\lambda\phi^4\right). \qquad (4.69)$$

Following the prescription outlined above, we introduce renormalized fields and couplings in $(4-\epsilon)$ dimensions

$$\phi = Z_\phi^{1/2}\phi_R \qquad\qquad m = Z_m m_R \qquad\qquad \lambda = Z_\lambda \lambda_R, \qquad (4.70)$$

and rewrite the action as

$$S = S_0 + S_I = \int d^{4-\epsilon}x\left(\mathcal{L}_0 + \mathcal{L}_I\right),$$

where,

$$\mathcal{L}_0 = \frac{1}{2}\partial_\mu\phi_R\,\partial^\mu\phi_R - \frac{1}{2}m_R^2\phi_R^2 - \frac{1}{4!}\lambda_R\phi_R^4, \qquad (4.71)$$

$$\mathcal{L}_I = \frac{(Z_\phi - 1)}{2}\partial_\mu\phi_R\,\partial^\mu\phi_R - \frac{(Z_\phi Z_m^2 - 1)}{2}m_R^2\phi_R^2$$
$$- \frac{(Z_\phi^2 Z_\lambda - 1)}{4!}\lambda_R\phi_R^4.$$

We shall treat all of S_I as perturbation as all terms in it are of order at least λ_R.

Notice that in $(4-\epsilon)$ dimensions the scalar field has dimension $[\phi] = M^{1-\frac{\epsilon}{2}}$ and consequently the coupling contant is no longer dimensionless. Indeed $[\lambda] = M^\epsilon$. It is more convenient to introduce a parameter μ of (mass) dimension one, $([\mu] = M)$, such that

$$\lambda_R = \mu^\epsilon g_R, \tag{4.72}$$

where g_R is a new dimensionless coupling constant.

Let us now calculate the two-point function $\left\langle \tilde{\phi}_R(p_1)\tilde{\phi}_R(p_2) \right\rangle$. To order λ_R, this is given by the sum of the following Feynman diagrams

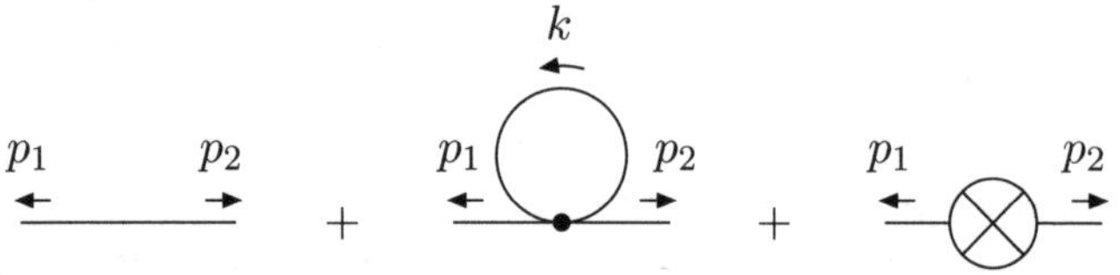

plus terms of $\mathcal{O}(\lambda_R^2)$. The first diagram above is the tree level contribution and is clearly finite, but both the second and the third diagrams give divergent contributions. More specifically, the second diagram is

$$\frac{i}{p_i^2 - m_R^2}\frac{i}{p_i^2 - m_R^2}(2\pi)^{4-\epsilon}\delta^{(4-\epsilon)}(p_1 + p_2)$$

$$\times \left(-\frac{ig_R\mu^\epsilon}{4!}\right) 4.3 \int \frac{d^{4-\epsilon}k}{(2\pi)^{4-\epsilon}}\frac{i}{(k^2 - m_R^2 + i\epsilon)}, \tag{4.73}$$

which diverges as $\dfrac{m_R^{2-\epsilon}}{(4\pi)^{2-\frac{\epsilon}{2}}}\dfrac{2}{\epsilon}$. The third diagram, on the other hand, contributes

$$\frac{i}{p_i^2 - m_R^2}\frac{i}{p_i^2 - m_R^2}(2\pi)^{4-\epsilon}\delta^{(4-\epsilon)}(p_1 + p_2)$$

$$\times i\left[\frac{1}{2}(\tilde{Z}_\phi - 1)p_1^2 - \frac{1}{2}(\tilde{Z}_\phi Z_m^2 - 1)m_R^2\right]. \tag{4.74}$$

Now we require that the divergences cancel. Since there is no divergent term dependent on p_1^2 in (4.73) but it is present in the m_R^2-dependent term, we find that

$$\tilde{Z}_\phi = 1, \qquad \tilde{Z}_m^2 - 1 = \frac{g_R}{16\pi^2\epsilon}. \qquad (4.75)$$

Let us notice that the above choices are not unique. This means that there is an ambiguity in the renormalization procedure. We shall choose counterterm to subtract only the divergent $1/\epsilon$, *i.e.* pole, term. This is known as the *minimal subtraction* scheme.

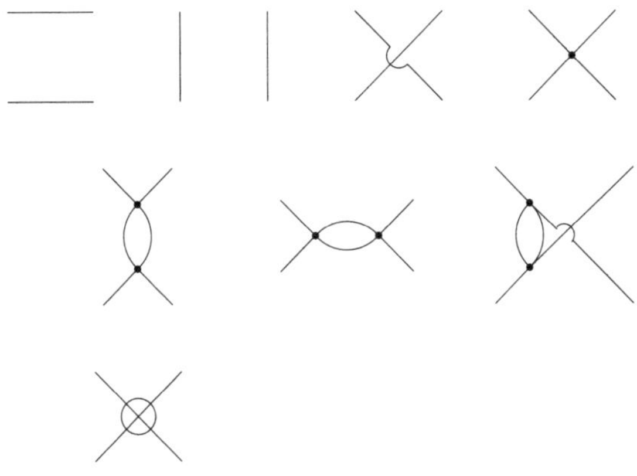

Feynman diagrams for the four-point function

in ϕ^4 field theory

Finally, in order to determine Z_λ, consider the four-point function

$$\left\langle \tilde{\phi}_R(p_1)\tilde{\phi}_R(p_2)\tilde{\phi}_R(p_3)\tilde{\phi}_R(p_4)\right\rangle.$$

The Feynman diagrams above contribute to this upto $\mathcal{O}(\lambda_R^2)$. The second and the third rows of the diagrams are divergent. Demanding that these divergences cancel determines

$$Z_\lambda = 1 + \frac{3g_R}{16\pi^2\epsilon}. \tag{4.76}$$

4.7   Lecture VII

Renormalization of QCD

In the last lecture, we learnt how to renormalize the ϕ^4 theory. We follow the same steps in renormalizing QCD.

First, start with the full gauge fixed action

$$S = \int d^{4-\epsilon}x\, \mathcal{L}\left(A_\mu^a, \chi^a, \bar{\chi}^a, \psi_s^i; g, \alpha, m_s\right). \tag{4.77}$$

Secondly, introduce renormalized fields and coupling constants:

$$\begin{aligned}
A_\mu^a &= \tilde{Z}_A^{1/2} A_{\mu R}^a, & \chi^a &= \tilde{Z}_\chi^{1/2}\chi_R^a, \\
\bar{\chi}^a &= \tilde{Z}_\chi^{1/2}\bar{\chi}_R^a, & \psi_s^i &= \tilde{Z}_{\psi_s}^{1/2}\psi_{sR}^i, \\
g &= Z_g g_R \mu^{\epsilon/2}, & & \\
\alpha &= Z_\alpha \alpha_R, & & \\
m_s &= Z_{m_s} m_{sR}.
\end{aligned} \tag{4.78}$$

A few remarks are in order. The function $\tilde{Z}_A$ is taken to be the same for all components A_μ. This is ensured by Lorentz invariance. Similarly, by SU(3) symmetry, all χ^a's are multiplied by the same $\tilde{Z}_\chi$. Further, we have set $\tilde{Z}_\chi = \tilde{Z}_{\bar{\chi}}$ for convenience. This can be done since they always appear together. Finally, we have separated the dimensionless coupling g_R by explicitly introducing appropriate power of μ.

Thirdly, write the Lagrangian $\mathcal{L}(A_\mu^a, \cdots)$ as a function of the renormalized fields and couplings, and as in the case of ϕ^4 theory, separate it into two parts. The first part dominates in the limit where the coupling goes to zero, and the second part is the 'interaction' including counterterms.

For example, from the quadratic terms in Lagrangian for the gauge fields, we have

$$-\frac{1}{2}\int \frac{d^{4-\epsilon}k}{(2\pi)^{4-\epsilon}}\, \tilde{A}_\mu^a(-k)\delta^{ab}\left(k^2\eta^{\mu\nu} - k^\mu k^\nu + \frac{1}{\alpha}k^\mu k^\nu\right)\tilde{A}_\nu^b(k)$$

$$= -\frac{1}{2}\int \frac{d^{4-\epsilon}k}{(2\pi)^{4-\epsilon}}\, \tilde{Z}_A \tilde{A}_{\mu R}^a(-k)\delta^{ab}$$
$$\left(k^2\eta^{\mu\nu} - \left(1 - \frac{Z_\alpha^{-1}}{\alpha_R}\right)k^\mu k^\nu\right)\tilde{A}_{\nu R}^b(k)$$

$$= -\frac{1}{2}\int \frac{d^{4-\epsilon}k}{(2\pi)^{4-\epsilon}}\, \tilde{A}_{\mu R}^a(-k)\delta^{ab}\left(k^2\eta^{\mu\nu} - \left(1 - \frac{1}{\alpha_R}\right)k^\mu k^\nu\right)\tilde{A}_{\nu R}^b(k)$$
$$- \frac{1}{2}\int \frac{d^{4-\epsilon}k}{(2\pi)^{4-\epsilon}}\, (\tilde{Z}_A - 1)\tilde{A}_{\mu R}^a(-k)\delta^{ab}\left(k^2\eta^{\mu\nu} - k^\mu k^\nu\right)\tilde{A}_{\nu R}^b(k)$$
$$- \frac{1}{2}\int \frac{d^{4-\epsilon}k}{(2\pi)^{4-\epsilon}}\, \frac{\tilde{Z}_A Z_\alpha^{-1} - 1}{\alpha_R}\, \tilde{A}_{\mu R}^a(-k)\delta^{ab}k^\mu k^\nu \tilde{A}_{\nu R}^b(k),$$

where the last two lines describe the counterterms $\mathcal{L}_{ct}$.

Next, pretend that we did not know the original fields and parameters and develop perturbation theory using renormalized fields and parameters. (To this end, include all the counterterms in the interaction part S_I.)

Finally, determine the functions Z's and $\tilde{Z}$'s by requiring that all Green's functions are finite in the limit $\epsilon \to 0$.

As an example, let us calculate the gluon two-point function to one-loop. We have the propagator,

$$k$$

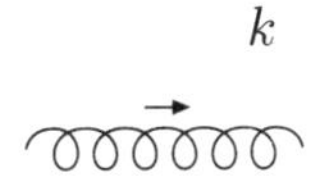

and radiative correction to it from gluon, fermion and ghost loops

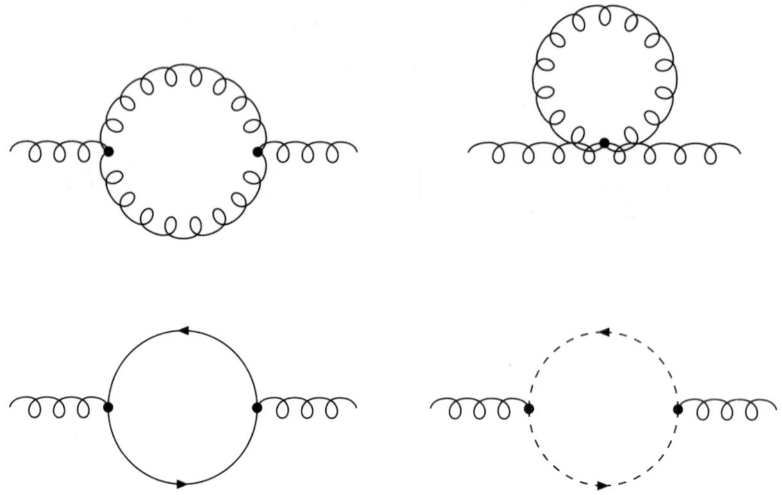

respectively. Finally, the counterterm

which equals

$$-\frac{i}{2}\left[(\tilde{Z}_A - 1)\left(k_1^2\eta^{\mu\nu} - k_1^\mu k_1^\nu\right) + \frac{(\tilde{Z}_A Z_\alpha^{-1} - 1)}{\alpha_R}k_1^\mu k_1^\nu\right]\delta^{ab}$$

Requiring that this Green function is finite at one-loop, we have

$$\tilde{Z}_A = 1 - \frac{g_R^2}{8\pi^2\epsilon}\left[\frac{4}{3}T_R N_f - \frac{1}{2}C_a\left(\frac{13}{3} - \alpha_R\right)\right], \tag{4.79}$$

where, N_f is the number of flavours, and $\tilde{Z}_A Z_\alpha^{-1} = 1$, that is

$$Z_\alpha = \tilde{Z}_A. \tag{4.80}$$

In the above, we have used the following convention for normalization.

$$\begin{aligned}
\text{Tr}\,(T^a T^b) &= T_R\, \delta^{ab}, \\
f^{acd} f^{bcd} &= C_a\, \delta^{ab}.
\end{aligned} \tag{4.81}$$

For instance, $C_a = 3$ for the group SU(3).

Similarly, from the ghost two-point function at one loop,

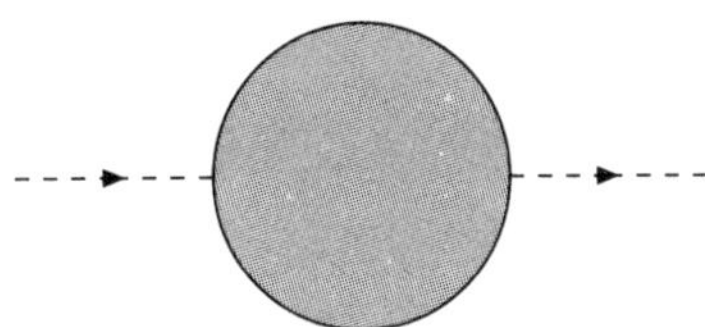

we have,

$$\tilde{Z}_\chi = 1 + \frac{g_R^2}{32\pi^2 \epsilon}\, C_a(3 - \alpha_R). \tag{4.82}$$

The corresponding expression for the fermions determine

$$\begin{aligned}
\tilde{Z}_{\psi_s} &= 1 - \frac{g_R^2}{8\pi^2 \epsilon}\, C_F \alpha_R, \\
\tilde{Z}_{\psi_s} Z_{m_s} &= 1 - \frac{g_R^2}{8\pi^2 \epsilon}\, C_F(3 + \alpha_R),
\end{aligned} \tag{4.83}$$

hence,

$$Z_{m_s} = 1 - \frac{3 g_R^2}{8\pi^2 \epsilon}\, C_F. \tag{4.84}$$

Here, C_F is defined as

$$(T^a T^b)_{ij} = C_F\, \delta_{ij}, \tag{4.85}$$

i.e. $C_F = 4/3$ for QCD.

Demanding that the gluon three-point function to one loop

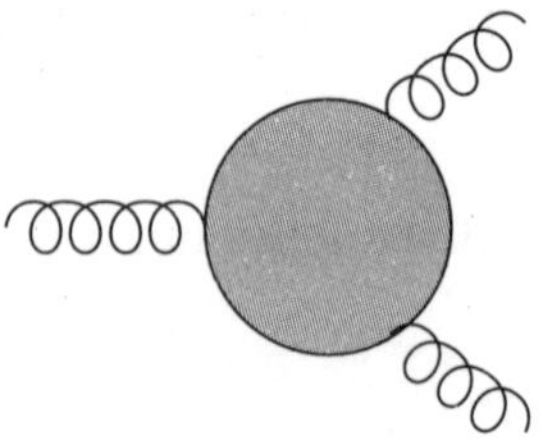

is finite, require that

$$\tilde{Z}_A^{3/2} Z_g = 1 - \frac{g_R^2}{8\pi^2\epsilon} C_a \left(-\frac{17}{2} + \frac{3\alpha_R}{4} + \frac{4}{3} T_R N_f \right).$$

Substituting the value of $\tilde{Z}_A$ from Eq.(4.79), we get

$$Z_g = 1 - \frac{g_R^2}{8\pi^2\epsilon} \frac{(11C_a - 4T_R N_f)}{6} + \mathcal{O}(g_R^4). \qquad (4.86)$$

Notice that while $\tilde{Z}_A$ depend on α_R, Z_g and Z_{m_s} are independent of it. The latter are related to physical quantities, which we do not expect to depend on the gauge choice.

Although we have determined all the counterterms, there are still some graphs left to be analyzed. For example, the interaction involving four gluons

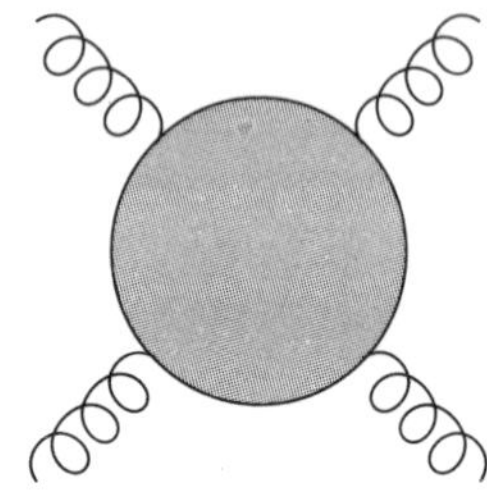

implies that

$$Z_g^2 \tilde{Z}_A^2 = 1 - \frac{g_R^2}{8\pi^2\epsilon} \left[\left(-\frac{2}{3} + \alpha_R \right) C_a + \frac{4}{3} T_R N_f \right]. \qquad (4.87)$$

Similarly, from the interactions of the gluon with the fermions or ghosts,

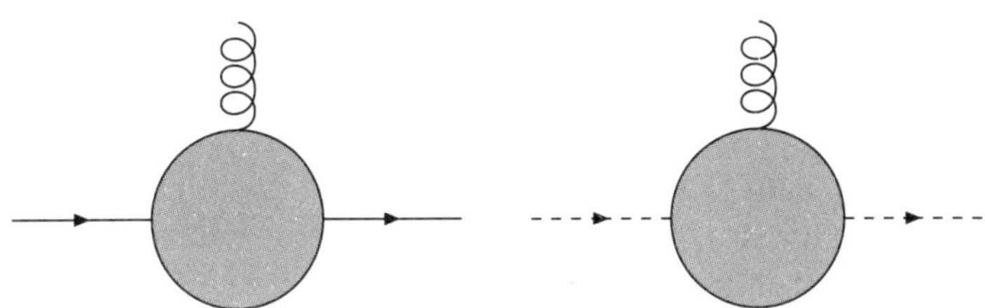

we get

$$\tilde{Z}_{\psi_s} Z_g \tilde{Z}_A^{1/2} = 1 - \frac{g_R^2}{8\pi^2\epsilon}\left[\left(\frac{3+\alpha_R}{4}\right)C_a + \alpha_R C_F\right]$$

$$\tilde{Z}_\chi Z_g \tilde{Z}_A^{1/2} = 1 - \frac{g_R^2}{16\pi^2\epsilon}C_a\alpha_R, \tag{4.88}$$

respectively.

> **Exercise:** The blobs in the diagrams above denote all graphs that contribute to that Green's function to the given order. In each case, determine these graphs explicitly.

> **Exercise:** Check that these relations are automatically satisfied to $\mathcal{O}(g_R^2)$. This shows that, at least to one loop order QCD is renormalizable.

Renormalization group

For a fixed ϵ, the renormalized theory has four sets of parameters: $g_R, \alpha_R, \{m_{sR}\}$ and μ. The original action, however, has only three sets of parameters: g, α and $\{m_s\}$. These are related by Eq.(4.78). All physical quantities, *e.g.* on-shell S-matrix elements, are functions of g, α and $\{m_s\}$. Therefore there is a redundancy in the description using the parameters $g_R, \alpha_R, \{m_{sR}\}$ and

μ. Put in another way, two different values of these sets of parameters may actually define the same theory. In order to specify the parameters, we could fix μ at a particular value μ_0 (say 1 GeV), and then specify g_R, α_R and $\{m_{sR}\}$. Suppose, we now choose a different value of μ, e.g. μ_0'. It is natural to ask the following question. What value of g_R, α_R and $\{m_{sR}\}$ should we choose so as to describe the same theory? In other words, how should g_R, α_R and $\{m_{sR}\}$ change with μ so that the theory they define remains the same? That is, g, α and $\{m_s\}$ remain the same. What we want is a set of differential equations for g_R, α_R and $\{m_{sR}\}$, and it easily follows from the required invariance of the independent parameters:

$$\mu\frac{dg}{d\mu} = 0 \;\Rightarrow\; \mu\frac{d}{d\mu}\left(Z_g(g_R)\,g_R\mu^{\epsilon/2}\right) = 0,$$

$$\mu\frac{d\alpha}{d\mu} = 0 \;\Rightarrow\; \mu\frac{d}{d\mu}\left(Z_\alpha(g_R)\,\alpha_R\right) = 0, \qquad (4.89)$$

$$\mu\frac{dm_s}{d\mu} = 0 \;\Rightarrow\; \mu\frac{d}{d\mu}\left(Z_{m_s}(g_R)\,m_{sR}\right) = 0.$$

Let us look at the first of these equations. Substituting, (see Eq.(4.86),

$$Z_g = 1 - \frac{1}{\epsilon}Ag_R^2$$

in Eq.(4.89), we find the variation of g_R with μ to be

$$\mu\frac{dg_R}{d\mu} = -\,Ag_R^3. \qquad (4.90)$$

The above equation is known as the β-function of QCD. When the coefficient A is positive[3], irrespective of its sign, g_R tends to zero as μ increases. This property is known as *asymptotic freedom*. So we finally see some light at the end of the tunnel. We can make QCD perturbative by taking μ large.

---

[3]The sign of A is determined by the factor $11C_a - 4T_RN_f$ (see Eq.(4.86)). Substituting $C_a = 3$ and $T_R = 1/2$, this is positive for $N_f \leq 16$.

Solving for g_R from differential Eq.(4.90),

$$g_R^2 = \frac{1}{C + 2A \ln \mu},$$

(4.91)

for some constant of integration C. This formula is approximate, because in determining g_R, we have neglected higher order terms. However, since $g_R \to 0$ as μ inceases, the approximation (4.91) becomes better and better.

One might wonder how it is possible to make the strong coupling problem of QCD into a weakly coupled one just by changing μ. Let us consider the following analogy. Suppose we have a function $f(x)$ with a known Taylor series expansion

$$f(x) = \sum_{n \geq 0} a_n x^n.$$

If we want the value the function at $x = 3$, the Taylor series above is not a good approximation for it. Now let us define a new variable y such that $x = y + 3y^2$. The Taylor series can be rewritten as

$$f(y) = \sum_{n \geq 0} b_n y^n,$$

where, the coefficients b_n can be calculated in terms of a_n. Moreover, upto a given order n, one only needs finitely many a_m's with m at most upto the same order n. At any given order, some of the original contribution is ignored. Since $x = 3$ corresponds to $y = 0.83$, one can hope to have a better convergence. This happens only if the new coefficients b_n's do not grow faster than the a_n's. This process can be repeated until this criterion is no longer satisfied and it is not profitable to do so. In exactly the same way, an expansion in $g_R(\mu_0)$ when traded in terms of another one in $g_R(\mu_0')$ may lead to a better perturbation theory.

Given any physical quantity in QCD, (like a cross section of some event), we can expand it is as

$$\sum_{n} C_n(\mu, \cdots) \, (g_R(\mu))^{2n}.$$

In order that perturbation theory works, we need to ensure that the coefficients $C_n(\mu, \cdots)$'s do not grow too fast as μ increases. So we need to ask if there are processes for which we can take μ to be large without increasing $C_n(\mu, \cdots)$. This may not possible for a given process.

4.8   Lecture VIII

The perturbation expansion of QCD in powers of $g_R(\mu)$ is valid for large μ, provided the coefficients of expansion are not large. Let us see what kind of process we can consider where the coefficients of expansion do not grow too fast as $\mu \to \infty$.

Consider some physical process involving external momenta p_1, p_2, etc. Suppose $Q(p_1, p_2, \cdots; \text{polarizations}; g_R(\mu); \{m_{sR}\}; \mu)$ be some physical quantity associated with this process that we want to compute. (A particular example of such a quantity would be a scattering cross-section.) Suppose Q has mass dimension d (*e.g.*, cross-section would have mass dimension -2). Let us ignore polarizations for the time being, then

$$Q(p_1, p_2, \cdots; g_R(\mu); \{m_{sR}(\mu)\}; \mu)$$
$$= (p_1 \cdot p_2)^{d/2} f\left(\{\frac{p_i}{\mu}\}; g_R(\mu); \frac{\{m_{sR}(\mu)\}}{\mu}\right),$$

where $f(\cdots)$ is a dimensionless quantity, which is a function of dimensionless parameters. We have pulled out dimensionful factor, in terms of $(p_1 \cdot p_2)^{d/2}$. We can do a perturbation expansion of f in powers of $g_R(\mu)$.

$$f\left(\left\{\frac{p_i}{\mu}\right\}; g_R(\mu); \left\{\frac{m_{sR}(\mu)}{\mu}\right\}\right)$$
$$= \sum_{n \geq 0} C_n\left(\left\{\frac{p_i}{\mu}\right\}; \left\{\frac{m_{sR}(\mu)}{\mu}\right\}\right) g_R^{2n}(\mu).$$

Notice that as $\mu \to \infty$, all the parameters $\{p_i/\mu\}$ and $\{m_{sR}\}/\mu$ become small. It may, however, happen that C_n's could be made of such functions that they grow even when the parameters become small. *e.g.*, $\ln|(p_i \cdot p_j/\mu^2)|$ or $\ln|(m_{sR}/\mu)|$. Such terms typically appear in the computation of C_n and these terms diverge when the parameters become small, *i.e.*, when $\mu \to \infty$. So first let us consider these terms. $\ln|(p_i \cdot p_j/\mu^2)|$ is small when $\mu \sim \sqrt{p_i \cdot p_j}$ and if $p_i \cdot p_j$ are of same order for all pairs i, j $(i \neq j)$.

So the lesson we learn from this is that perturbation theory makes sense if we confine ourselves to those processes for which $p_i \cdot p_j$, $(i \neq j)$ are large. This is the conclusion we obtained from $\ln|(p_i \cdot p_j/\mu^2)|$ but we also have terms of the form $\ln|(m_{sR}/\mu)|$. Since we have set $\mu \sim \sqrt{p_i \cdot p_j}$, the second kind of terms can be written as $\ln|(m_{sR}(\mu)/\sqrt{p_i \cdot p_j})$. We will now look at the various possible values of this term.

1. If $m_{sR}(\sqrt{p_i \cdot p_j}) \gg \sqrt{p_i \cdot p_j}$ then $m_{sR}(\mu)/\sqrt{p_i \cdot p_j}$ is large. This case can be analysed by taking $m_{sR}(\mu) \to \infty$ limit. In this limit, the corresponding quark is infinitely massive and hence we can omit that quark from all our calculations, including that of the β-function. Let us consider an example. Suppose $m_t \gg \sqrt{p_1 \cdot p_2} \gg m_b$ then we can pretend that we have only five quarks u, d, s, c, and b. This is an intuitive argument but we can show using decoupling of the heavy particle in the field theory that it does not contribute the C_n's.

2. If $m_{sR}(\sqrt{p_i \cdot p_j}) < \sqrt{p_i \cdot p_j}$ then we would like to ask whether C_n's diverge in the limit $m_{sR}(\sqrt{p_i \cdot p_j})/\sqrt{p_i \cdot p_j} \to 0$. *E.g.*, suppose $m_t \sim \sqrt{p_1 \cdot p_2}$ then, $m_u/\sqrt{p_1 \cdot p_2} \sim m_u/m_t$ is very small and the question is what happens to C_n's in this case? Only those processes for which C_n's have finite limit as $m_{sR} \to 0$ can be analysed by perturbative QCD.

Thus we can see that the processes with large external momenta and with small m_{sR} can be analysed by perturbative QCD if and only if C_n's are finite in the limit $m_{sR} \to 0$. What kind of processes give finite C_n's in this limit? There are two ways of approaching this problem.

1. *Rigorous approach:* We should only search among gauge invariant processes.

2. *Intuitive approach:* Start with the gauge fixed action and pretend that we never heard of gauge invariance. In other words, treat gauge fixed action as ordinary field theory without any gauge symmetry and calculate all kinds of scattering processes, like $q\bar{q}$ pair production.

We will take the intuitive approach, *i.e.*, we will calculate "physical processes" in the gauge fixed theory, like on-shell S-matrix elements and cross sections. We will see that the theory is clever enough to tell us when we are asking a wrong question. That is, if we ask right questions it qives sensible answers but if we ask wrong questions we would, in most cases, end up getting answers that makes no sense.

Let us examine the process

$$e^+ e^- \longrightarrow q_s^i \bar{q}_s^i.$$

Since electrons do not have strong interaction, the process is mediated by photon. This, therefore, is a QED process. To leading order the diagram is the following.

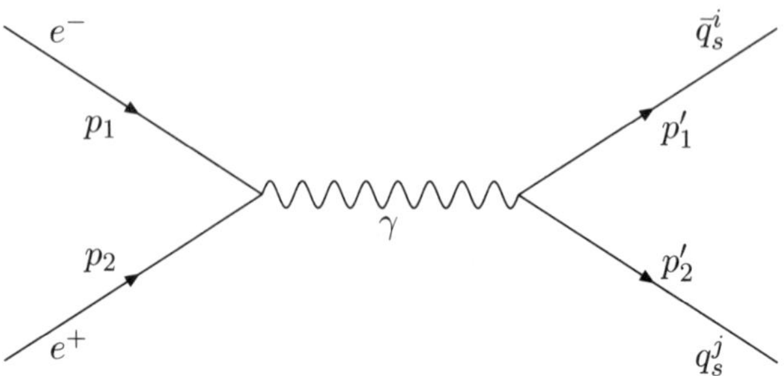

We can easily calculate the cross section for this process. Leading corrections to this process come from QCD. We get QCD corrections because outgoing fermions are quarks, which have QCD interactions. Since QCD coupling is large compared to QED coupling, the leading loop order correction to this process are from

QCD.

One loop QCD corrections

QCD corrections can come from the part of the diagram which involves the $q\bar{q}\gamma$ vertex and not from the part containing $e^+e^-\gamma$ vertex. We can then leave out the e^+e^- part and concentrate only on the quark end of the diagram.

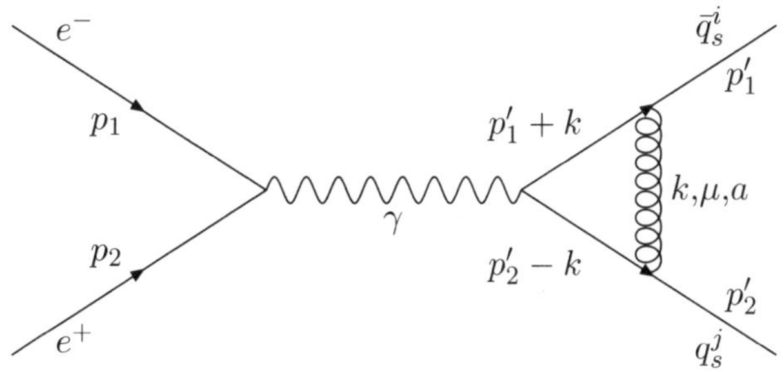

The amplitude is given by

$$\mathcal{M} = \int \frac{d^4k}{(2\pi)^4} \frac{i}{k^2 + i\varepsilon} \bar{u}(p_2')\gamma^\mu \frac{(\slashed{p}_2' - \slashed{k}) + m_{sR}}{(p_2' - k)^2 - m_{sR}^2 + i\varepsilon}$$
$$\gamma^\rho \frac{-(\slashed{p}_1' + \slashed{k}) + m_{sR}}{(p_1' + k)^2 + m_{sR}^2 + i\varepsilon} \gamma_\mu v(p_1')(T^a)_{jk}(T^a)_{ki} \quad (4.92)$$

We will isolate the factor which is likely to diverge and keep other terms out of our computation. The amplitude can be written as

$$\int \frac{d^4k}{(2\pi)^4} \frac{i}{k^2 + i\varepsilon} \frac{1}{(p_2' - k)^2 - m_{sR}^2 + i\varepsilon} \frac{1}{(p_1' + k)^2 + m_{sR}^2 + i\varepsilon} N(k),$$
$$(4.93)$$

where we have explicitly displayed the divergent terms in the amplitude, $N(k)$ contains terms from the numerator which do not contribute to the divergence. Recall, we have alreay taken care of UV divergences by regularization and renormalization. The divergences we are looking at occur when m_{sR} goes to zero or equivalently p_1', p_2' become large.

$$\begin{aligned}
p_1'^2 = p_2'^2 &= m_{sR}^2, \\
(p_1' + k)^2 - m_{sR}^2 &= 2p_1' \cdot k + k^2, \\
(p_2' - k)^2 &= -2p_2' \cdot k + k^2.
\end{aligned}$$

First, let us consider a limit in which all components of k are small. In that case the divergent part of the integral is

$$\int \frac{d^4k}{(2\pi)^4} \frac{i}{k^2 + i\varepsilon} \frac{1}{2p_1' \cdot k + i\varepsilon} \frac{1}{-2p_2' \cdot k + i\varepsilon} \sim \int \frac{d^4k}{(2\pi)^4} \frac{1}{k^4}. \tag{4.94}$$

It is worth mentioning here that the numerator $N(k)$ does not vanish as $k \to 0$. This integral is logarithmically divergent. This divergence comes from small k and hence is called soft or infrared divergence. We encounter this divergence because we are not asking the right question. $e^+e^- \to q\bar{q}$ is not a gauge invariant process. Nevertheless it still makes sense to look at this process as we can try to infer something about the gauge invariant objects.

Let us now consider the limit $m_{sR} \to 0$. We will work in the centre of mass frame. In this frame, momenta p_1' and p_2' can be written as

$$p_1' = p'(1,0,0,1), \quad p_2' = p'(1,0,0,-1).$$

We will use the light cone variables. In the light cone frame, components of any vector a^μ are

$$a^\pm = a^0 \pm a^3, \quad \vec{a}_\perp = (a^1, a^2).$$

In terms of these variables, inner product of two vectors is

$$a \cdot b = \frac{1}{2}(a^+ b^- + b^+ a^-) - \vec{a}_\perp \cdot \vec{b}_\perp.$$

The formulas for Dirac gamma matrices also change in these variables and they are

$$(\gamma^+)^2 = 0, \quad (\gamma^-)^2 = 0, \quad \{\gamma^+, \gamma^-\} = 4, \quad \{\gamma^\pm, \gamma_\perp\} = 0.$$

Let us get back to our problem, we have chosen the centre of mass frame. The light cone variables in this frame become

$$p_1'^+ = 2p', \qquad p_1'^- = 0 = \vec{p}_{1\perp},$$
$$p_2'^- = 2p', \qquad p_2'^+ = 0 = \vec{p}_{2\perp}. \qquad (4.95)$$

In these variables, the amplitude is

$$\mathcal{M} \sim \int \frac{dk^+ dk^- d^2 k_\perp}{(2\pi)^4} \frac{N(k)}{k^+ k^- - k_\perp^2 + i\varepsilon}$$
$$\times \frac{1}{\left((p_1'^+ + k^+)k^- - \vec{k}_\perp^2 + i\varepsilon\right)\left(-(p_2'^- - k^-)k^+ - \vec{k}_\perp^2 + i\varepsilon\right)}.$$

This amplitude has a different kind of divergence. To see this let us consider the change of variables

$$k^- = \lambda \vec{k}_\perp^2,$$

and use k^+, λ and $\vec{k}_\perp$ as independent variables. It is easy to see that in these variables the integration measure becomes

$$dk^+ \, dk^- \, d^2 k_\perp = dk^+ \, d\lambda \, \vec{k}_\perp^2 \, d^2 k_\perp.$$

Consider the region of integration where k^+ and λ are finite and $|\vec{k}_\perp|$ is small, then the amplitude is

$$\mathcal{M} \sim \int \frac{dk^+ d\lambda \vec{k}_\perp^2 \, d^2 k_\perp}{(2\pi)^4} \frac{N(k)}{(\lambda k^+ - 1)\vec{k}_\perp^2 + i\varepsilon}$$
$$\times \frac{1}{\left(((p_1'^+ + k^+)\lambda - 1)\vec{k}_\perp^2 + i\varepsilon\right)\left((-p_2'^- k^+) + o(\vec{k}_\perp^2)\right)}.$$

This amplitude is logarithmically divergent as $|\vec{k}_\perp| \to 0$. This is called collinear divergence or mass singularity. It is called mass singularity because it occurs when we take the limit $m_{sR} \to 0$. It is called collinear divergence because $|\vec{k}_\perp| \sim 0$ and $k^- \sim 0$, so only nonvanishing component is k^+. This means k is nearly parallel to p_1', so the divergence comes from the region where gluon momentum is collinear with quark momentum. This amplitude will have another divergence when k is collinear with p_2'.

As mentioned earlier, $e^+e^- \to q\bar{q}$ process suffers from all these divergences because we are not asking the right kind of question. It is, however, still useful to go through this analysis because there is a closely related process, to which we will turn to momentarily, for which we can ask the relevant questions.

Consider the total cross section for the process

$$e^+e^- \longrightarrow \text{anything},$$

by 'anything', we really mean strongly interacting particles. That is, we will sum over cross sections for

$$e^+e^- \longrightarrow q\bar{q}, \quad e^+e^- \longrightarrow q\bar{q}g, \quad e^+e^- \longrightarrow q\bar{q}gg, \dots$$

and we will sum over final state quantum numbers and integrate over momenta.

We have already looked at $e^+e^- \to q\bar{q}$. The next term that has the lowest order QCD coupling is $e^+e^- \to q\bar{q}g$.

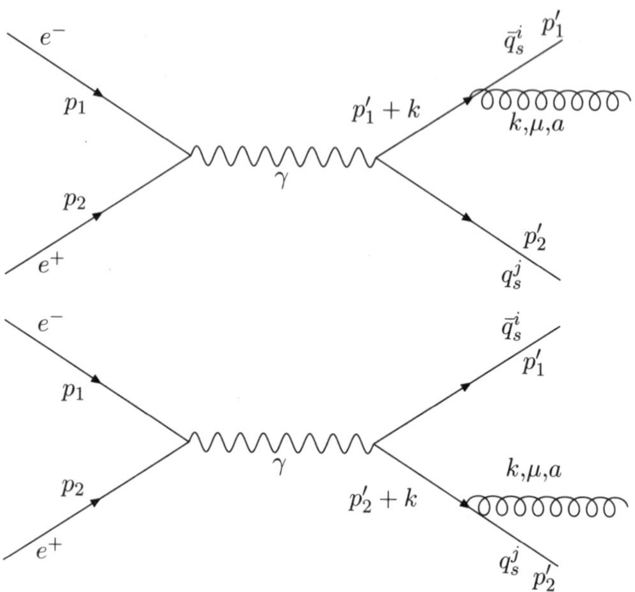

Since the final states in both these diagrams are same, we should sum the amplitudes. One of the terms in the integrand of the cross section is

$$\frac{1}{(p_1' + k)^2 + i\varepsilon} \frac{1}{(p_2' + k)^2 - i\varepsilon}.$$

This term comes as the cross term when we square the modulus of the sum of two amplitudes. The negative sign in the $i\varepsilon$ prescription is due to the fact that the second term is complex conjugated. The integral that we need to perform is

$$\int \frac{d^4k}{(2\pi)^4} \frac{1}{(p_1' + k)^2 + i\varepsilon} \frac{1}{(p_2' + k)^2 - i\varepsilon}. \tag{4.96}$$

This integral is divergent when any of the following conditions are met.

1. $|\vec{k}_\perp|$ is small,

2. k is parallel to p_1',

3. k is parallel to p_2'.

To this order we can regularise these integrals by introducing a small mass for the quarks as well as the gluons.

When we sum all the diagrams, we find that the infrared divergences of the real graphs are cancelled with that of the virtual graphs and collinear divergences of the real graphs are cancelled with the collinear divergences of the virtual graphs[4]. In the end we can take the masses of the quarks and gluons to zero. Thus the

---

[4]To this order in perturbation theory real graphs are those in which in there is a gluon in the final state. Virtual graphs, on the other hand, do not have gluons in the final state.

final result is that in the total cross section the divergences cancel in the $m_{sR} \to 0$, $m_g \to 0$ limit. This result holds to all orders in perturbation theory.

Some remarks

1. We cannot use gluon masses as an infrared regulator because it is not a gauge invariant regulator and it leads to a breakdown of renormalizability. In practice, we should use dimensional regularization. That is, work in $D = 4 + \epsilon$ dimensions and then take the limit $D \to 4$ at the end of the manipulations. (More precisely, $D = 4 + \epsilon$ for infrared regularization and $D = 4 - \epsilon$ for ultraviolet regularization.)

2. Virtual graphs have amplitudes which are proportional to g_R^2 whereas real graphs have amplitudes which are proportional to g_R. The cross setions, therefore, are proportional to g_R^4 for virtual graphs and g_R^2 for real graphs. A natural question that arises at this point is how do we hope to cancel the divergences between them? The correct procedure is to take sum of all the diagrams upto a given order in perturbation theory with a fixed number of particles in the final state and add their amplitudes and then take the square of their absolute value. In the case of $q\bar{q}$ final state, upto one loop level we have a tree graph as well as one loop QCD graphs with virtual gluons. Summing these amplitudes and then squaring them gives us terms proportional to $(1 + g_R^2 + g_R^4)$ and the term proportional to g_R^2 from this cancels with the g_R^2 term from the real graphs. At g_R^4 level there are other contributions from real as well as virtual graphs.

4.9   Lecture IX

Consider the scattering process $e^-(p_1) + e^+(p_2) \to X$ where X denotes the strongly interacting particles produced in this process. Here, all the final states are summed over (such as spin, colour, flavour, phase space etc) and p_1, p_2 are the momenta of e^- and e^+ respectively. The center of mass energy is defined as $s = (p_1 + p_2)^2$.

In perturbative QCD, this cross section can be expanded in powers of the square of strong coupling constant $g_R(\mu)$, where R refers to the fact that the coupling constant is renormalised and the scale μ is the renormalisation scale.

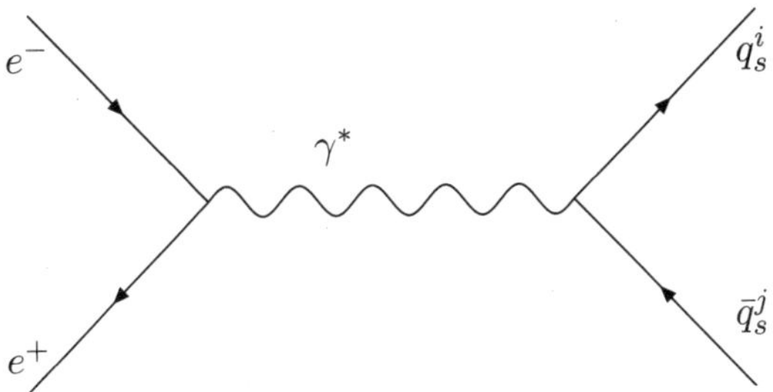

The cross-section is

$$\sigma(e^- + e^+ \to X) = \frac{1}{s}\sum_n C_n\left(\frac{\sqrt{s}}{\mu}, \left\{\frac{m_q}{\mu}\right\}\right)(g_R(\mu))^{2n}. \quad (4.97)$$

We want $g_R(\mu)$ small, which requires that μ should be large. To avoid $\sqrt{s}/\mu$ being small and give a large contribution to C_n, we are forced to choose μ of the order of $\sqrt{s}$. If we choose $\mu = \sqrt{s}$,

$$\sigma = \frac{1}{s}\sum_n C_n\left(1, \left\{\frac{m_q}{\mu}\right\}\right)(g_R(\sqrt{s}))^{2n}. \quad (4.98)$$

In order to be able to use perturbation theory, $C_n\left(1, \left\{\frac{m_q}{\mu}\right\}\right)$ must tend to finite values as $\{m_q/\mu\} \to 0$.

It turns out that the coefficients C_n for the above process are finite in the limit when quark mass become zero since all the collinear divergences cancel. Also these coefficients are free of any soft singularities. So we can choose μ to be large and apply perturbative QCD, *i.e.*, we can compute the total inclusive cross section for the process $e^-(p_1) + e^+(p_2) \to X$ using perturbative QCD. It is sensible to ask the question: what is the cross-section for $e^-(p_1) + e^+(p_2) \to X$, where we sum over final states, *i.e.*, inclusive cross section, and choose μ to be large? Detecting a quark or an anti-quark in the final state is not sensible as the result of the cross section diverges when we compute the cross section beyond leading order (say, to order $g_R^2(\mu)$).

To leading order in strong coupling constant, namely to $(g_R(\mu))^0$, we find

$$R = \frac{\sigma(e^- e^+ \to X)}{\sigma(e^- e^+ \to \mu^- \mu^+)} = 3 \sum_s Q_s^2, \qquad (4.99)$$

where Q_s is the charge of the quark of flavour s. Beyond leading order, QCD corrections to this process come from (a) virtual corrections such as vertex correction and self energy corrections and (b) real gluon emission from the external quark and antiquark lines as shown in the figure. The computation of these diagrams is straightforward and the result in the limit $\{m_q/\mu\} \to 0$ is found to be

$$R = 3 \sum_s Q_s^2 \left(1 + \frac{3}{4} C_F \frac{g_R^2(\sqrt{s})}{4\pi} \right). \qquad (4.100)$$

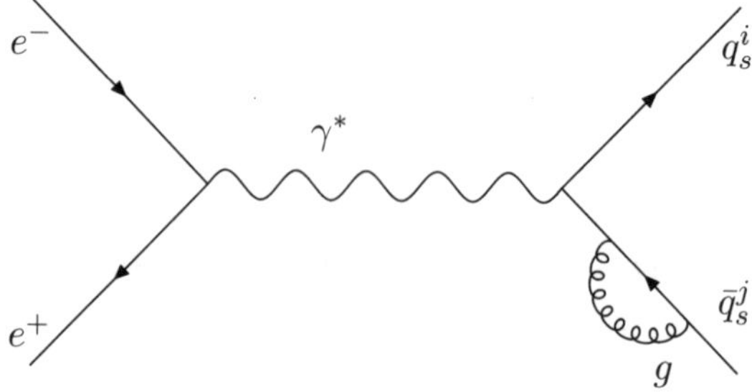

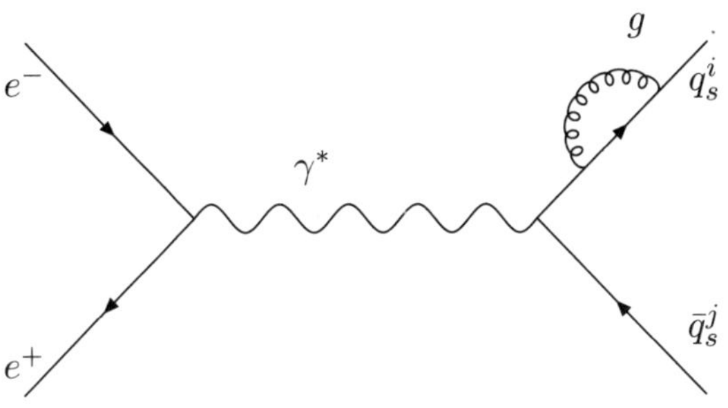

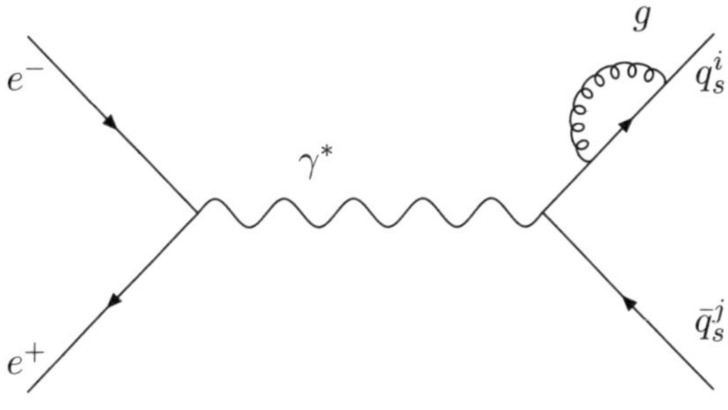

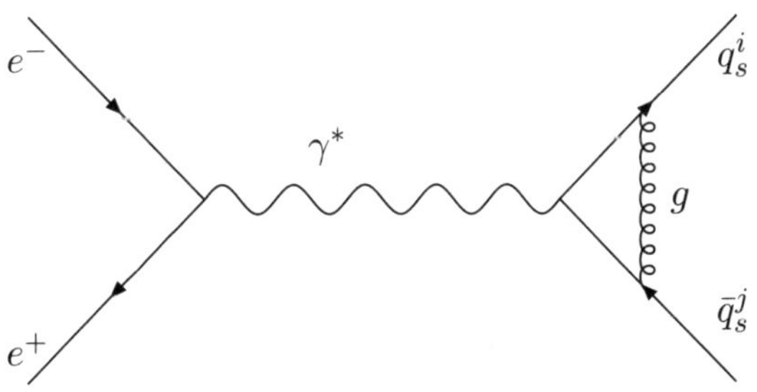

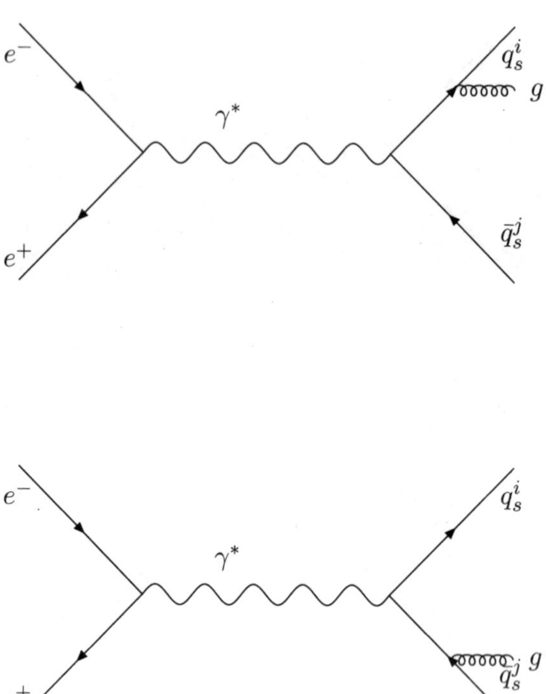

It is worth mentioning that the above computation involves careful handling of soft divergence coming from region where the gluon 4-momentum tends to zero and collinear divergence coming from the region where the momentum of the gluon is parallel to the quark or antiquark momentum. As is clear from the above equation, the final result is finite and free of any soft and collinear singularities. This is due to fact that the soft and collinear divergences appearing in the intermediate stages cancel among various diagrams leaving the final inclusive cross section finite.

The above exercise shows that the singularities coming from one set of diagrams cancel against the singularities coming from another set of diagrams. This happens when we compute inclusive cross-section in which all the final states are summed. Instead we can also ask: what is the cross section for producing a jet at an angle θ and another jet at an angle $\pi - \theta$ with total energy between $\sqrt{s} - \Delta E$ and $\sqrt{s}$? Here a jet of energy E along an axis A is a set of particles moving within a solid angle $\Delta\omega$ of A carrying the total energy E. It is still a sensible question because the singularities coming from soft and collinear divergences do not appear in the final result for such cross sections. Notice that the above question is about a cross-section with hadrons in the final states with specific properties. It turns out the result for such cross-sections are of the form

$$\left(1 + \cos^2(\theta)\right) \frac{3}{16\pi} \Delta\omega \, \sigma_{total}$$

at the tree level. The higher order corrections involve terms such as $\log\left(\Delta E/\sqrt{s}\right)$ and $\log \Delta\omega$.

The other processes which can be studied in perturbative QCD are the deep inelastic scattering (DIS) $e^- + N \to e^- + X$) and the Drell-Yan process $(p + \bar{p} \to \mu^- \mu^+ + X)$. Let us consider the DIS. This is a scattering process in which, an electron scatters off a proton or a nucleon producing a debris of hadrons, denoted by X, in the final state. Here also, we sum over all the final state hadrons. A typical Feynman diagram for this process is given in the figure.

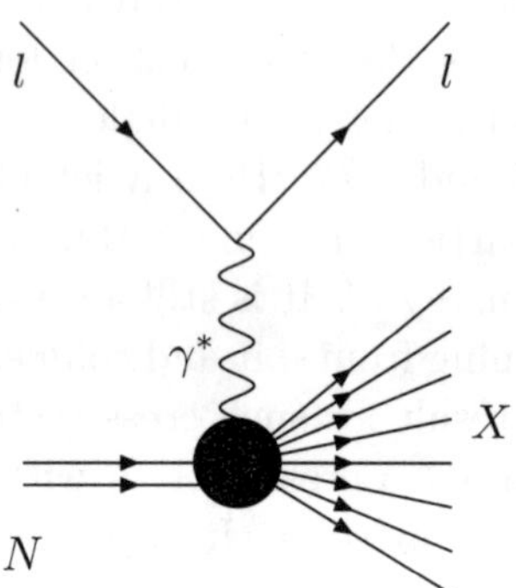

Let p_e, p'_e be the momenta of initial and final electrons and $q = p_e - p'_e$ be the momentum transfer through the intermediate photon. The invariants of the process are $q^2 = -Q^2 = (p_e - p'_e)^2 < 0$ and $p \cdot q$. Here $p^2 = m_N^2$ is a parameter of the theory. Let us define a dimensionless variable $w = -2p \cdot q/q^2$. The limit defined by $Q^2 \to \infty$, $p \cdot q \to \infty$ with w finite is called the Bjorken limit. In this limit, the cross section is found to be a function of this dimensionless variable only. This goes under the name *Bjorken scaling*. To study this process in perturbative QCD, one usually uses two approaches: (a) the *operator product expansion* (OPE) and (b) an intuitive approach called the *parton model*. In the OPE approach, one uses standard operator product expansion techniques to study the cross section. Notice that the DIS cross section is proportional to the hadron matrix elements of products of certain currents. Translating the Bjorken limit $Q^2 \to \infty$, $p \cdot q \to \infty$ in terms of the light cone limits in coordinate space, one can expand the product of these currents in terms of certain gauge invariant local operators with appropriate (singular) coefficient functions. These coefficient functions are computable order by order in perturbative QCD, although the matrix elements of these local operators are non-perturbative. This way one can study the DIS and also show the scaling behaviour at leading order perturbation theory. In the parton model, one views the nucleon as a collection of partons such as quarks, antiquarks, gluons, moving parallel to the nucleon. The transverse momentum is assumed to be small. One defines $f_i(x)dx$, the probability that the ith parton carries the longitudinal momentum between xp and $(x + dx)p$ with $0 \le x \le 1$. Energy momentum conservation demands that

$$\sum_i \int_0^1 dx \, x \, f_i(x) = 1. \tag{4.101}$$

In this model, the hadronic cross-section can be expanded in terms of partonic cross-section with appropriate parton probability densities $f_i(x)$:

$$\sigma(e^- N \to e^- X) = \sum_i \int_0^1 dx \, f_i(x) \, \sigma(e^- + i \to e^- + X). \tag{4.102}$$

Using this, we find that in the lab frame,

$$\frac{d\sigma}{d\Omega dE'} = \frac{\alpha^2 \cos^2(\theta/2)}{4E_e^2 \sin^4(\theta/2)} \left(W_2 + 2W_1 \tan^2(\theta/2) \right), \tag{4.103}$$

where,

$$W_2 \;=\; \frac{1}{\nu\omega} \sum_l Q_l^2 \, f_i\left(\frac{1}{\omega}\right),$$

$$W_1 \;=\; \frac{1}{2m_N} \sum_l Q_l^2 \, f_i\left(\frac{1}{\omega}\right),$$

for $l = u, u, d$. To leading order, we see that νW_2 and W_1 are independent of $-q^2$. This is the Bjorken scaling. Beyond the leading order, the scaling behaviour is expected to disappear due to virtual gluon correction as well as real gluon emission. The natural choice of the scale which enters the coupling constant involved is $\sqrt{-q^2}$. Therefore the coupling constant should be small in the large $-q^2$ limit (Bjorken limit) and that is why one can use perturbation theory to compute the corrections. As expected, the soft divergences cancel among various diagrams. But it turns out that the collinear divergences in the limit $-q^2 \to \infty$ do not disappear — terms such as $m^2/-q^2 \to 0$ give logarithmic divergences. But these collinear divergences factorise. Hence they can be absorbed into $f_i(x)$ by a simple redefinition $f_i(x) \to f_i(x, \mu^2)$. This leads to scaling violation.

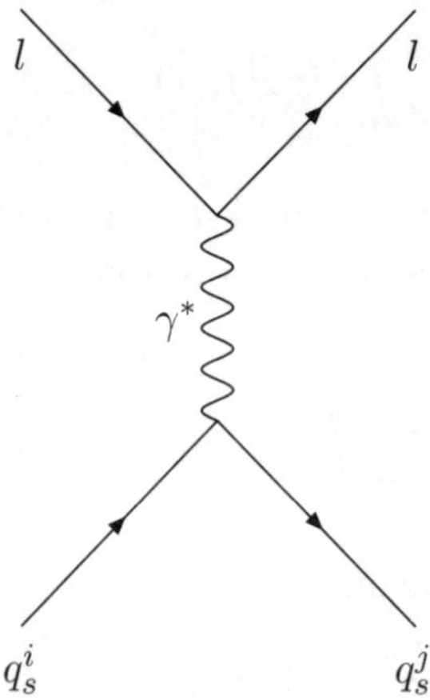

4.10   Lecture X

In this lecture and the next[5], we shall consider the process $e^+ + e^- \to q + \bar{q}$ beyond the leading order in perturbative QCD. In fact, we shall see that some nonperturbative results can be obtained by summing certain contributions from all orders in perturbation theory.

For example, the next to leading order perturbative QCD contribution to this process involves a vertex correction to the photon-quark-antiquark vertex.

---

[5]These last two lectures were follow up ones and were given immediately after the SERC school.

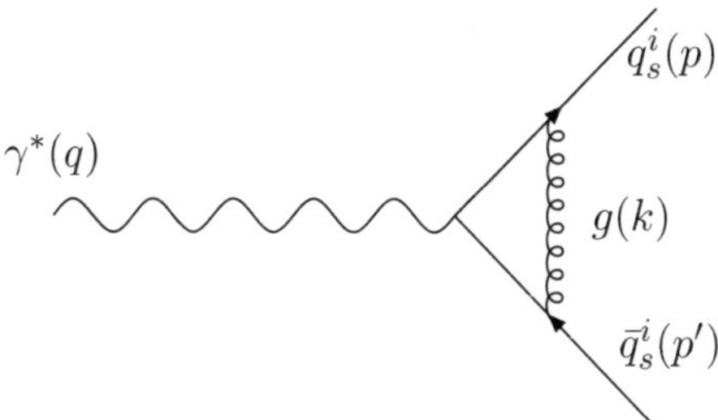

We have already seen that the vertex correction due to gluon exchange between quark and antiquark lines is divergent due to the ultraviolet singularity coming from the large momentum flow inside the loop. In addition, the loop integral contains collinear divergence when the mass of the quark and antiquark is much smaller than the energy scale associated with the process. That is, when the energy scale $q^2 \gg \Lambda^2$, $m^2 \to 0$ is a good approximation. But this approximation leads to a collinear divergence. So the natural question one would like to ask is whether the cross section of the process $e^+ + e^- \to q + \bar{q}$ is infinite. The answer to this question is that this cross section is zero. It turns out that if we include all the higher order corrections systematically, then the final result would look like the exponential of the leading singular contribution coming from the vertex contribution: $e^{-g^2 \times \infty} \sim 0$. It means that the probability of detecting a quark and an antiquark alone in $e^+ e^-$ collision is identically zero. If m is finite, Q is finite and gluon mass is zero, one ends up with an IR divergence (usually called *soft* divergence) but no collinear divergence. When $Q^2 \to \infty$ with an IR cutoff, say finite ΔE, then one encounters collinear divergence. To study this situation one should introduce the infrared cutoff in a gauge invariant way. Dimensional regularisation is gauge invariant regularisation to study large Q^2 behaviour of the processes in perturbation theory. In QED one uses massive photon to regulate IR divergence. The other regularisation which is used involves the introduction of an additional scalar field. The spontaneous breaking of the gauge invariance gives masses to gluons. Keeping the vacuum expectation value small and treating it as the IR cutoff regulates the IR divergence.

Consider the following generalised vertex $\gamma q\bar{q}$:

$$\bar{u}(p)\, \epsilon_\mu \Gamma^\mu\, v(p'),$$

where, $\Gamma_\mu = \Gamma_\mu\,(p, p', m, \mu, g, m, m_{IR})$ with $p^2 = m^2$, $p'^2 = m^2$.

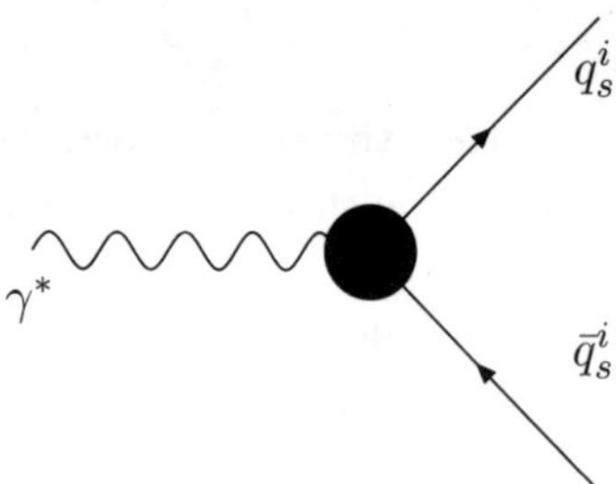

Let us choose the following kinematics for the momenta q of the photon, p of the quark and p' of the antiquark:

$$q \;=\; (Q\cosh\beta,\, 0,\, 0,\, Q\sinh\beta)$$

$$p \;=\; \left(\frac{Q}{2}\cosh\beta + \sqrt{\frac{Q^2}{4} - m^2}\,\sinh\beta,\, 0,\, 0,\right.$$

$$\left.\sqrt{\frac{Q^2}{4} - m^2}\,\cosh\beta + \frac{Q}{2}\sinh\beta\right).$$

In the limit $Q \gg m$

$$p = \frac{1}{2}\,Qe^\beta(1, 0, 0, 1) + \cdots,$$

where the dots refer to higher orders in m/Q which are ignored in this approximation. In the above, β is velocity factor and is frame-dependent. Different choices of β corresponds to different choices of frames. We also choose to work with light-cone coordinates defined by:

$$a^\mu = (a^+, a^-, a_\perp),$$

where,

$$a^\pm = a^0 \pm a^3, \qquad a_\perp = (a^1, a^2).$$

The scalar product of two four vectors is

$$a \cdot b = \frac{1}{2}(a^+ b^- + a^- b^+) - a_\perp \cdot b_\perp.$$

So in the approximation we are working, the momenta of the quark and the antiquark become:

$$p^+ \approx Q\, e^\beta, \qquad p^- \approx 0, \qquad p_\perp = 0,$$
$$p'^- \approx Q\, e^{-\beta}, \qquad p'^+ \approx 0, \qquad p'_\perp = 0.$$

The Dirac gamma matrices in this coordinate frame satisfy

$$(\gamma^+)^2 = 0, \qquad (\gamma^-)^2 = 0,$$
$$\{\gamma^+, \gamma^-\} = 4, \qquad \{\gamma^\pm, \gamma_\perp\} = 0.$$

The equations of motion for the quark and anti-quark fields are

$$\bar{u}(p)\gamma^- = 0, \qquad \gamma^+ v(p') = 0,$$

where we make a large Q approximation which leads to $\not{p} \approx \frac{1}{2}p^+\gamma^-$ and similarly $\not{p}' \approx \frac{1}{2}p'^-\gamma^+$. We also choose to work in the axial gauge $n \cdot A = 0$, where n is an arbitrary spacelike vector. We parametrise n as $n = (0, n_\perp, n^3)$, with $n_\perp$ proportional to the transverse polarisation vector $\epsilon_\perp$ of the photon. We can also choose $n_\perp = 0$.

The gluon propagator $-iN_{ab}^{\mu\nu}(k)$ in this gauge is found to be:

$$\delta_{ab}\left(\frac{-i}{k^2 + i\epsilon}\right)\left[\eta^{\mu\nu} - \frac{k^\mu n^\nu + k^\nu n^\mu}{k \cdot n} + \frac{k^\mu k^\nu}{(k \cdot n)^2}n^2\right], \qquad (4.104)$$

where, the singular denominator $k \cdot n$ is regulated using the following prescription:

$$\frac{1}{(k \cdot n)^\alpha} = \frac{1}{2}\lim_{\epsilon \to 0}\left(\frac{1}{(k \cdot n + i\epsilon)^\alpha} + \frac{1}{(k \cdot n - i\epsilon)^\alpha}\right)$$

for $\alpha = 1, 2$. Notice that the ghosts decouple in this gauge. In order to regulate the UV divergences, we use dimensional regularisation and follow minimal subtraction procedure to remove the divergences. For external quarks and antiquarks, we follow on-shell renormalisation prescription.

Let us first study the quark (antiquark) propagator $S(p^+, p^-, p_\perp)$. To leading order, this look like

$$S(p^+, p^-, p_\perp) = \frac{i}{\not{p} - m}$$

with $p^2 = p^+ p^- - p_\perp^2$. In general, due to radiative corrections, the propagator takes the following form:

$$S(p^+, p^-, p_\perp) = \frac{iZ(p^+)}{\not{p} - m},$$

and similarly for the antiquark:

$$S(p'^+, p'^-, p'_\perp) = \frac{i\tilde{Z}(p^-)}{\not{p}' - m},$$

where $\tilde{Z}(p^-) = Z(m^2/p^-)$. Here, both $Z, \tilde{Z}$ are renormalisation group (RG) invariant, although gauge dependent.

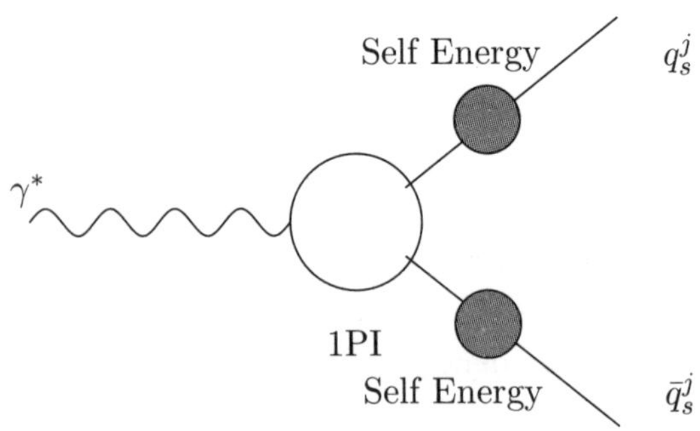

The Sudakov form factor for the photon-quark-antiquark vertex is defined as

$$Z^{\frac{1}{2}}\left(p^{+}, \mu, g, m, m_{d}\right) \bar{u}(p)\, \epsilon_{\mu} V^{\mu}\left(p^{+}, p^{\prime -}, \mu, g, m, m_{d}\right)$$
$$Z^{\frac{1}{2}}\left(p^{\prime -}, \mu, g, m, m_{d}\right) v(p'),$$

where m_{d} is any IR regulator. We find that

$$\epsilon_{\mu} V^{\mu} \approx \gamma_{\perp} \cdot \epsilon_{\perp} V(p^{+}, p^{\prime -}, \mu, g, m, m_{d}).$$

Now the form factor becomes,

$$\bar{u}(p)\, \gamma_{\perp} \cdot \epsilon_{\perp}\, v(p')\, \Gamma(p^{+}, p^{\prime -}, \mu, g, m, m_{d}),$$

where

$$\Gamma(p^{+}, p^{\prime -}, \mu, g, m, m_{g}) \;=\; V(p^{+}, p^{\prime -}, \mu, g, m, m_{d}) Z^{\frac{1}{2}}\left(p^{\prime -}, \mu, g, m, m_{d}\right)$$
$$Z^{\frac{1}{2}}\left(p^{+}, \mu, g, m, m_{d}\right).$$

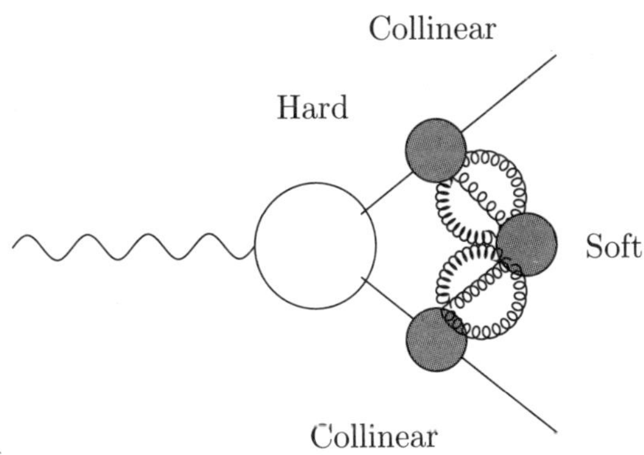

The computation of Γ involves loop integrals with loop momentum, say, k. We can split the domain of integration as follows:

1. the hard region, namely $k^\mu > Q$, for all μ;

2. the soft region, namely $k^\mu \approx \lambda Q$, $\lambda \ll 1$ for all μ;

3. the collinear (to p) region, namely $k^+ \approx Q$, $k^- \approx \lambda^2 Q$, $k_\perp \approx \lambda Q$;

4. the collinear (to p') region, namely $k^- \approx Q$, $k^+ \approx \lambda^2 Q$, $k_\perp \approx \lambda Q$.

Using the standard power counting techniques in the axial gauge one can separate hard, soft and collinear regions of the loop integrals to all orders in perturbation theory. How small k^- is, determines the degree of collinearity and how small $k_\perp$ is, determines softness. Noting that

$$\int \frac{d|k_\perp|}{|k_\perp|} \int \frac{dk^-}{k^-} (\cdots) \approx \log^2(Q),$$

we find that the typical form of the perturbation series looks like

$$\Gamma = \sum_{n=0}^{\infty} g^{2n} \left[a_{2n}^{(n)} \log^{2n}\left(\frac{Q}{\mu}\right) + a_{2n-1}^{(n)} \log^{2n-1}\left(\frac{Q}{\mu}\right) + \cdots \right],$$

where the first non-trivial term in the series is called the leading log contribution. The sum of the leading logs leads to

$$\exp\left(-Cg^2 \log^2\left(\frac{Q}{\mu}\right)\right),$$

where the constant C is calculable perturbatively and is greater than zero. Hence, in the large Q limit, the sum of leading order contributions vanishes exponentially. This does not necessarily mean that $\Gamma = 0$ in the large Q limit. To show that it happens, one has to include all the non-leading terms in the sum as well. Moreover, the leading order behaviour comes purely from the renormalisation constants Z and $\tilde{Z}$ which are unphysical.

4.11 Lecture XI

In this lecture, we shall argue in favour of some exact results which are valid beyond perturbation theory. To this end, we shall work with the functions $\log V$, $\log \sqrt{Z}$ and $\log \sqrt{\tilde{Z}}$. Let us define,

$$\chi\left(p^+, p'^-, \mu, g, m, m_g\right) = \frac{\partial \log V}{\partial \log p^+}$$

$$\chi'\left(p^+, p'^-, \mu, g, m, m_g\right) = \frac{\partial \log V}{\partial \log p'^-}.$$

One can prove the following:

Theorem: The contributions to χ and χ' come only from hard loops, hence they do not contain any infrared or collinear divergences.

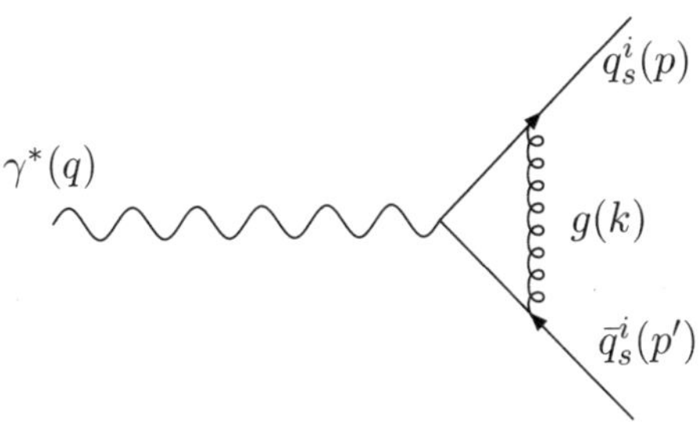

Consider the correction to the photon-quark-antiquark vertex due to a gluon exchange. The amplitude for this process can be written as the integral (over k) of

$$\bar{u}(p)iN_{\mu\nu}\gamma^\mu\left[\frac{(\slashed{p}-\slashed{k}+m)\,\epsilon\cdot\gamma\,(-\slashed{p}'-\slashed{k}+m)}{((p-k)^2-m^2+i\epsilon)\,((p'+k)^2-m^2+i\epsilon)}\right]\gamma^\nu v(p').$$

Let us study this amplitude in the limit $k^\mu \to 0$. Using the equation of motion, we find that the numerator of the expression inside the square bracket is

$$(2p^\mu)\,\epsilon\cdot\gamma\,(2p'^\nu) = 4\,p^+ p'^-\,\epsilon\cdot\gamma.$$

The same result may be obtained by dropping m and noting that $\not{p} \approx p^+\gamma^-$, $\not{p}' \approx p'^-\gamma^+$ and using $\{\gamma^\pm,\gamma^\pm\} = 0$, $\{\gamma^+,\gamma^-\} = 4$. Likewise, the denominator simplifies to

$$-\frac{1}{4(p\cdot k)(p'\cdot k)} = -\frac{1}{4p^+p'^-k^+k^-}.$$

From the above, we find that the integrand goes as $\sim N_{+-}/k^+k^-$ and since $N_{+-} \approx 1/k^2$, the vertex function behaves as

$$\int d^4k\,\frac{1}{k^2k^+k^-}.$$

This is logarithmically divergent. This proves that, to order g^2, $\partial \log V/\partial \log p^+$ and $\partial \log V/\partial \log p'^-$ are free of soft and collinear divergences.

Now recall that $\Gamma\left(p^+,p'^-\right) = Z^{\frac{1}{2}}(p^+)\tilde{Z}^{\frac{1}{2}}(p'^-)V(p^+,p'^-)$. Differentiating with respect to p^+ and p'^- respectively,

$$\frac{\partial \log \Gamma}{\partial \log p^+} = A(p^+) + \chi(p^+,p'^-),$$

$$\frac{\partial \log \Gamma}{\partial \log p'^-} = B(p'^-) + \chi'(p^+,p'^-),$$

where

$$A(p^+) = \frac{\partial \log \sqrt{Z}}{\partial \log p^+}, \qquad B(p'^-) = \frac{\partial \log \sqrt{\tilde{Z}}}{\partial \log p'^-}.$$

Lorentz invariance demands that $\Gamma\left(p^+,p'^-\right)$ is a function of the product $p^+p'^-$, i.e., $\Gamma = \Gamma\left(p^+p'^-\right)$. Therefore, $\frac{\partial \log \Gamma}{\partial \log p^+} = \frac{\partial \log \Gamma}{\partial \log p'^-}$ and thus:

$$\begin{aligned}
A(p^+) - B(p'^-) &= \chi'(p^+,p'^-) - \chi(p^+,p'^-) \\
&= \chi''(p^+,p'^-,\mu,g) \\
&= \chi''\left(p^+,p'^-,\lambda\mu,\bar{g}(\lambda\mu)\right) \\
&= \chi''\left(p^+/\lambda,p'^-/\lambda,\mu,\bar{g}(\lambda\mu)\right),
\end{aligned}$$

where, in the last two steps, we have used the RG equations and scaling arguments respectively. Now, using the fact that the coupling constant evolves by the RG equation

$$\mu \frac{d\bar{g}(\mu)}{d\mu} = \beta\left(\bar{g}(\mu)\right)$$

with the boundary condition $\bar{g}(\mu) = g$, we arrive at

$$\frac{\partial \chi''}{\partial \log p^+} + \frac{\partial \chi''}{\partial \log p'^-} - \beta(\bar{g}) \frac{\partial \chi''}{\partial \bar{g}} = 0.$$

This, in turn, implies that

$$\frac{\partial A}{\partial \log p^+} - \beta(\bar{g}) \frac{\partial A}{\partial \bar{g}} = \frac{\partial B}{\partial \log p'^-} - \beta(\bar{g}) \frac{\partial B}{\partial \bar{g}}.$$

Notice that the LHS of the above equation is function of p^+ alone, while the RHS depends only on p'^-. This is possible if they are independent of both p^+ and p'^-. In other words,

$$\frac{\partial A}{\partial \log p^+} - \beta(\bar{g}) \frac{\partial A}{\partial \bar{g}} = C(\mu, \bar{g}, m, m_g),$$

$$\frac{\partial B}{\partial \log p'^-} - \beta(\bar{g}) \frac{\partial B}{\partial \bar{g}} = C(\mu, \bar{g}, m, m_g),$$

where C is a constant independent of p^+ and p'^-.

The solution to the first order differential equations above can be obtained as follows. First, we use scaling arguments to write

$$-\lambda \frac{\partial}{\partial \lambda} \left(A \left(\frac{p^+}{\lambda}, \mu, \bar{g}(\lambda\mu), m, m_g \right) \right) = C\left(\bar{g}(\lambda\mu), m, m_g\right).$$

Next, integrating with respect to the scale parameter λ,

$$-\int dA \left(\frac{p^+}{\lambda}, \mu, \bar{g}(\lambda\mu), m, m_g \right) = \int_1^{\frac{p^+}{\mu}} \frac{d\lambda}{\lambda} C\left(\mu, \bar{g}(\lambda\mu), m, m_g\right),$$

we find:

$$\begin{aligned}
A\left(p^+, \mu, \bar{g}(\mu), m, m_g\right) &= A\left(\mu, \mu, \bar{g}(p^+), m, m_g\right) \\
&\quad + \int_\mu^{p^+} \frac{dx}{x} C\left(\mu, \bar{g}(x), m, m_g\right).
\end{aligned}$$

It follows that

$$
Z^{\frac{1}{2}}(p^+) \;\; = \;\; Z^{\frac{1}{2}}(\mu) \, \exp\left[\int_\mu^{p^+} \frac{dy}{y} \int_\mu^y \frac{dx}{x} \, C\left(\mu, \bar{g}(x), m, m_g\right) \right.
$$

$$
\left. + \int_\mu^{p^+} \frac{dx}{x} \, A\left(\mu, \mu, \bar{g}(x), m, m_g\right) \right].
$$

Moreover, in case $\bar{g}$ is a constant,

$$
Z^{\frac{1}{2}}(p^+) \;\; = \;\; Z^{\frac{1}{2}}(\mu) \, \exp\left[\frac{1}{2} C(\mu, g, m, m_g) \, \log^2\left(\frac{p^+}{\mu}\right) \right.
$$

$$
\left. + A(\mu, \mu, g, m, m_g) \, \log\left(\frac{p^+}{\mu}\right) \right], \quad (4.105)
$$

which vanishes as $p^+ \to \infty$ $(C < 0)$. We can find a similar result for $\tilde{Z}(p'^-)$. It can be shown that the scale dependent coupling constant does not affect our conclusion as they can only generate non-leading terms such as $\log\log(p^+)$.

Finally, we can find V using

$$
\frac{\partial \log V}{\partial \log Q} = (\chi + \chi')\bigg|_{p^+ = p'^- = Q}.
$$

Therefore,

$$
\frac{\partial \log V}{\partial \log p^+} + \frac{\partial \log V}{\partial \log p'^-} \;\; = \;\; \left(\chi + \chi'\right)(Q, \mu, g)
$$

$$
= \;\; \left(\chi + \chi'\right)(Q, \lambda\mu, \bar{g}(\lambda\mu))
$$

$$
= \;\; \left(\chi + \chi'\right)(Q/\lambda\mu, \bar{g}(\lambda\mu))
$$

$$
= \;\; \left(\chi + \chi'\right)(\mu, \mu, \bar{g}(Q)),
$$

where, again, we have used RG and scaling arguments as before. Integrating, we find

$$
V(Q) = V(\mu) \, \exp\left[\int_\mu^Q \frac{dx}{x} \left(\chi + \chi'\right)(\mu, \mu, \bar{g}(x)) \right].
$$

If the coupling constant does not depend on the scale,

$$V(Q) \to \exp\left[\left(\chi + \chi'\right)(\mu, \mu, \bar{g}) \, \log\left(\frac{Q}{\mu}\right)\right]. \tag{4.106}$$

Once again, the scale dependent coupling produces only terms which have less leading logarithms.

Therefore the Eqs.(4.106), (4.105) and the analogue of the latter for $\tilde{Z}$ are exact leading order results. Corrections to these involve less dominant terms such as $\log \log p^+$, etc.

Bibliography

[1] T.P. Cheng and L.F. Li, *Gauge theory of elementary particle physics*, Clarendon Press, Oxford (1984).

[2] T. Muta, *Foundations of quantum chromodynamics: an introduction to perturbative methods in gauge theories*, World Sci. Lect. Notes Phys. **5**, 1 (1987).

[3] M.E. Peskin and D.V. Schroeder, *An introduction to quantum field theory*, Addison-Wesley (1995).

[4] P. Ramond, *Field theory: a modern primer*, Front. Phys. **74**, 1, Addison-Wesley (1989).

[5] G. Sterman, *An introduction to quantum field theory*, Cambridge University Press (1993).

[6] S. Weinberg, *The quantum theory of fields, Vol. 1: Foundations, Vol. 2: Modern applications*, Cambridge University Press (1995).

The above are some text books on quantum field theory that treats perturbative QCD. For more details on the subject of the lectures X and XI, see:

[7] J.C. Collins, *Algorithm to compute corrections to the Sudakov form-factor*, Phys. Rev. **D22** (1980) 1478.

[8] A. Sen, *Asymptotic behavior of the Sudakov form-factor in QCD*, Phys. Rev. **D24** (1981) 3281.

Chapter 5

Quark-Gluon Plasma

Rajiv V. Gavai

5.1 Introduction

Lattice QCD *predicts* a phase transition to a hitherto unseen state of matter called quark-gluon plasma (QGP). A few microseconds after the big bang, our universe was most likely filled with QGP. Amazingly, attempts to produce QGP in the laboratory are currently going on in the relativistic heavy ion collider (RHIC) in the USA. In the following, we introduce the concept of QGP and then move on to its description in terms of the underlying field theory, namely quantum chromodynamics (QCD). Lattice QCD has yielded the most reliable predictions for QGP. After introducing the basics, the elementary aspects of heavy ion physics is covered with some examples of signatures of QGP.

Many thanks are due to Sharmistha Mukhopadhyay for making a preliminary draft, to Mohan Shinde for additional typing assistance and to Rajendra Pawar for drawing the figures. These notes are based on lectures were given at the SERC school in HRI. I would like to thank the organizers, especially Debashis Ghoshal for doing a great job of the organization and more importantly, for coaxing me into writing these notes up and being patient with me.

5.1.1   Confinement & why quark-gluon plasma is important

Quantum chromodynamics (QCD) is now the widely accepted theory of strong interactions which describes the theory of interactions of quarks and gluons, the basic building blocks of protons and neutrons. Unlike the electroweak theory, the coupling of QCD, α_s can have large values in many interesting physical applications, although QCD has been tested extensively and successfully in experiments for small α_s. The property of asymptotic freedom of QCD states that for sufficiently large Q^2 (or small distances) the running coupling $\alpha_s(Q^2)$ is small. I refer you to Ashoke Sen's lectures in this volume for further details on perturbative QCD and its successes[1].

However, QCD also has to explain how the entire spectrum of observed hadrons arises and what their masses, decay constants etc. are. Even for certain electroweak decays, which naively should be computable using perturbative techniques, one needs hadronic matrix elements where the corresponding α_s is at low Q^2 and therefore large. The phenomenon of confinement of quarks or gluons is a qualitative prediction expected from QCD: Indeed, it is otherwise a mystery as to why no free quarks or gluons have been observed in any experiments in spite of vigourous searches. Usual perturbation theory fails for all these situations. New tools or techniques are needed if the beloved Standard Model is to account for the observed physics in its entirety.

Gauge theories formulated on discrete space-time lattice have the potential to handle all the questions above. Over the past two decades or so, tremendous progress has been made in establishing the answers to the questions above. QCD on a discrete space-time lattice postdicts confinement, hadron masses and various weak decay matrix elements. A true test of a theory and/or a new technique is in confronting its predictions with the experiments successfully. Lattice techniques have yielded a *non-perturbative*

---

[1]We shall attempt to provide references in the text. A brief list of useful references and guide to further reading is given at the end.

predictions of QCD : A phase transition to a new state of strongly interacting matter, called Quark-Gluon Plasma (QGP). It turns out that such a state can be, and may have been, produced in heavy ion collisions in CERN and very recently in Relativistic Heavy Ion collider (RHIC) at the Brookhaven National Loboratory (BNL), New York. If so, this would constitute the observation a deconfined state of quarks and gluons, and thus the first experimental proof of confinement at lower temperatures and densities by implication. The mystical nature of quarks and gluons, which are 'felt' but not 'seen' in most experiments makes the confinement hypothesis quite a bit unconvincing/unreal. Lattice QCD not only demonstrates it to be an in-built feature of the theory, it also predicts that confinement can be overcome if extreme conditions are created in the laboratory or the early universe.

These lectures are devoted to explaining

- how QCD provides the basic, fundamental information of QGP from first principle, and

- how this new state QGP could be produced and detected in the laboratory.

As you will see the first part is theoretically well developed and one can ask innovative questions to obtain detailed information on QGP. As I will outline, however, there are quite a few conceptual questions as well. Some are technical but some are still stumbling blocks. No satisfactory treatment of QCD at finite density (or equivalently at finite baryonic chemical potential) has so far emerged, depriving us of a first principles approach to investigate exciting speculations as colour superconductivity or the strange quark stars. The second part is rewarding for the connection it makes with the real world but is theoretically less developed. In fact, there are many challenges here to even develop the right framework or at least, to establish the existence of QGP experimentally in the face of somewhat weaker theoretical setting.

5.1.2   A simple model for quark-gluon plasma: the MIT bag model

As already outlined above, one needs new non-perturbative techniques to extract information on the hadron spectrum and, consequently, to deal with hadronic world under extreme conditions. Before we turn to lattice QCD, let us obtain some physical insight by considering simple model of hadrons which exploit the basic features of QCD under these conditions. *MIT bag model* is one such model which incorporates both confinement of quarks and gluons and the property of asymptotic freedom of QCD. It does so by treating the ordinary vacuum as a medium in which hadron exists as a bag (or a bubble). As shown in Fig. 5.1, a colour neutral baryon in this model has (almost) free coloured quarks inside it which cannot escape out due to the bag shown as a boundary. Thus the bag incorporates the physics of confinement while the free quarks inside account for asymptotic freedom.

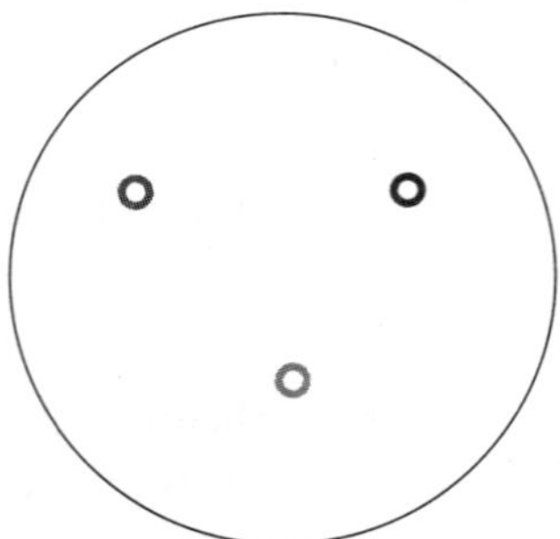

Figure 5.1: A baryon in the bag model.

Assuming the vacuum to have an energy density B, the energy of such a hadron is given by,

$$E_H = \frac{4}{3}\pi R^3 B + \frac{C}{R},\tag{5.1}$$

where C/R is the contribution coming from the kinetic energy estimated by using the uncertainty principle and the first term is the energy needed to create the bag. A stable hadron H will result when

$$\frac{\partial E_H}{\partial R} \;=\; 4\pi R^2 B - \frac{C}{R^2} = 0,$$

$$\text{yielding } R_H \;=\; \left(\frac{C}{4\pi B}\right)^{\frac{1}{4}}.$$

This, in turn, implies that the mass of the hadron is given by

$$M_H = \frac{4}{3}\pi R_H^3 B + \frac{4\pi B R_H^4}{R_H} = \frac{16\pi}{3} B R_H^3. \tag{5.2}$$

Note that $\dfrac{1}{4\pi R^2}\dfrac{\partial E_H}{\partial R}$ is the radial pressure on the bag surface, being the force per unit area of it's surface. So for a static bag the pressure $P = -B + (C/4\pi R_H^4) = 0$. Thus, the vacuum pressure B balances the kinetic energy pressure $(C/4\pi R_H^4)$ to create a stable hadron. A more detailed calculation, taking the appropriate full kinetic energy term, enables to fix the bag constant B from the experimental data on hadron masses.

Imagine now a box of volume V having some hadrons. Increase its density of hadrons or increase the temperature T of the box. As the bags (hadrons) begin to overlap, one can envisage a state of matter in the box where the mobility of the quarks will extend beyond the size of an individual bag. This is the state we will call quark-gluon plasma in which colour can flow over distance scales much larger than typical hadron size.

5.1.3 μ-T phase diagram

The above qualitative physical picture can be made more quantitative. Recall that the partition function of a system,

$$Z = \mathrm{Tr}\left[\exp\left(-\frac{\hat{H} - \mu\hat{N}}{T}\right)\right]$$

gives the entire thermodynamics. Here, $\hat{H}$ is the hamiltonian of the system and $\hat{N}$ is its some conserved number, for us here the baryon number. μ is the corresponding chemical potential. All physical observables can be obtained as appropriate derivatives of the partition function. *E.g.*, the energy density of the system is given by

$$\epsilon \equiv \frac{\langle E \rangle}{V} \;\; = \;\; \frac{1}{VZ}\mathrm{Tr}\left[\hat{H}\exp\left(-\frac{\hat{H}-\mu\hat{N}}{T}\right)\right]$$

$$= \;\; \frac{T^2}{V}\frac{\partial \ln Z}{\partial T} + \mu n,$$

where,

$$n \equiv \frac{\langle N \rangle}{V} \;\; = \;\; \frac{1}{VZ}\mathrm{Tr}\left[\hat{N}\exp\left(-\frac{\hat{H}-\mu\hat{N}}{T}\right)\right]$$

$$= \;\; \frac{T}{V}\frac{\partial \ln Z}{\partial \mu}$$

is the (baryon) number density.

The phase of our box above at low T (or low density) can be approximated by a gas of noninteracting hadrons (ideal gas of hadrons), whereas the high T phase (or the high density phase) can be assumed to be an ideal gas of quarks and gluons, but with the vacuum pressure. For noninteracting fermions or bosons, one can easily evaluate the (log of the) partition function to be

$$\ln Z = \frac{gV}{6\pi^2 T}\int_0^\infty \frac{dp\, p^4}{\sqrt{p^2+m^2}}\left[\frac{1}{e^{\frac{E-\mu}{T}}+\eta}+\frac{1}{e^{\frac{E+\mu}{T}}+\eta}\right], \quad (5.3)$$

where $E = \sqrt{p^2+m^2}$, $\eta = \pm 1$ for fermions and bosons respectively and g denotes the degeneracy due to spin polarizations. For $m = 0$, the integrals can be evaluated easily:

$$T(\ln Z)_f \;\; = \;\; \frac{g_f V}{12}\left[\frac{7}{30}\pi^2 T^4 + \mu^2 T^2 + \frac{\mu^4}{2\pi^4}\right],$$

$$T(\ln Z)_B \;\; = \;\; \frac{g_B V}{90}\pi^2 T^4 \qquad\qquad (5.4)$$

Problem 1: Derive Eq.(5.4) from Eq.(5.3).

Using the relation $P = T \ln Z/V$, one can write down the pressure in the two phases. Since we expect these two phases to be the relevant ones at the two ends of T (or μ), one may invoke the Gibbs principle to select the favoured phase of maximum pressure. Let us consider some special cases to illustrate this and obtain the μ-T phase diagram for our simple model of hadrons.

Case I $T \neq 0$, $\mu = 0$.

The hadronic phase has the usual pions, kaons, protons, ρ-mesons, etc. with masses given by $m_\pi = 140$ MeV, $m_K = 495$ and $m_p = 940$ MeV, $m_\rho = 770$ MeV. If the temperatures of interest are sufficiently low, as we will shortly see is the case, then the hadronic phase can be approximated by three pions since the heavier particles will not contribute due to a typical suppression factor $\exp(-E/T) \leq \exp(-M/T)$. We can simplify further by assuming the pions to be massless bosons so that one can use the analytic solutions (5.4); this assumption can be trivially removed by doing the integrals (5.3) numerically. The pressure of the hadronic phase is then given by

$$P_H = \frac{\pi^2}{30} T^4.$$

For the quark-gluon phase, we only need to get the correct degrees of freedom, since gluons are massless and only light quarks will be relevant. Note that quarks with masses $m_q < \Lambda_{QCD}$ are light quarks whereas $m_q > \Lambda_{QCD}$ are heavy, Λ_{QCD} is known from experiments to be ~ 200–250 MeV. One mostly deals with light quarks in QGP as the Boltzmann suppression factor of $\exp(-m_q/T)$ is substantial for the charm, bottom or top quark. Indeed, the strange quark with a mass of ~100-150 MeV is already a borderline case, as shown below. Assuming therefore massless quarks of three colours and two flavours, *i.e.*, $N_c = 3, N_f = 2$, one gets for quarks, $g_f = 2 \times 3 \times 2 = 12$, and for gluons, $g_B = 2 \times (N_c^2 - 1) = 16$, where the extra factor of 2 in each case is due to the possible spin polarizations. Using these values with (5.4), one obtains the pressure of the quark-gluon phase as :

$$P_{q+g} = \frac{37}{90}\pi^2 T^4 - B. \tag{5.5}$$

Note the presence of the bag pressure term above which reduces the pressure of the quarks and gluons and keeps them from flying apart.

The thermodynamically favoured state of minimum thermodynamic potential or maximum pressure for low temperatures is the hadronic phase since $P_H > 0, P_{q+g} < 0$ for it. On the other hand, when T is large, $P_{q+g} > P_H$ and the quark-gluon phase is preferred by the Gibbs criterion. The phase transition will occur at a temperature T_c found by equating $P_H(T_c) = P_{q+g}(T_c)$. One finds $T_c = (45B/17\pi^2)^{\frac{1}{4}} = 0.72\ B^{\frac{1}{4}}$. Since the nucleon mass $M_p = \frac{16}{3}\pi R^3 B = 1$ GeV, and its radius $R_p = 1$ fm $(= 200\text{MeV})^{-1}$, one can estimate $B^{1/4} \simeq 3^{-1/4} \times 200$ MeV $\simeq 150$ MeV. This, in turn, implies a $T_c \simeq 100 - 140$ MeV where the higher value is obtained if one uses a higher average value $\bar{M}_p$ to take care of the presence of various higher resonances. Note that $m_K/T_c, m_\rho/T_c, m_N/T_c >> 1$ and therefore can be *a posteriori* ignored in obtaining the pressure of hadronic phase and thus $T_c(\mu = 0)$. It is easy to compute the energy densities $\mathcal{E}_H(T_c)$ and $\mathcal{E}_{q+g}(T_c)$ of the two phases at the transition point T_c. Since the fermionic and bosonic energy densities are give by

$$\mathcal{E}_f(T, \mu = 0) \;=\; \frac{g_f}{12}\frac{7}{10}\pi^2 T^4, \tag{5.6}$$

$$\mathcal{E}_B(T, \mu = 0) \;=\; \frac{g_B}{30}\pi^2 T^4. \tag{5.7}$$

The energy densities of the two phase at T_c are found to be

$$\mathcal{E}_{p+q}(T_c) \;=\; \frac{7}{10}\pi^2 T_c^2 + \frac{8}{15}\pi^2 T_c^4 + B,$$

$$\mathcal{E}_H(T_c) \;=\; \frac{\pi^2 T_c^4}{10} \simeq T_c^4. \tag{5.8}$$

The pressure is continuous at T_c by construction, but as is evident above, the energy density jumps discontinuously. The latent heat at this first order phase transition is

$$L = \mathcal{E}_{p+q}(T_c) - \mathcal{E}_H(T_c) = \frac{68\pi^2}{45} T_c^4 \qquad (5.9)$$

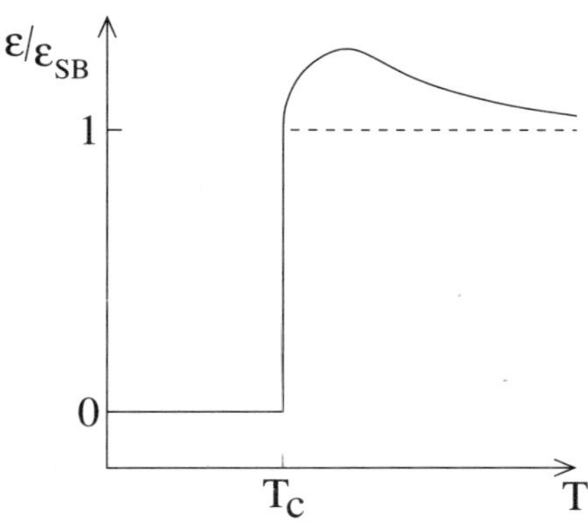

Figure 5.2: The energy density vs. T in bag model for $\mu = 0$.

In summary, the features of the quark-hadron transition in this case are

1. a first order phase transition;

2. a huge latent heat: $\dfrac{L}{T_c^4} \simeq 15$ or $L = 4B$, *i.e.* a strong discontinuity in the energy density;

3. as shown in Fig.5.2, $\mathcal{E} \to \mathcal{E}_{SB}$ from above in the $T \to \infty$ limit, where $\mathcal{E}_{SB}$ is the ideal gas or Stefan-Boltzmann limit.

Case II $T = 0$, $\mu \neq 0$.

Let us now consider the limiting case of $T = 0$ with nonzero μ. As μ increases, the net baryon number increases. A sensible approximation to the hadronic phase is to assume dominance of protons and neutrons. Using again the ideal gas expressions in this $T \to 0$ limit,

$$P_H = \frac{M^4}{6\pi^2}\left[\frac{\mu}{M}\left(\frac{\mu^2}{M^2}\sqrt{\frac{\mu^2}{M^2}-1}-\frac{5}{2}\right)\right.$$

$$\left.+\frac{3}{2}\ln\left\{\frac{\mu}{M}+\sqrt{\frac{\mu^2}{M^2}-1}\right\}\right],$$

$$n_B = \frac{2}{3\pi^2}\left(\mu^2-M^2\right)^{\frac{3}{2}}$$

$$\text{and}\quad \mathcal{E} = -p+sT+\mu n = \mu n_B - p_B, \qquad (5.10)$$

where, n_B is the baryon density and s is the entropy density.

Problem 2: Derive Eq.(5.10) from Eq.(5.3).

Using the expressions (5.4) for the ideal quark gas, one has

$$P_q = \frac{\mu_q^4}{2\pi^2}-B, \quad n_B = \frac{n_q}{3}=\frac{2\mu_q^3}{3\pi^2}, \quad \text{and}\ \mathcal{E}_q = \frac{3\mu_q^4}{2\pi^2}+B. \quad (5.11)$$

Since a baryon is made of three quarks, the baryon density is one third of the quark density. Accordingly, the baryon chemical potential is related to the quark chemical potential by $\mu = 3\mu_q$. The same method as in **case I** above can be used to determine the thermodynamically favoured phase at a given μ (hadronic for low μ and quark for high μ) and the chemical potential at the transition. Thus at the phase transition $P_q = P_H$ whereas the energy density may in general be discontinuous at μ_c. Since latent heat $L = \mathcal{E}_q(\mu_c) - \mathcal{E}_H(\mu_c) \geq 0$, one obtains the relation $\left(3\mu_q^4/2\pi^2\right) + B - \mu n_B + P_H \geq 0$. Using the continuity of pressure $P_H(\mu_c) = P_q(\mu_c) = \frac{\mu_{qc}^4}{2\pi^2} - B$, this inequality becomes $\left(2\mu_{qc}^4/\pi^2\right) - \mu_c n_B \geq 0$. Finally, substituting for n_B from (5.10) and using $\mu_{qc} = \mu_c/3$, it can be converted as an inequality for μ_c:

$$\mu_c\left[\left(\frac{\mu_c}{3}\right)^3 - (\mu_c^2 - M^2)\right] \geq 0,$$

which implies

$$M \leq \mu_c \leq \frac{3M}{2\sqrt{2}}. \tag{5.12}$$

Substituting these limits in (5.10) and (5.11), one finds that for the lower limit, $\mu_c = M$, $P_H = 0 = P_q$, leading to $B = \frac{1}{2\pi^2}\left(\frac{M}{3}\right)^4$. For the upper limit, $\mu_c = \frac{3}{2\sqrt{2}}M$,

$$P_q = \frac{1}{2\pi^2}\frac{M^4}{64} - B, \qquad \text{and } P_B = \frac{M^4}{6\pi^2}\left[-\frac{33}{64} + \frac{3}{4}\ln 2\right]. \tag{5.13}$$

From the equality of these two, one has $B = \frac{M^4}{8\pi^2}\left(\frac{3}{4} - \ln 2\right)$. Thus a solution for μ_c exists provided,

$$\frac{1}{2\pi^2}\left(\frac{M}{3}\right)^4 \leq B \leq \left[\frac{3}{4} - \ln 2\right]\frac{M^4}{8\pi^2}, \tag{5.14}$$

which is a very narrow range indeed. However, it does include the standard $B^{1/4}$ value we usually have: $0.158M \leq B^{\frac{1}{4}} \leq 0.164M$.

In summary, the features of the quark-hadron transition in this case are the following.

1. A first order phase transition, depending on value of B. For the highest value of B, the latent heat $L = 0$, leading to a second order transition;

2. For the lowest value of B, one has the strongest first order transition with $L = 4B$ again;

3. As $\mu \to \infty$, $\mathcal{E} \to \mathcal{E}_{SB}$ from above.

Note that $n_B = 0.15\,\text{fm}^{-3}$ is the cold nuclear matter density. This corresponds to $\mu \simeq 0.78$ GeV (or $\mu_q \simeq 0.26$ GeV). The bag model, on the other hand, suggests the highest possible $\mu_c \simeq 1.06M = 1$ GeV, which is a bit too low.

Case III $T \neq 0$, $\mu \neq 0$.

This general case is not analytically tractable even for the simplifying approximations we made above. However, the same method works in this case as well. The only key difference is that the equation $P_{q+g}(T_c, \mu_{qc}) = p_H(T_c, \mu_c) = P_{\text{meson}}(T_c) + P_{\text{nucleon}}(T_c, \mu_c)$ for a given pair (T_c, μ_c) has to be solved numerically. This the leads to the popular T-μ phase diagram of strongly interacting matter, shown in Fig.5.3.

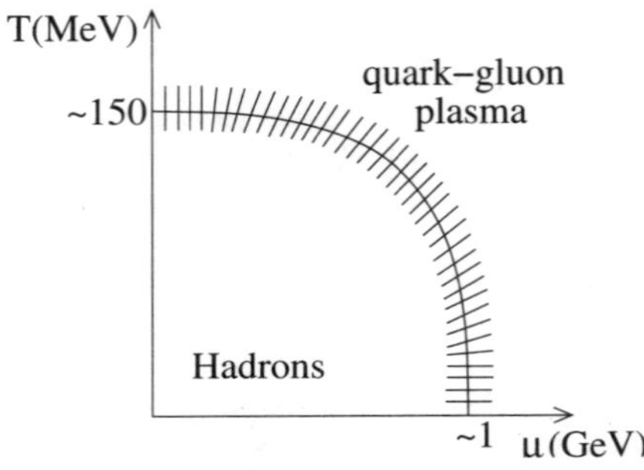

Figure 5.3: The $T - \mu$ phase diagram for bag model.

Simple though it may be for illustrative purposes, the bag model phase diagram is clearly not very satisfactory, although unfortunately it is heavily used in phenomenological analyses of heavy ion data and the cosmology of the early universe. Its major drawback is that the phase transition line appears by construction. Since one assumes the hadronic phase and the ideal quark-gluon phase to be that of the *same matter*, one is able to apply the Gibbs criterion. There is no strong theoretical basis for this assumption other than the naive picture we presented earlier. On a somewhat technical level, it is too simplistic to assume an ideal gas picture or a lack of interaction (apart from B). This can, of course, be improved by incorporating more hadrons and resonances as well. Similarly, the ideal quark-gluon gas can be improved upon by including weak perturbative QCD interactions.

One can thus take recourse to better models or insist that one must simply use quantum chromodynamics (QCD), the known theory of strong interactions to obtain this phase diagram. Due to a lack of time, we will not dwell on any other model but follow the latter option in the next lecture.

5.2 QGP from QCD

5.2.1 Path integral formalism

Let us begin by recalling that the partition function

$$Z_{QCD} = \text{Tr} \, \exp\left[-\frac{\hat{H}_{QCD} - \mu_B \hat{N}_B}{T} \right]$$

contains all the information we are looking for. All that one has to do is to choose the appropriate complete set of states and compute the trace. For QCD, we can begin from the gauge invariant Lagrangian $\mathcal{L}_{QCD}$, construct $\hat{H}_{QCD}$ and obtain Z. Unfortunately, this turns out to be too complicated and intractable. First of all, one has to worry about gauge dependence : $\hat{H}_{QCD}$ has to be defined in a gauge specific manner. Then, the trace must be carried over gauge independent physical states, by appropriate projectors. A neater solution is to reformulate the problem as a functional integral of the gauge invariant euclidian Lagrangian. One can then exploit its similarity with $T = 0$ field theory and employ essentially the same techniques to obtain thermal expectation value $\langle \theta \rangle$ of a given observable θ.

For simplicity, let us demonstrate the idea of reformulation with a quantum mechanical example: Set $\mu = 0$ and consider the hamiltonian $\hat{H} = \frac{\hat{p}^2}{2m} + V(\hat{x})$ in stead of $\hat{H}_{QCD}$. Employing the eigenstates of the position operator, $\hat{x}|x\rangle = x|x\rangle$ to evaluate the trace, the partition function can be written as:

$$Z = \text{Tr} \, e^{-\beta \hat{H}} = \int \langle x| \, e^{-\beta \hat{H}} \, |x\rangle. \tag{5.15}$$

Let $\beta = n\epsilon$ with ϵ small and n large : $\lim \epsilon \to 0$. The operator in (5.15) can then be written as a product of n terms involving $e^{-\epsilon \hat{H}}$. Inserting the completeness relation below,

$$\int dx |x\rangle\langle x| = 1, \qquad \langle x|x'\rangle = \delta(x - x') \tag{5.16}$$

$(n-1)$ times in between the terms of this product for x_i, $i = 1, \cdots n-1$ and defining $x_0 = x_n = x$ in (5.15), Z becomes

$$
\begin{aligned}
Z &= \int \prod_{i=1}^{n} dx_i \, \langle x_{i-1}| \, e^{-\epsilon \hat{H}} \, |x_i\rangle \\
&= \int \prod_{i=1}^{n} dx_i \, \langle x_{i-1}| \, e^{-\epsilon V(\hat{x})/2} \, e^{-\epsilon \hat{p}^2/2m} \, e^{-\epsilon V(\hat{x})/2} |x_i\rangle .
\end{aligned}
$$

(A symmetric (Weyl) ordering of operators has been used above.) Let $|p\rangle$ denote the momentum eigenket, $\hat{p}|p\rangle = p|p\rangle$. Using the corresponding completeness and orthonormality relations,

$$
\langle p|p'\rangle = \delta(p-p'), \qquad \int dp \, |p\rangle\langle p| = 1,
$$

$2n$ times, we obtain

$$
\begin{aligned}
Z &= \int \prod_{i=1}^{n} dx_i \, dp_i \, dp'_i \, \langle x_{i-1}| \, p'_i \rangle \, e^{-\epsilon(V(x_{i-1})+V(x_i))/2} \\
&\qquad \times \langle p_i|x_i\rangle \, \langle p'_i| \, e^{-\epsilon \hat{p}^2/2m} \, |p_i\rangle \\
&= \int \prod_{i=1}^{n} dx_i \, dp_i \exp\left[-\epsilon\left(\frac{p_i^2}{2m} + \frac{V(x_{i-1})+V(x_i)}{2} \right. \right. \\
&\qquad\qquad\qquad \left.\left. -ip_i \frac{x_i - x_{i-1}}{2} \right)\right].
\end{aligned}
\tag{5.17}
$$

Finally, taking the continuum limit $\epsilon \to 0$, one sees that $x_i \to x(t)$, $p_i \to p(t)$ and $\frac{1}{\epsilon}(x_i - x_{i-1}) \to \dot{x}(t)$. Thus Z is given by

$$
Z = \int [dp] \int_{x(0)=x(\beta)} [dx] \; e^{\int_0^\beta dt[ip(t)\dot{x}(t) - H[x(t),p(t)]]}
\tag{5.18}
$$

Remarks :

1. The integrals over x and p, $\int dx$ and $\int dp$, are to be thought of as functional integrals. At each t, all possible values of these functions are allowed.

2. The integration over p is unrestricted but that over x has periodic boundary condition $x(0) = x(\beta)$. This arises due to the trace in Eq.(5.15).

3. There are no operators in Eq.(5.18) anymore. All variables are classical.

One can choose to do the p_i integration before sending $\epsilon \to 0$. For our hamiltonian this amounts to simply completing the square and then doing the gaussian integrals in p_i:

$$\int dp_i \ \exp\left(-\epsilon\frac{p_i^2}{2m} + ip_i(x_i - x_{i-1})\right)$$
$$= \quad \sqrt{\frac{m}{2\pi\epsilon}} \ \exp\left(-\frac{m}{2\epsilon}(x_i - x_{i-1})^2\right).$$

Therefore, in the limit $\epsilon \to 0$, the partition function is

$$Z \quad = \quad \sum_{\substack{x(0)=x(\beta) \\ \text{paths}}} e^{-S}, \tag{5.19}$$

$$\text{where,} \quad S \quad = \quad \int_0^\beta d\tau \left(\frac{m\dot{x}^2}{2} + V(x)\right). \tag{5.20}$$

Remarks :

1. Z is now like an Euclidean path integral, being the sum over all possible classical paths of the action S with a Boltzmann weight e^{-S}.

2. $\beta \to \infty \Leftrightarrow T \to 0$, $\quad Z$ reduces to usual zero temperature path integral.

As is usually done in field theory, one can define a partition function with a source $j(\tau)$:

$$Z(\beta, j) = \int \mathcal{D}x e^{-S(\beta)+\int_0^\beta j(\tau)x(\tau)d\tau}$$

Differentiate twice with respect to j and set $j = 0$ to obtain the propagator:

$$\frac{1}{Z(\beta)}\frac{\delta^2 Z(\beta, j)}{\delta j(\tau_1)\delta j(\tau_2)}\Big|_{j=0} = \frac{1}{Z(\beta)}\int \mathcal{D}x \ x(\tau_1)x(\tau_2) \ e^{-S} \tag{5.21}$$

Recall that $\langle \hat{A} \rangle_\beta = \operatorname{Tr} \hat{A} e^{-\beta \hat{H}} / \operatorname{Tr} e^{-\beta \hat{H}} = \frac{1}{Z(\beta)} \operatorname{Tr} \hat{A} \exp(-\beta \hat{H})$ (since $\mu = 0$ for us here). We will now show how this is related to the propagator in (5.21). Since $\hat{x}(t) = e^{i\hat{H}t} \hat{x} e^{-i\hat{H}t}$, one can write $\hat{x}(-i\tau) = e^{\hat{H}\tau} \hat{x} e^{-\hat{H}\tau}$.

Defining

$$T\left(\hat{x}(-i\tau_1)\hat{x}(-i\tau_2)\right) = \begin{cases} \hat{x}(-i\tau_1)\,\hat{x}(-i\tau_2), & \text{for} \quad \tau_1 > \tau_2, \\ \hat{x}(-i\tau_2)\,\hat{x}(-i\tau_1), & \text{for} \quad \tau_2 > \tau_1. \end{cases}$$

one can show that

    i) $\langle T(\hat{x}(-i\tau_1)\hat{x}(-i\tau_2))\rangle_\beta$ is the RHS of Eq.(5.21);

   ii) $\langle T(\hat{x}(-i\beta)\hat{x}(-i\tau))\rangle_\beta = \langle T|\hat{x}(0)\hat{x}(-i\tau)\rangle_\beta$;

  iii) $\Delta(\tau - \beta) = \Delta(\tau)$,   where $\Delta(\tau) = \langle T(\hat{x}(-i\tau)\hat{x}(0))\rangle_\beta$.

Problem 3: Show the above.

Problem 4: Choose $V(x) = \frac{1}{2}\omega^2 x^2$ corresponding to a harmonic oscillator and set $m = 1$ to make a connection to field theory.

$$\begin{aligned} Z(\beta, j) &= \int_{\text{pbc}} \mathcal{D}x \, \exp\left(-\int_0^\beta d\tau \left[\frac{1}{2}\dot{x}^2 + \frac{1}{2}\omega^2 x^2 - j(\tau)x(\tau)\right]\right) \\ &= \int_{\text{pbc}} \mathcal{D}x \, \exp\left(-\int_0^\beta d\tau \frac{1}{2}x(0)\left[-\frac{d^2}{d\tau^2} + \omega^2\right]x(\tau) \right. \\ &\qquad\qquad \left. + \int_0^\beta j(\tau)x(\tau)d\tau\right) \\ &= Z(\beta)\exp\left(\frac{1}{2}\int d\tau d\tau' j(\tau)K(\tau,\tau')j(\tau')\right), \end{aligned}$$

where $K(\tau, \tau')$ is the inverse of $\left(-\frac{d^2}{d\tau^2} + \omega^2\right)$, i.e.,

$$\left(-\frac{d^2}{d\tau^2} + \omega^2\right) K(\tau, \tau') = \delta(\tau - \tau') \tag{5.22}$$

Using the definition of Δ, one obtains $K(\tau, \tau') = \Delta_F(\tau - \tau')$. The subscript F stands for harmonic oscillator (or equivalently Free

Field Theory). Solving (5.22), subject to the boundary condition $\Delta(\tau - \beta) = \Delta(\beta)$ leads to

$$\Delta_F(\tau) = \frac{1}{2\omega}[1 + n(\omega)]e^{-\omega\tau} + \frac{1}{2\omega}n(\omega)e^{\omega\tau}, \qquad (5.23)$$

with $1/n(\omega) = \exp(\beta|\omega|) - 1$ corresponding to the Bose-Einstein distribution.

5.2.2   Spectral function

The spectral function is a useful tool for extracting physical information in a variety of situations including those involving deviations from thermal equilibrium. Let us see how it arises in our simple example.

Let us define two new two-point functions

$$\begin{aligned}
D^>(t,t') &= \langle\,\hat{x}(t)\,\hat{x}(t')\,\rangle_\beta \\
D^<(t,t') &= \langle\hat{x}(t')\hat{x}(t)\rangle_\beta = D^>(t,t')
\end{aligned}$$

Here time t is complex in general. Insert the complete set of eigenvectors of $\hat{H}$ in between $\hat{x}(t)$ and $\hat{x}(t')$

$$D^>(t,t') = \frac{1}{Z(\beta)} \sum_n \langle n|e^{-\beta\hat{H}}\hat{x}(t)\hat{x}(t')|n\rangle, \qquad (5.24)$$

to obtain

$$\begin{aligned}
D^>(t,t') &= \frac{1}{Z(\beta)} \sum_{n,m} e^{-\beta E_n} \langle n|e^{i\hat{H}t}\hat{x}(0)e^{-i\hat{H}t}|m\rangle \\
&\qquad \times \langle m|e^{i\hat{H}t'}\hat{x}(0)e^{-i\hat{H}t'}|n\rangle \qquad (5.25) \\
&= \frac{1}{Z(\beta)} \sum_{n,m} e^{-\beta E_n}e^{iE_n(t-t')}|\langle n|\hat{x}(0)|m\rangle|^2 e^{iE_m(t-t')}
\end{aligned}$$

For $\beta \le \mathrm{Im}(t - t') \le 0$, $D^>(t,t')$ is well behaved and defined whereas for $\beta \ge \mathrm{Im}(t - t') \ge 0$, the same is true of $D^<(t,t')$.

Further,

$$D^>(t,t') \;=\; \frac{1}{Z}\,\mathrm{Tr}\,e^{-\beta\hat{H}}\hat{x}(t)\hat{x}(t')$$

$$=\; \frac{1}{Z}\mathrm{Tr}\,\hat{x}(t+i\beta)e^{-\beta\hat{H}}\hat{x}(t')$$

$$=\; \frac{1}{Z}\mathrm{Tr}\,e^{-\beta\hat{H}}\hat{x}(t')\hat{x}(t+i\beta)$$

$$=\; D^<(t+i\beta,t'),$$

which is called the Kubo-Martin-Schwinger relation.

Exploiting translation invariance, t' can be set to zero without loss of generality. Defining now, $D^{<,>}(k_0) = \int\limits_{-\infty}^{\infty} e^{-ik_0 t}dt\; D^{<,>}$ and using KMS relation one deduces

$$D^<(k_0) = e^{-\beta k_0}D^>(k_0) \;. \tag{5.26}$$

The spectral function can be defined in terms of these as

$$\rho(k_0) \;=\; D^>(k_0) - D^<(k_0) \tag{5.27}$$

$$=\; \int\limits_{-\infty}^{\infty} e^{-ik_0 t}\,\langle\, [\hat{x}(t),\hat{x}(0)\,]\,\rangle_\beta dt \;.$$

From (5.26) and (5.27) one obtains $D^>(k_0) = (1+f(k_0))\rho(k_0)$ and $\quad D^<(k_0) = f(k_0)\rho(k_0)\quad$ with $f(k_0) = 1/\left(e^{\beta k_0}-1\right)$. Furthermore, using (5.26) and (5.27) one can show that

$i)\quad \rho(k_0) = -\rho(-k_0)$

$ii)\quad \epsilon(k_0)\rho(k_0) > 0,\quad \epsilon(k_0)$ is the sign function

$iii)\quad \int\limits_{-\infty}^{\infty} dk_0\,\epsilon(k_0)\rho(k_0) = 1.$

When $\hat{x}(t)$ is the position operator for harmonic oscillator, one can use

$$\hat{x}(t) = \frac{1}{\sqrt{2\omega}}(ae^{-i\omega t} + a^\dagger e^{i\omega t})$$

to show $\rho_F = 2\pi\epsilon\delta(k_0^2 - \omega^2)$ which obeys all the three properties i) , ii), iii) above. Moreover, ρ_F is temperature dependent.

5.2.3   Neutral scalar fields

We can use the experience gained in the previous subsection to make a leap to neutral scalar field theory, defined by the lagrangian density:

$$\mathcal{L} = \frac{1}{2}\partial_\mu\phi\partial^\mu\phi - \frac{1}{2}m^2\phi^2 - U(\phi), \qquad (5.28)$$

where the potential $U(\phi) = g\phi^3 + \lambda\phi^4$. Since $\pi = \dot{\phi} = \frac{\partial\mathcal{L}}{\partial\dot{\phi}}$, the hamiltonian density $\mathcal{H} = \pi\dot{\phi} - \mathcal{L}$ can be written as

$$\mathcal{H} = \frac{\pi^2}{2} + \frac{1}{2}\nabla\phi^2 + \frac{1}{2}m^2\phi^2 + U(\phi). \qquad (5.29)$$

Note now the similarity with the harmonic oscillator example considered above: $\pi \leftrightarrow p$, $V(x) \leftrightarrow$ the rest. The partition function for the neutral scalar field theory is thus

$$Z_{NS} = \int_{\phi(0,\vec{x})=\phi(\beta,\vec{x})} \mathcal{D}\phi \, \exp\left(-\int_0^\beta d\tau \int d^3x \, \mathcal{L}(\phi, \partial_\mu\phi)\right), \qquad (5.30)$$

where the euclidean lagrangian density is

$$\mathcal{L}(\phi, \partial_\mu\phi) = \left[\left(\frac{\partial\phi}{\partial\tau}\right)^2 + \nabla\phi^2 + m^2\phi^2 + U(\phi)\right]. \qquad (5.31)$$

Here we have let $t \to -i\tau$ giving rise to $\dot{\phi}(t)^2 \to -\dot{\phi}(\tau)^2$. For $U(\phi) = 0$, one has free scalar field theory corresponding to free bosons. Since τ direction is compact and since $\phi(0,\vec{x}) = \phi(\beta,\vec{x})$, p_0 is discrete, with allowed values given by $p_0 = 2\pi nT$. By integrating $\frac{\partial\phi}{\partial\tau}$, $\nabla\phi$ by parts and using periodic b.c. of the fields ϕ, one gets

$$Z = \int \mathcal{D}\phi \, \exp\left(-\int_0^\beta d\tau \int d^3x \frac{1}{2}\phi D\phi\right),$$

$$\text{where} \quad D = \frac{\partial^2}{\partial\tau^2} + \nabla^2 + m^2$$

Therefore the gaussian functional integration can be done formally:

$$Z = \text{const} \times \det^{-\frac{1}{2}} D,$$
$$\text{or, } \ln Z = -\frac{1}{2} \ln \det D + C$$
$$= -\frac{1}{2} \ln \text{Tr} D + C,$$

where C is a constant.

Problem 5: This problem is an explicit calculation of $\ln \det D$.

Show that

$$\ln Z = V \int \frac{d^3 p}{(2\pi)^3} \left[-\frac{1}{2}\beta\omega - \ln(1 - e^{-\beta\omega}) \right],$$

where $\omega^2 = p^2 + m^2$.

Using Fourier transform of $\phi(\tau, \vec{x})$ evaluate $E(T = 0)$ and show it to be infinite. The vacuum energy can, however, be explicitly subtracted using the freedom to choose the scale for energy.

Perturbative expansion: Unlike above, the potential $U(\phi)$ is not zero in general. For $U(\phi) \neq 0$, *i.e.* for a general $\mathcal{L}$ one needs a method of computation of Z or some approximation scheme. If the coupling is sufficiently small, one can use a perturbative expansion in it. Denoting the free action density above as S_0, which is quadratic in field ϕ, and the interaction part as S_I, the partition function can be expanded as below.

$$Z = \int \mathcal{D}\phi \; e^{S_0 + S_I} = \int \mathcal{D}\phi \, e^{S_0} \sum_{l=0}^{\infty} \frac{1}{l!} S_I^l$$
$$= \left(\int \mathcal{D}\phi e^{S_0} \right) \times \left[1 + \frac{1}{\int \mathcal{D}\phi e^{S_0}} \sum_{l=1}^{\infty} \int \mathcal{D}\phi e^{S_0} S_I^l \right].$$

Or, equivalently:

$$\ln Z = \ln Z_0 + \ln Z_I.$$

The first term $\ln Z_0$ has been evaluated above. The second term can be expanded as a power series. For simplicity, let us consider, $-S_I = U(\phi) = \lambda\phi^4$. The expansion then is governed by powers of λ. The lowest order term is

$$\ln Z_1 = -\frac{\lambda}{\int \mathcal{D}\phi \, e^{S_0}} \int \mathcal{D}\phi e^{S_0} \left(\int_0^\beta d\tau \int d^3x \, \phi^4(\vec{x},\tau) \right). \quad (5.32)$$

In order to evaluate it, let us write down the Fourier expansion of the field ϕ in momentum space.

$$\phi(\vec{x},\tau) = \left(\frac{1}{VT}\right)^{\frac{1}{2}} \sum_n \sum_p e^{i(\vec{p}\cdot\vec{x}+\omega_n\tau)}\phi_n(p) \,, \quad (5.33)$$

where we have assumed a finite spatial volume V, thus discretizing the three momenta as well in addition to the energy (which, recall, has to be discrete due to $T \neq 0$). The action S_0 is then transformed to

$$S_0 = -\frac{1}{2T^2} \sum_{n,\vec{p}} (\omega_n{}^2 + \vec{p}^2 + m^2)\phi_n(p)\phi_{-n}(-p) \,, \quad (5.34)$$

and correspondingly $\mathcal{D}\phi \, e^{-S_0}$ is:

$$\prod_{n,p} \int d\phi_n(p) \exp\left[-\frac{1}{2T^2}(\omega_n{}^2 + \vec{p}^2 + m^2)\phi_n(p)\phi_{-n}(-p)\right]. \quad (5.35)$$

Note that the periodic boundary condition, $\phi(\beta,\vec{x}) = \phi(0,\vec{x})$ was used above in writing S_0. The Fourier transform of the four fields in the numerator yields sums over n_i, $i = 1,2,3$ and of $\exp(\sum i\omega_{n_j} t)$ and similar sums over $\vec{p}_i$ multiplied by $\prod_{j=1}^4 \phi_{n_j}(p_j)$. The integrations over $\vec{x}$ and τ lead to Kronecker δ-constraints which eliminate some sums over energy and momenta. One finally obtains (in the limit $V \to \infty$),

$$\ln Z_1 = -3\lambda\beta V \left[T\sum_n \int \frac{d^3p}{(2\pi)^3}\mathcal{D}_0(\omega,\vec{\omega})\right]^2, \quad (5.36)$$

where $\mathcal{D}_0(\omega,\vec{\omega}) = ({\omega_n}^2 + \vec{p}^2 + m^2)^{-1}$ is the free boson propagator. Diagrammatically, the above equation can be represented as shown in Fig.(5.4).

$$\ln Z_1 = 3$$ 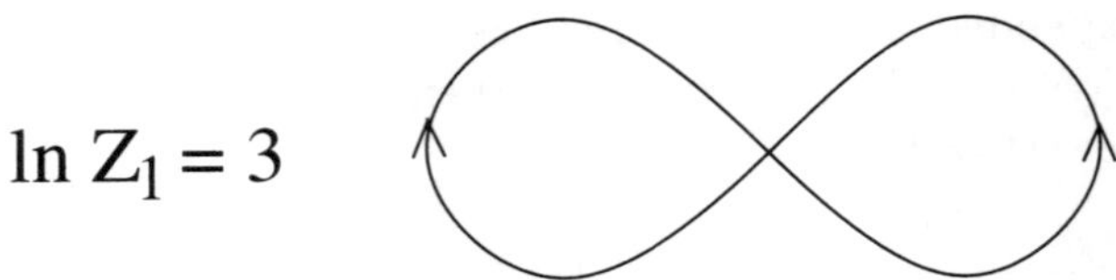

Figure 5.4: Diagrammatic representation of Eq.(5.36).

The closed loop in the figure denotes a factor of energy-summation and momentum integrations, the vertex stands for a factor of $-\lambda$, and the lines denote the free boson propagators.

As can be guessed from Fig.5.4, one can devise Feynman rules similar to $T = 0$ and use them for higher order calculations. Let us define general propagator $\mathcal{D}(x_1,\tau_1;x_2,\tau_2) = \langle \phi(x_1,\tau_1)\ \phi(x_2,\tau_2)\rangle$, Its Fourier transform is $\mathcal{D}(\omega_n,\vec{p}) = \beta^2 \langle \phi_n(\vec{p})\phi_{-n}(-\vec{p})\rangle$. Since the full action $S = S_0 + S_I$ contains $\beta^2 \phi_n(p)\phi_{-n}(-p)$ with $\mathcal{D}_0 = (\omega_n^2 + p^2 + m^2)$ as coefficient, one sees that

$$\mathcal{D}(\omega_n,\vec{p}) = -2\frac{\delta}{\delta\mathcal{D}_0^{-1}}\ln Z = 2\mathcal{D}_0^2\frac{\delta}{\delta\mathcal{D}_0}\ln Z. \qquad (5.37)$$

Define self energy $\Pi(\omega_n,\vec{p})$ by

$$\mathcal{D}(\omega_n,\vec{p}) = \frac{1}{\omega_n^2 + p^2 + m^2 + \Pi(\omega_n,\vec{p})}$$
$$= \frac{1}{\mathcal{D}_0^{-1} + \Pi} = \frac{\mathcal{D}_0}{1 + \Pi\mathcal{D}_0}.$$

Then (5.37) and (5.38) together imply that

$$(1 + \Pi\mathcal{D}_0)^{-1} = 2\mathcal{D}_0\frac{\delta}{\delta\mathcal{D}_0}\ln Z = 1 + 2\mathcal{D}_0\frac{\delta}{\delta\mathcal{D}_0}\ln Z_I \qquad (5.38)$$

Using (5.38) the self energy Π can be computed in powers of λ : $\Pi = \sum_{l=1}^{\infty}\Pi_l$. The first and second order terms Π_1 and Π_2

satisfy

$$1 - \mathcal{D}_0 \Pi_1 \;=\; 1 + 2\mathcal{D}_0 \frac{\delta}{\delta \mathcal{D}_0} \ln Z_1$$

$$-\mathcal{D}_0 \Pi_2 + \mathcal{D}_0 \Pi_1 \mathcal{D}_0 \Pi_1 \;=\; 2\mathcal{D}_0 \frac{\delta}{\delta \mathcal{D}_0} \ln Z_2 \;\;.$$

Recall that diagrammatically, $\ln Z_1$ is as shown in Fig. 5.4. Taking its derivative with respect to $\mathcal{D}_0$ will break one loop, leading to the diagram in Fig. 5.5, which stands for

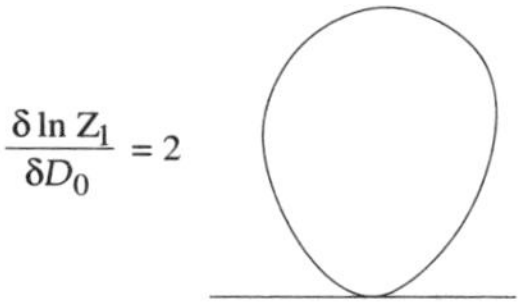

$$\frac{\delta \ln Z_1}{\delta D_0} = 2$$

Figure 5.5: Diagrammatic representation of Π_1.

$$\Pi_1 = 12\lambda T \sum_n \int \frac{d^3 p}{(2\pi)^3} \frac{1}{\omega_n^2 + \omega^2} \;\;. \tag{5.39}$$

Problem 6: Show that the evaluation of Eq.(5.39) leads to

$$\Pi_1 = 12\lambda \int \frac{d^4 p}{(2\pi)^4} \frac{1}{p_4^2 + p^2 + m^2} + 12\lambda \int \frac{d^3 p}{(2\pi)^3} \frac{1}{\omega} \frac{1}{e^{\beta\omega} - 1} \;,$$

where the first term is temperature independent and quadratically divergent.

The above is the canonical divergence we are familiar in scalar field theories. The usual $T = 0$ couterterm can be employed here too to eliminate it: add $\frac{1}{2}\delta m^2 \phi^2$. The remaining finite part $\Pi_1^{ren} \propto \lambda T^2$ on dimensional grounds alone. Its being nonzero for $T \neq 0$ is an important feature of $T \neq 0$ field theory, used in various applications to cosmology etc.

Seeing the connection of Π_1 with $\ln Z_1$, it is no surprise that it contributes to pressure

$$P = -\frac{T}{V}\ln Z = T^4\left(\frac{\pi^2}{90} - \frac{\lambda}{48} + \cdots\right).$$

The pressure *decreases* due to the lowest order correction at $O(\lambda)$. Let us make some general remarks, based on this example. It is true in general that (i) no new ultra-violet divergences arise at $T \neq 0$ $or(\mu \neq 0)$; the only divergences to be encountered are the $T = 0$ ones which can be handled in the usual way and (ii)the interactions change pressure, energy , etc. in a manner similar to above and related to Π_1, Π_2 etc.

5.2.4   Infra-red problems

As noted above, there are no additional ultra-violate divergences at finite temperature or density. However, the finite temeprature field theory has severe infra-red problems in general. In order to illustrate these in our simple example considered in this chapter, let us consider the massless case: $m = 0$. As we will argue below, the resultant infra-red divergences, when treated adequately, change the naive expectation for the next correction to the pressure P_2. Instead of being of $O(\lambda^2)$ it turns out to be of $O(\lambda^{3/2})$, making the λ^2 term the next one in the perturbative expansion.

Qualitatively, one can understand this by inspecting $\ln Z_2$ diagrammatically. Fig. 5.6 shows the two diagrams which contribute to it.

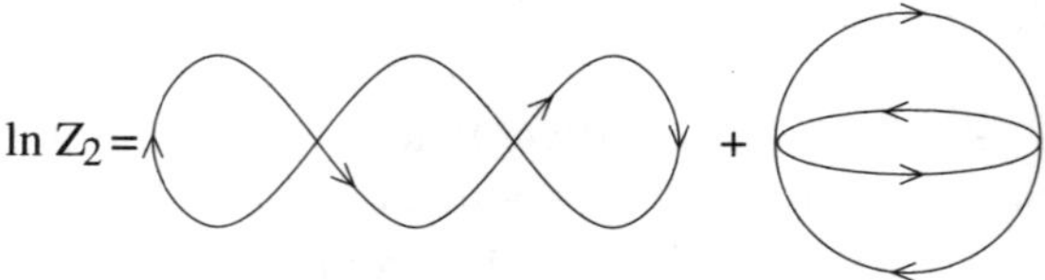

Figure 5.6: Diagrammatic representation of $\ln Z_2$.

When $\omega_n = 0$ in the middle loop of the first diagram, it becomes $\propto \Pi_1^2 \, dp \, p^{-2}$ (since the middle loop has two propagators and a $p^2 dp$ from the 3-momentum integration) and is thus divergent. This divergent contribution is absent at $T = 0$ since $\Pi_1 \propto \lambda T^2$. The nonzero Π_1 acts like an effective thermal mass. Although we chose $m_{bare} = 0$, the non-vanishing Π_1 at $T \neq 0$ yields $m_{1-loop} \propto \sqrt{\lambda} T$. Thus, if we could exploit the existence of the 1-loop thermal mass in the propagators in Π_2, the infra-red divergence encountered above can be tackled. At the order one is computing, the replacement of bare propagators by the 1-loop ones is, of course, permitted. This infra-red problem keeps coming at higher orders. In fact at $O(\lambda^N)$ the divergent terms are $\propto dp \, p^{-2(N-1)}$, making it more and more severe. It turns out that the offending terms can be resummed to all orders which leads to.

$$P = \frac{\pi^2 T^4}{90} \left[1 - \frac{15}{8} \left(\frac{\lambda}{\pi^2} \right) + \frac{15}{2} \left(\frac{\lambda}{\pi^2} \right)^{\frac{3}{2}} + \cdots \right].$$

In order to understand the origin of the $\lambda^{\frac{3}{2}}$ term better, let us imagine redrawing the diagram in Fig. 5.6 and draw the loops at the edges much smaller than that in the middle. The smaller loops can be then seen as decorations of the central scalar propagator loop. We know that $\ln Z = \ln Z_0 + \ln Z_1 + \ln Z_2 \ldots$ is an infinite series with increasing order in λ, or equivalently with diagrams increasing in number of loops. Now, we saw above that

$$\ln Z_2 \sim \Pi_1^2 \, \frac{d^3 p}{(p_0^2 + \vec{p}^{\,2})} \simeq (-\Pi_1 \, \mathcal{D}_0)^2 d^3 p.$$

At the next order, one has to add one more bubble on the big loop at the center (one more *petal* on the flower). Mathematically, this means one multiplies the above expression by an extra $(-\Pi_1 \, \mathcal{D}_0)$. Taking all such diagrams into account, leads to a sum: $T \sum_{n=2}^{\infty} \int \frac{d^3 p}{(2\pi)^3} \sum_N \frac{[-\mathcal{D}_0 \pi_1]^N}{N}$ where the combinational factor N actually arises as a product of $(N-1)!/2$ (number of ways of arranging the 'petals') and $1/N!$ from the expansion. This series, which is diagarammatically expressed as shown in Fig. 5.7, can be summed to

$$\sum_n \int \frac{d^3p}{(2\pi)^3} \ln[1 + \Pi_1 \mathcal{D}_0] - \Pi_1 \mathcal{D}_0.$$

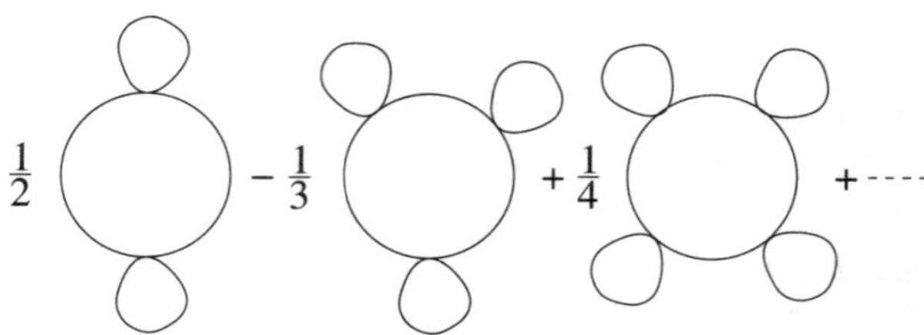

Figure 5.7: Diagrammatic representation of diagrams from $\ln Z_N$.

Note that $\Pi_1 = \Pi_1^{vac} + \Pi_1^{T \neq 0}$, with a quadratically $(c\Lambda^2)$ divergent contribution in the first term which has to be cancelled by a counter term δm^2. A careful evaluation including the counter terms modifies the summed result for the series as below.

$$\text{Series} \;\rightarrow\; \sum_n \int \frac{d^3p}{(2\pi)^3} \left\{ \ln[1 + \Pi_1^{T \neq 0} \mathcal{D}_0] - \Pi_1^{T \neq 0} \mathcal{D}_0 \right\}$$

$$= \sum_n \int \frac{p^2 dp}{2\pi^2} \left\{ \ln\left[1 + \frac{\lambda T^2}{\omega_n^2 + p^2}\right] - \frac{\lambda T^2}{\omega_n^2 + p^2} \right\}$$

$$= \lambda^{\frac{3}{2}} T^3 \sum_n \int x^2 dx \left[\ln\left[1 + \frac{1}{x^2 + y_n^2}\right] - \frac{1}{x^2 + y_n^2} \right].$$

In the above, we have changed variables: $x = p/\sqrt{\lambda}T$ and $y_n = \omega_n/\sqrt{\lambda}T$.

An alternative, perhaps better, way to see how the $\lambda^{3/2}$ comes about is from the Π of Fig.5.5. At the Nth order, $N - 1$ loops will contribute with the same number of external legs. This can be achieved by adding the diagram successively on the loop to decorate its loop by 'petals'.

Since in the massless ($m = 0$) limit,

$$
\begin{aligned}
\Pi_1 &= 12\lambda T \sum_n \int \frac{d^3p}{(2\pi)^3} \frac{1}{\omega_n^2 + p^2} \\
&\equiv 12\lambda T \sum_n \int \frac{d^3p}{(2\pi)^3} \mathcal{D}_0(\omega, \vec{p}),
\end{aligned}
$$

the graph at the N th order is

$$
12\lambda T \sum_n \int \frac{d^3p}{(2\pi)^3} \left[\frac{-\Pi_1(\omega_1 p)}{\omega_n^2 + p^2} \right]^N \frac{1}{\omega_n^2 + p^2}.
$$

Sum over all N can be done by recognising that this is just an expansion like $(k^2 + k_0^2)^{-1} = k^{-2} \left[-k_0^2/k^2\right]^N$. Hence the expression for the full resummed propagator becomes $\Pi = 12\lambda T \sum_n \int d^3p \frac{1}{\omega_n^2 + p^2 + \Pi_1}$. Using $\Pi_1 = \lambda T^2$ and taking $n = 0$ with $p \to 0$, one has

$$
\Pi \to \lambda T \int \frac{p^2 dp}{p^2 + \lambda T^2} \simeq \sqrt{\lambda} T \left[\tan^{-1} \frac{p}{\sqrt{\lambda} T} \right]_0^\infty,
$$

giving back the non-analytic behaviour in coupling.

Similar infra-red problems also occur in QED and QCD, with in fact a similar solution in form of a resummation of a class of diagrams to all orders. Note the alternating sign in the expression above. Even this is repeated in gauge theories, leading to poorer convergence of the series for P, ϵ etc. in QCD.

Finally, renormalization group equations lead us in the scalar field theory to $\bar{\lambda}_R = 2\pi^2 \left[9 \ln(\Lambda/M)\right]^{-1}$ at leading order. The renormalized coupling $\bar{\lambda}_R \to 0$ as $M/\Lambda \to 0$. M can be taken as a constant c times the temperature T and the corresponding $\bar{\lambda}$ can be used in P above. As $(T/\Lambda) \to 0$,

$$
P = P_{\text{ideal}} \left[1 - \frac{15}{8} \left(\frac{2}{9 \ln(\Lambda/cT)} \right) + \frac{15}{2} \left(\frac{2}{9 \ln(\Lambda/cT)} \right)^{\frac{3}{2}} + \cdots \right],
$$

where the ideal gas pressure P_{ideal} for this case of scalar field theory is $\pi^2 T^4/90$.

5.3   QGP from QCD II

The treatment of the previous lecture can be generalized to QCD to obtain Z_{QCD} in terms of its basic fields, namely, quark and gluon fields. The generalization, however, has to face many key technical problems. These are related to the gauge invariance the theory must have, the fermionic nature of the quarks and the non-abelian nature of the gluons. We will begin by considering the simpler example of the $U(1)$ gauge theory first to address the problems related to the first issue and then gradually move on to the quark and gluon fields.

5.3.1   Quark and gluon fields

The $U(1)$ gauge theory of the photon field $A_\mu(x)$ is defined by the Lagrangian

$$L = -\frac{1}{4} \int d^4x \, F_{\mu\nu} F^{\mu\nu}, \tag{5.40}$$

where the field tensor $F_{\mu\nu}(x) = \partial_\mu A_\nu(x) - \partial_\nu A_\mu(x)$. Let us note that an $A \to eA$ scaling changes $\mathcal{L} \to \mathcal{L}/e^2$. We will prefer the latter form for the non-abelian gauge theories later. As is well known, the theory has a gauge invariance:

$$A'_\mu(x) = A_\mu(x) + \partial_\mu \Lambda(x), \tag{5.41}$$

where $\Lambda(x)$ is an arbitrary scalar function. Thus all four $A_\mu(x)$ are not independent of degrees of freedom. In fact, there are only two, as expected of a massless vector field. Choosing the $A_0(x) = 0$ gauge, the lagrangian above can be written as

$$\begin{aligned} L &= \frac{1}{2} \int d^4x \, ((\partial_0 \vec{A})^2 - \vec{B}^2) \\ &= \frac{1}{2} \int d^4x \, (E^2 - B^2) \end{aligned}$$

L now has a residual gauge invariance,

$$A'_i(\vec{x}, t) = A_i(\vec{x}, t) + \partial_i \Lambda(\vec{x}), \tag{5.42}$$

where the permissible gauge functions are time-independent. Using the standard method, one finds the canonical momenta from the L above.

$$\Pi_i = \frac{\partial \mathcal{L}}{\partial \dot{A}_i} = \dot{A}_i = E_i. \tag{5.43}$$

Using these the hamiltonian density can be obtained as

$$\mathcal{H} = \Pi_i E_i - \mathcal{L}$$
$$\text{and} \quad H = \int d^3x\, \mathcal{H} = \frac{1}{2} \int d^3x\, (E^2 + B^2).$$

It may be commented that one need not have fixed the gauge $A_0(x) = 0$. This would have lead to a gauge field dependent hamiltonian density

$$\mathcal{H} = \frac{1}{2}(E^2 + B^2) + E \cdot \nabla A_0. \tag{5.44}$$

Integrating by parts and using the Maxwells equation, one could still obtain the same hamiltonian in the pure photonic theory (without sources).

Having identified the fields A_i, $i = 1$, 2 and 3, and the corresponding canonical momenta E_i, one can impose the canonical commutation

$$[\, E_j(\vec{x}, t),\, A_j(\vec{x}, t)\,] = i\,\delta_{ij}\,\delta(\vec{x} - \vec{y}),$$

in order to define the quantum theory and then $Z = \mathrm{Tr}\,\exp(-\beta H)$ defines its partition function. Of course, the trace must be taken over only physical states (or equivalently gauge invariant states)

Problem 7: Show that $[H, \nabla \cdot E] = 0$, *i.e.* H and $\nabla \cdot E$ can be simultaneously diagonalized.

Exploiting the property above, the trace over physical states can be defined as follows. Gauss' law states that the divergence of the electric field is given by the charge density. Thus any eigenstate $|\psi\rangle$ of the hamiltonian satisfies

$$\nabla \cdot E |\psi\rangle = \rho(\vec{x})|\psi\rangle, \tag{5.45}$$

which can be used to select the gauge invariant states. So a projection operator P_ρ into gauge invariant states is given by a δ-functional for the Gauss' law :

$$P_\rho = \int \mathcal{D}\theta(\vec{x}) \, \exp\left[-i \int d^3x (\nabla.E - \rho(\vec{x}))\theta(x)\right]. \qquad (5.46)$$

Since $P_\rho|\psi\rangle = |\psi^{\mathrm{phys}}\rangle$, the constraint of summing over only physical states in Z can now be easily handled: $Z = \mathrm{Tr}\, P_\rho \exp(-\beta H)$, with trace now on *all* possible states. Note we have used the commutation relation above as well as the relation $P_\rho^2 = P_\rho$ to write Z. As in the harmonic oscillator example above, the partition function can now be seen to be

$$Z = \int \mathcal{D}\vec{A}(\vec{x}) \, \langle \vec{A}|e^{-\beta H} P_\rho|\vec{A}\rangle. \qquad (5.47)$$

Proceeding as in the simple quantum mechanical example, β can be subdivided to obtain a product of factors $\exp(-\epsilon H)$. Since the projection operator can be raised to Nth power trivially, it can be associated with each such factor. The $|E\rangle\langle E|$ and $|A\rangle\langle A|$ completeness relations can be inserted each time step and the canonical momenta E can be eliminated in favour of $\dot{A}$. The final result is

$$Z = \int \mathcal{D}\theta\, \mathcal{D}A_i \, \exp\left[\int_0^\beta d^3x\, d\tau \{\frac{1}{2}(\vec{B}^2 + (\dot{A} - \nabla\theta)^2) + i\theta(\vec{x},t)\rho(\vec{x})\}\right]. $$
$$(5.48)$$

We have used the relation

$$\exp\left(i \int \theta \, \nabla \cdot E d^3x\right) = \exp\left(-i \int E \cdot \nabla\theta d^3x\right),$$

provided $\theta(x) = 0$ as $r \to \infty$. Renaming $\theta(\vec{x},t)$ as $A_0(\vec{x},t)$, the partition function can be written in an explicitly gauge invariant fashion:

$$Z = \int_{pbc} \mathcal{D}A_\mu(x) \, \exp\left(-\int_0^\beta d\tau \int d^3x \, [\mathcal{L} + i\rho(x)A_0(x)]\right). \qquad (5.49)$$

Again it is the trace in the definition of Z which results in the periodic boundary conditions (*pbc*) on the gauge fields:

$$A_\mu(\vec{x}, 0) = A_\mu(\vec{x}, \beta) \qquad \forall \quad \mu = 1, 2, 3, 4. \tag{5.50}$$

For $\rho = 0$, *i.e.*, the sourceless theory, $Z_{U(1)}$ is, as may have been anticipated, the functional integral of $\exp(-\int_0^\beta d\tau \int d^3 x \mathcal{L})$ over the gauge fields $A_\mu(x)$. Note that all four components of the vector potential appear above and Z is explicitly gauge invariant. The full set of allowed gauge trasformation has to maintain the *pbc*, $A_\mu(\vec{x}, 0) = A_\mu(\vec{x}, \beta)$.

> **Problem 8:** Find out all independent gauge transformation subject to the boundary condition *pbc* above.

It is straightforward to see that in the limit $T \to 0$, $\beta \to \infty$ the partition function reduces to the usual generating functional of the $U(1)$ gauge theory. We will employ this simple example to illustrate one important concept below.

Let us now consider sources in the $U(1)$ theory and choose $\rho = \delta(\vec{x} - \vec{x}_0)$, which corresponds to a point source. Substituting in (5.48), one has

$$Z(\rho) = \int \mathcal{D} A_\mu \, e^{-S_E} \, L(\vec{x}_0), \tag{5.51}$$

where we have defined

$$L(\vec{x}_0) = \exp\left[i \int_0^\beta d\tau A_0(\vec{x}_0, \tau) \right]. \tag{5.52}$$

Using (5.51) and (5.48) and the definition of thermal expectation value, one obtains,

$$\langle L(\vec{x}_0) \rangle_\beta = \frac{Z(\rho)}{Z(\rho = 0)}. \tag{5.53}$$

Problem 9: Show that L above is gauge invariant.

One can further show that $-T \ln\langle L(\vec{x})\rangle = (F_q(\vec{x}_0) - F_0)$, where $F_q(\vec{x}_0)$ is the free energy of a point charge (or an abelian quark) and F_0 is the free energy of vacuum. Taking now a pair of opposite charges, *i.e.*,

$$\rho(x) = \delta(\vec{x} - \vec{x}_0) - \delta(\vec{x} - \vec{y}_0) \ ,$$

one can similarly define $\langle L(\vec{x}_0) L^\dagger(\vec{y}_0)\rangle$ which is related to the free energy of the pair of charges, $F_{q\bar{q}}(\vec{x}_0 - \vec{y}_0)$, in the same manner.

We will now generalize the discussion above to a non-abelian gauge theory as a first step of going towards QCD. Recall that an $SU(N)$ gauge theory describes a theory of $N^2 - 1$ gluons. An important distinction the $SU(N)$ theory has over the $U(1)$ example is that it is a fully interacting theory since the gluons have self-coupling. The case of $N = 3$ colours corresponds to QCD (without quarks). In general, the same considerations as $U(1)$ go through except that the electric field $E \to E^a$ with colour index a which runs from 1 to $(N^2 - 1)$ for an $SU(N)$ gauge theory. Similarly, for the magnetic field $B \to B^a$ and the charge density $\rho \to \rho^a$ etc. The non-abelian version of the Gauss' law is $D \cdot E = \rho$, where we have used the generators of the $SU(N)$ group in the fundamental representation, T^a, to define matrices $E = E^a T^a$, $\rho = \rho^a T^a$ etc. and $D_{ij} = \delta_{ij}\nabla + (T^b)_{ij} A^b$ with i, j running from 1 to N. It is easy to check that the Gauss's law does not commute with H in this case. So trace in Z has to average over all a for ρ^a. In fact, it would otherwise be itself gauge dependent. Except for this complication, all the above considerations go through to yield the partition function of the Yang-Mills theory,

$$Z_{YM} = \int\limits_{A_\mu(\vec{x},0) = A_\mu(\vec{x},\beta)} \mathcal{D}A_\mu(\vec{x},\tau) \exp\left[-\int\limits_0^\beta d\tau \int d^3x\, \mathcal{L}_{YM}\right] . \quad (5.54)$$

In view of the deceptively simple looking equation above, let us remind ourselves that the fields A_μ are actually $N \times N$ matrices and many complicated interaction terms are hidden in (5.54). The free energy of a static quark is given again by $-T \ln Z = (F_q(\vec{x}_0) -$

F_0) but with L given by

$$L(\vec{x}_0) = \frac{1}{N} \operatorname{Tr} T \exp\left[i \int_0^\beta d\tau\, A_0^a\, (\vec{x}_0, \tau)\, T^a \right]. \tag{5.55}$$

L is called the thermal Wilson loop or the Polyakov loop. If all quarks are confined, one expects the free energy of a single quark to be infinite, $F_q(\vec{x}_0) = \infty$, leading to $\langle L(\vec{x}) \rangle = 0$, whereas the existence of a free quark with finite free energy (amounting to its deconfinement) implies $\langle L \rangle \neq 0$. The free energy of a pair of quark antiquark pair, located respectively at $\vec{x}_0$ and $\vec{y}_0$, is given by the correlation function $\langle L(\vec{x}_0)L^\dagger(\vec{y}_0)\rangle$, as before.

In order to reach the final goal of writing down the partition function of full QCD, one needs to introduce quarks in (5.54). Bringing in the quarks necessitates handling of another technical problem. Quarks obey Fermi-Dirac statistics, and the corresponding operators obey anticommutation relations. Introduction of the sum over all possible paths for them has to take these properties into account. I will not deal with the details of this here but simply state that one uses anticommuting Grassmann variables to define the fermionic path integrals. Details about the Grassmann variables, and their calculus can be found in standard references.

The Lagrangian density for the free quarks (or fermions) is given by

$$\mathcal{L} = \bar{\psi}(i\gamma^0 \frac{\partial}{\partial t} + i\gamma.\nabla - m)\psi. \tag{5.56}$$

One can easily work out the canonical momentum to be $\pi_\psi = \partial\mathcal{L}/\partial\dot{\psi} = i\psi^\dagger$, thus suggesting to use $\psi, \bar{\psi} = \psi^\dagger\gamma_0$ as independent variables in the path integrals. The hamiltonian density is given by

$$H = \Pi_\psi \dot{\psi} - \mathcal{L} = \bar{\psi}(-i\gamma \cdot \nabla + m)\psi. \tag{5.57}$$

Using it, and the expression for the number density, the partition function $Z = \text{Tr} \exp\left(-(\hat{H} - \mu\hat{N})/T\right)$ can be written as

$$Z = \int_{apbc} \mathcal{D}\bar{\psi}\,\mathcal{D}\psi \, \exp\left(\int_0^\beta d\tau \int d^3x\, \bar{\psi}(\gamma^0 \frac{\partial}{\partial\tau} + i\gamma \cdot \nabla - m + \mu\gamma^0)\psi\right),$$

(5.58)

where *apbc* denotes antiperiodic boundary conditions on the fields:

$$\bar{\psi}(\vec{x},0) = -\bar{\psi}(\vec{x},\beta) \qquad \text{and} \qquad \psi(\vec{x},0) = -\psi(\vec{x},\beta).$$

These arise due to the trace again but are antiperiodic for the fermionic fields. This leads to the quantized energy sums in this case to run over

$$\omega_n = (2n+1)\pi T, \qquad -\infty < n < \infty.$$

Note that $\omega_n \neq 0$ for any n and the lowest value is $\omega_0 = \pi T$. This should be contrasted with the bosonic case discussed earlier, where $\omega_0 = 0$. Defining $i\gamma^\mu \to \gamma^\mu$ for $\mu = 1, 2, 3$, one has $\{\gamma^\mu, \gamma^\nu\} = 2\delta^{\mu\nu}$ as may be expected in a Euclidean theory. The resultant quark partition function is

$$Z = \int \mathcal{D}\psi \, \mathcal{D}\bar{\psi} \, \exp\left[-\int_0^\beta d\tau \int d^3x\, \bar{\psi}(\gamma \cdot \partial + m + \mu\gamma^0)\psi\right].$$

(5.59)

The integration over Grassmann variables can be done to obtain $Z = \det D$ or alternatively, $\ln Z = \text{Tr} \ln D$, where D stands for the Dirac matrix sandwitched in $\bar{\psi}$ and ψ in (5.59). For the free fermion case, the trace can be evaluated by going to the Fourier space and one obtains

$$\ln Z = 2V \int \frac{d^3p}{(2\pi)^3} \left[\beta\omega + \ln(1 + e^{-\beta(\omega-\mu)}) + \ln(1 + e^{-\beta(\omega+\mu)})\right],$$

(5.60)

where $\omega^2 = p^2 + m^2$.

Problem 10: Obtain Eq.(5.60) from Eq.(5.59).

Remarks:

- The energy density obtained from (5.60) diverges quartically. As for bosons, this zero point energy needs to be subtracted. This can be easily done by using the freedom to shift the origin of the energy density scale and thus by defining it to be identically zero at $T = 0$.

- Actually there also exists a quadratic μ-dependent divergence in the energy density as well as the number density in the free fermion theory. This can be attributed to the zero point number subtraction. However, unlike the case above, the prescription to eliminate it cannot be related to any physical principle but appears to be arbitrary.

Putting all the elements above together, and recognising that gauge interaction can be introduced in the fermionic action in (5.59) by substituting the covariant derivative D^a defined earlier in place of the ordinary derivative, the partition function for QCD is

$$Z_{QCD} = \int_{bc} \mathcal{D}A_\nu \, \mathcal{D}\psi \, \mathcal{D}\bar{\psi} \exp\left[-S(A_\nu, \, \psi, \, \bar{\psi}; \, g, \, m_f, \, \mu_f, \, T)\right],$$

$$(5.61)$$

where bc denotes collectively the boundary conditions on the quark (antiperiodic) and the gluon fields (periodic) :

$$
\begin{aligned}
A_\nu(x, 0) &= A_\nu(x, 1/T) \qquad \text{(pbc)}, \\
\psi(x, 0) &= -\psi(x, 1/T) \qquad \text{(apbc)}, \\
\bar{\psi}(x, 0) &= -\bar{\psi}(x, 1/T) \qquad \text{(apbc)}.
\end{aligned}
$$

The QCD action S is given by

$$S = \int_0^{1/T} dt \int_{-\infty}^{\infty} d^3x \left(\frac{1}{4} F^a_{\mu\nu} F^a_{\mu\nu} + \sum_f [\bar{\psi}(\slashed{D} + m_f)\psi - \mu_f \, \bar{\psi}_f \, \gamma_0 \, \psi_f] \right),$$

$$(5.62)$$

and thermal expectation value of a physical observable θ is given by

$$\langle \theta \rangle = \frac{1}{Z} \int \mathcal{D}A_\nu \, \mathcal{D}\psi \, \mathcal{D}\bar{\psi} \, \theta \, e^{-S}.$$

$$(5.63)$$

The index f runs over flavours: up, down, strange, charm etc. and the corresponding mass and chemical potentail are denoted by m_f and μ_f respectively.

Comments :

1. The (baryonic) chemical potential μ appears to couple like a constant electromagnetic potential, $A_\nu = (1,0,0,0)$. This is a helpful mnemonic in diagramatics but one should be weary of attaching a physical significance to it since the action does not have any corresponding gauge invariance.

2. Integrating the fermions, the flavour contribution is seen to be $\ln Z \propto \sum_{f=1}^{6} \det \mathcal{D}_f$. Ignoring heavy quarks in Z, since the masses of charm, bottom and top quarks are very large compared to the temprature scales we will need (*a posteriori*), Z is determined by u, d and s quarks. If $m = m_u = m_d = m_s$, then the QCD partition function has $SU(3)$ flavour symmetry.

5.3.2 Symmetries and order parameters

It is useful to study the symmetries of the QCD partition function (5.61) in various limits and define corresponding order parameters to check whether i) the vaccum respects a given symmetry or breaks it spontaneously and ii) increasing the temperature/density has any effects on the breaking.

In the limit of $m_q \to \infty$ for all the flavours, the quarks are effectively frozen. In this quenched approximation to QCD, the partition function reduces to that of an $SU(3)$ gauge theory. The fermionic boundary conditions then become irrelevant, resulting in an extra global $Z(3)$ summetry in that case. For a general $SU(N)$ gauge theory, the origin of the extra $Z(N)$ symmetry can be understood as below. Under $V \in SU(N)$, the gauge fields transform as

$$A'_\mu = V A_\mu V^{-1} + iV \partial_\mu V^{-1},$$

$$\text{along with} \quad A'_\mu(\vec{x},0) = A'_\mu(\vec{x},\beta),$$

where $A_\mu = A_\mu^a t^a$ are $N \times N$ matrices. The periodic boundary condition restricts the set of allowed gauge transformations to

$$V_\mu(\vec{x}, 0) = z V_\mu(\vec{x}, \beta), \qquad z \in Z(N). \tag{5.64}$$

Problem 11: Derive the above condition.

One would have, of course, expected the set of allowed gauge transformations to be periodic at $T \neq 0$ but it turns out to have a further global $Z(N)$ symmetry. It is straightforward to see that under the $Z(N)$, $L(\vec{x}_0) \to L'(\vec{x}_0)$, where

$$
\begin{aligned}
L'(\vec{x}_0) &= \frac{1}{N} \operatorname{Tr} V(\vec{x}_0, 0) \, T \exp \left(\int_0^\beta A_0^a t^a d\tau \right) V^\dagger(\vec{x}_0, \beta) \\
&= z L(\vec{x}_0).
\end{aligned}
$$

If this $Z(N)$ global symmetry is exact, then the above transformation property implies the thermal average of the Polyakov loop or thermal Wilson loop, $\langle L(\vec{x}_0) \rangle = 0$. On the other hand, if it is broken spontaneously by the vaccum, $\langle L(\vec{x}_0) \rangle \neq 0$, which is a signal of deconfinement as we saw earlier, since

$$
\begin{aligned}
\langle L \rangle &= \exp\left[-(F_q(x)/T)\right], \\
F_q(x) = \infty \quad &\Longleftrightarrow \quad \langle L \rangle = 0 \quad \text{confinement}, \\
F_q(x) < \infty \quad &\Longleftrightarrow \quad \langle L \rangle \neq 0 \quad \text{deconfinement}.
\end{aligned}
$$

Thus a deconfinement phase transition with $\langle L \rangle$ as its order parameter is synonymous to the spontaneous breaking of the global $Z(N)$ symmetry the theory has in the quenched approximation. Since $\langle L(x) L^\dagger(y) \rangle$ is the correlation function of static quarks, its behaviour can tell us about the nature of the phases, it being a measure of the free energy of the quark and antiquark pair. In a confining phase, we expect $V_{Q\bar{Q}} = \sigma x$, and hence, $\langle LL^\dagger \rangle \sim \exp(-\sigma|r|/T)$, where r is the distance between the $Q\bar{Q}$ pair. With increasing temperature T, the tension $\sigma(T)$ of the string connecting them decreases, leading to a a decrease in the correlation function as well.

The deconfinement phase transition between the confining phase, where we exist and no free quarks are observed and the phase at high temperatures, where free quarks can exist, appears to be similar to that in spin models with $Z(N)$ symmetry. The order parameter for the former, $\langle L \rangle$, plays the role of magnetization M which is the order parameter for the latter. If for a given number of colours, N, the deconfinement phase transition is second order one can appeal to universality and expect to obtain predictions for various *critical exponents*:

$$
\begin{aligned}
\langle L \rangle &\sim |T - T_c|^\beta &&\text{as } T \to T_c^\dagger, \\
\chi &\sim \sum \langle LL^\dagger \rangle_c \sim |T - T_c|^\gamma &&\text{as } T \to T_c^-, \\
C_v &\sim |T - T_c|^{-\alpha} &&\text{as } T \to T_c, \\
\xi &\sim T/\sigma \sim |T - T_c|^{-\nu} &&\text{as } T \to T_c^-.
\end{aligned}
$$

In addition to the order parameter which vanishes in the manner prescribed above, its susceptibilty χ, the specific heat C_v and the correlation length ξ diverge at T_c as per the indices of the corresponding observables of the $Z(N)$ spin/Potts models. In case of a first order phase transition, one observes a discontinuous turn-on for the order parameter, and the dimensionality of space (three in our case) governs the divergences above.

If the fermion masses are finite, then the $Z(N)$ global symmetry no longer exists: with $\psi(\vec{x}, 0) = -\psi(\vec{x}, \beta)$, one obviously has $V(\vec{x}, 0) = V(\vec{x}, \beta)$ only and $\langle L \rangle$ is no longer required to be zero at any T. It is thus strictly not an order parameter for confinement/deconfinement in full QCD. One may, however, continue to use the limiting case as a guide inside the $m - T$ phase diagram. For large enough quark masses, one can imagine them as having been tuned down from infinity. If the deconfinement phase transition is first order at $m = \infty$, such a tuning down will result in a line of first order phase transition in the $m - T$ phase diagram. The strngth of the transition, the latent heat, will decrease as the quark mass decreases. For the second order case, however, the transition ceases to exist inside the $m - T$ diagram as a result of such a tuning.

In the opposite limit of vanishing quark masses, the QCD lagrangian has another extra symmetry. Its expected $SU(3)$ flavour symmetry of for equal u, d and s quark masses is enhanced to $SU(3)_L \otimes SU(3)_R \otimes U(1)_A \otimes U(1)_B$ symmetry when all these quarks are massless. Here the subscripts L and R denote left and right moving quarks obtained by the projection operators $P = (1 \pm \gamma_5)/2$. Physically, this is the statement that for free massless spin half particles, spin projections are intrinsic properties of the particles which cannot be changed by any Lorentz transformations.

Problem 12: Show this.

QCD thus has for equal quark masses: $m_f = m$,

1. a $U(1)_B$ symmetry corresponding to the conservation of baryon number,

2. an $SU(N_f)$ flavour symmetry leading to the conserved quantum numbers (by strong interaction) such as isopsin and strangeness.

As $m \to 0$, these are enhanced to $SU(N_f) \otimes SU(N_f)$ chiral symmetry and $U(1) \otimes U(1)$ symmetry. If the interactions respect the symmetry, *i.e.*, the vaccum is chirally symmetric, then the order parameter for this symmetry, the *chiral condensate*, $\langle \bar{\psi}\psi \rangle$, must vanish. In addition, one expects the hadrons to be parity degenerate since the L- and R-quarks do not transform into each other. Since this is not seen experimentally, one is lead to postulate that the chiral symmetry is broken dynamically by the QCD vacuum and $\langle \bar{\psi}\psi \rangle \neq 0$. A consequence of such a breaking of a continuous symmetry is the prediction of massless Goldstone bosons. Since m_π^2, $m_K^2 \ll m_N^2$, these can be identifed as the Goldstone bosons. These do have masses though : $m_\pi^2 \neq 0$ and $m_K^2 > m_\pi^2$. Even this can be understood as due to the small explicit breaking of the chiral symmetry by nonzero up, down and strange quark masses. It is natural to expect that the magnitude of the quark masses will govern the Goldstone masses. One can show that $m_\pi^2 \propto m_u + m_d$ and $m_K^2 \propto m_s$, assuming $m_u, m_d < m_s$. The quark masses are, in fact, obtained using these relations.

At finite temperature, one naturally expects $\langle \bar\psi\psi \rangle$ to be a function of T. If at $T \geq T_{CH}$, $\langle \bar\psi\psi \rangle = 0$, one may have a chiral symmetry restoring phase transition (χPT). As for the deconfinement phase transition above, the chiral condensate is a strict order parameter for the χPT only in the massless limit for the quarks. Depending the number of flavours, *i.e.*, the chiral symmetry, this transition may be a first order one. It is expected to be so for $N_f \geq 4$. Again as one tunes up the masses from zero, one ought to expect a line of first order chiral phase transitions in that case and no chiral transition for nonzero quark masses in the second order chiral transition for massless quarks. The m-T diagram thus generically has lines of first order deconfinement and chiral transitions coming into from $m = \infty$ and zero respectively. The real world has small unequal up and down quark masses in the range of a few MeV and a moderately heavy strange quark of about hundred MeV. For both analytical as well as numerical methods, it turns out easier to obtain information for the real world from either limits discussed above. The questions one therefore has to clarify are the nature of the transitions in the limiting cases, the flow of the lines inside the diagram, whether they merge or cross, and finally which line is close to the real world, and thus whether one has chiral, deconfining or both phase transitions at high temperatures/densities. In addition to the order parameters above, one can also investigate various other physical quantites, such as the energy density ϵ, the pressure P, the number density n, the quark number susceptibilities χ, etc. These can be obtained from the partition function (5.61) by employing the canonical formulæ, *e.g.*, $\epsilon = \frac{T^2}{V} \frac{\partial \ln Z}{\partial T}$.

5.3.3   Free energy in pertubation theory

As must be clear from the discussion of the previous section, one needs to evaluate $\ln Z$ and/or the thermal expectation values of the physical observables and the order parameters to extract the

information on QCD phase diagram. One very straightforward
and familiar method is to use the small coupling, g, expansion.
This perturbative approach works along the lines we discussed
earlier, provided one takes adequate care of the additional technical problems of gauge invariance and fermions. The pressure in
this approximation can be written down order by order as

$$
\begin{aligned}
P/T^4 &= P_0(1 + P_1 + P_2 + P_3 + P_4 + \cdots) \\
\text{with } P_0 &= \text{ideal gas terms} = p_{q+g}, \\
P_1 &= -g^2 \frac{5N_f + 12}{72}, \\
P_2 &= +\frac{g^3}{12\pi} c_2, \\
P_3 &= g^4 \ln g^2 \, c_3, \\
P_4 &= -g^4 \, c_4 \\
P_5 &= +g^5 \, c_5.
\end{aligned}
$$

Here the c_i's are numerical constants which have been evaluated
and their exact values can be found in the literature cited at the
end. The leading term is that of an ideal gas of appropriate number of quarks and gluons, as expected. The series is not a simple
g^2 expansion like at zero temperature but has *all* powers of g and
even log g. The g^3 and g^5 terms arise due to the nontrivial mass
the electric gluons have : $m_{el} \propto gT$, with a proportionality constant decided by the number of colours and flavours. These terms
also come with positive sign, making the series alternating in sign
between successive terms. A natural consequence would be its expected convergence for very small values of the coupling g. Since
QCD is an asymptotically free theory, its coupling can be expected
to be small for asymptotically large temperatures/densities:

$$
g^2(T) = \frac{1}{2b_0 \, \ln(T/\Lambda_T)} - \left(\frac{2b_1}{2b_0^3}\right) \frac{\ln\ln(T/\Lambda_T)}{\ln(T/\Lambda_T)^2},
$$

where b_0 and b_1 are the first two coefficients of the QCD β-
function. For $T >> \Lambda_T$, $g^2(T)$ small and use of perturbation
theory is justified. This yields P_{ideal} as the pressure for $T \to \infty$,

which was the *ansatz* we used in the Bag model case. The P_i's
provide corrections to it, and the region of validity of the weak
coupling expansion is also where the *ansatz* can be used safely.
Various criteria of stability of a series can be used to figure it
out. Naively, one expects the difference in the absolute values of
$p(g^{n+1})$ and $p(g^n)$ should be small. When they become comparable, the alternating sign of these terms would induce instabilites.
Putting in the known values of the constants c_i's, it turns out that
typically $(P - P_{ideal})/P_{ideal} \sim 1\text{--}2\%$ for a stable series, leading to
an estimate of $T \sim 10^8$ MeV or so up to which the series is valid.
This is, of course, way above any temperature of interest for any
applications in the heavy ion physics or even the early Universe
(due to the dominance of other physics at such high temperatures,
such as electroweak symmetry breaking etc.).

5.3.4   Hierarchy of scales in QCD at finite temperature

In principle, the pressure, P, and any other thermodynamic quantity could be computed to any order in g using the diagrammatic
approach discussed above. One simply has to draw all the allowed
diagrams, obtain the expressions for them using the Feynman rules
and evaluate them. However, as was first shown by Linde, an in-
surmountable barrier occurs at $O(g^6)$ due to the severe infra-red
problems of QCD, which are in fact general for any non-Abelian
gauge theory at nonzero temperature. Consider the $(l + 1)$ loop
diagram shown in Fig. 5.8. Note this is possible only due to the
self-couplings of gluons, and thus occurs exclusively in non-abelian
gauge theories. Since it has $2l$ three-gluon vertices, one gets a factor $g^{2l}p^{2l}$. Its $3l$ propagators yield $(p^2 + m^2)^{-3l}$, where we have
introduced an infra-red cut off m, anticipating problems for small
momenta $p \rightarrow 0$. The presence of $(l + 1)$ loops implies $(l + 1)$ integrals, and a factor, $[T \int d^3p]^{l+1}$. Putting all the factors together,
the diagram is proportional to $g^{2l}\, T^{l+1} \int d^{3l+3}p \cdot p^{2l}/(p^2 + m^2)^{3l}$.
Dimensionally, it can be worked out to be T^4. Since no extra ultra-
violet divergences arise due to nonzero temperature, one can take

$\sim T$ as the upper limit for the integrals. It can be shown that for $l \leq 2$, the contribution of the diagram is $g^{2l}T^4$. For $l = 3$, it is of $\int dp/p$ type and is $g^6\, T^4 \ln\left(\frac{T}{m}\right)$. For $l > 3$, it is $g^6\, T^4 (g^2 T/m)^{l-3}$, as can be seen already on dimensional grounds. We know that the electric gluons acquire a Debye mass gT but at that order the magnetic ones do not. If they do so at the next order, *i.e.*, if $m \sim O(g^2 T)$, then for $l \geq 3$ all diagrams contribute at the same order, $O(g^6)$. This is, of course, a breakdown of perturbative method where one usually expects the higher order diagrams to contribute only at higher order of g. Of course, this is not entirely a new phenomenon, since the perturbative expansion of the last section already had terms coming from an infinite set of diagrams which give rise to the terms with odd powers in g. The new feature, however, is that the all the diagrams in the infinite set contribute at the *same* order, making a perturbative computation unfeasible.

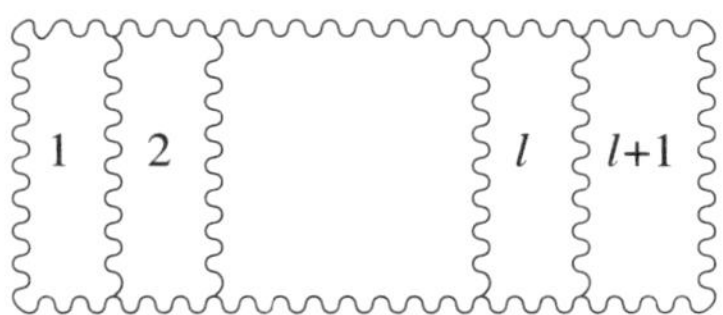

Figure 5.8: A $2l$-th order diagram.

Nevertheless, for sufficiently small coupling, $g \ll 1$, one has a hierarchy of scales at finite temperature: $T > gT$ (electric Debye mass) $> g^2 T$ (magnetic Debye mass). Note that the temperaures at which this is really valid are very large, $T \sim 10^8$ MeV, since the running coupling $\alpha_s(T) \ll 1/4\pi$. One can imagine integrating out the degrees of freedom at the successive scales and obtain a sequence of effective theories at high temperatures. Thus integrating out fields up to the scale T, the usual $\mathcal{L}_{QCD}(T \neq 0)$ goes over to a three (spatial) dimensional theory, $\mathcal{L}_{3dgauge+higgs}$, while integrating up to the scale gT, yields a three dimensional gauge theory, $\mathcal{L}_{3dgauge}$. This is the essence of the (perturbative) dimensional reduction program, which has also been extended to obtain the pressure for temperatures where naive perturbation theory fails.

5.4   QGP and lattice QCD

In our goal to obtain predictions for quark-gluon plasma from the basic underlying theory, QCD, using its partition function,

$$
\mathcal{Z}_{QCD} = \int_{bc} \mathcal{D}\psi \, \mathcal{D}\bar{\psi} \, \mathcal{D}A_\mu \, \exp\left(- \int\limits_0^\beta d\tau \int d^3x \, \mathcal{L}_{QCD} \right),
$$

the next step is to introduce a space-time lattice as a tool for computations. In the equation above, bc denote collectively the boundary conditions on the fields due to the trace in $\mathcal{Z}$ as before: antiperiodic for the quark fields $\bar{\psi}$, ψ and periodic for the gauge fields A_μ.

Just as ordinary integration can be done by taking a limit of sums on discrete points in the range of integration, lattice QCD could be thought of as an attempt to do the functional integrals in Z by discretizing the space-time over which the fields are defined. Indeed, the formally but imprecisely defined functional integrals in $\mathcal{Z}$ can actually be given a precise meaning using the lattice. The lattice need not be regular but is usually chosen to be so. Let N_σ and N_β denote the number of points in the spatial and temporal (equivalently inverse temperature) direction and let a_σ, a_β denote the corresponding lattice spacings. Then the volume V and temperature T are given by $V = N_\sigma^3 \, a_\sigma^3$ and $T^{-1} = \beta = N_\beta a_\beta$. One, of course, needs to take a continuum limit at the end of the calculation, somewhat like the ordinary integration where the interval between the points has to shrink to zero. The number of lattice points $N_\sigma, N_\beta \to \infty$ as $a_\sigma, a_\beta \to 0$ in this limit, keeping the volume and the temperature fixed. In addition, we need to take $V \to \infty$ for getting QCD thermodynamics.

Recall that we needed a regularization to compute momentum integrals in perturbation theory. A regulator is essential in quantum field theory. Space-time lattice provides one such regularization. It restricts the range of allowed momenta to $-\pi/a \leq p_i \leq \pi/a$. So lattice can be used for doing calculations in perturbation theory as well. But its great strength lies in being able to handle large couplings. In fact, large g^2 is in a sense trivial on lattice, with easy demonstration of, *e.g.*, confinement of quarks in this limit. The problem here is the $a \to 0$ limit which has to be taken since there is no experimental evidence for a discrete space-time.

A space-time lattice breaks Lorentz symmetry. Nevertheless, its great virtue is the ability to define a gauge invariant theory and perform all computations in a gauge invariant way, not always feasible with many other regularizations. One should try to define the quark and gluon fields and their action so as to preserve as many remaining continuum symmetries as possible. As we shall see shortly, one crucial symmetry is the chiral symmetry which turns out to be very nontrivial on the lattice.

5.4.1 Lattice fermions

Let us first consider free fremions, defined by the usual Lagrangian:

$$\mathcal{L}_{\text{Euclidian}} = \bar{\psi}(\slashed{\partial} + m)\psi., \tag{5.65}$$

Here we have already chosen the four-dimensional euclidean space and correspondingly gamma matrices satisfy $\{\gamma_\mu, \gamma_\nu\} = 2\delta_{\mu\nu}$ along with the relation $\gamma_\mu^\dagger = \gamma_\mu$. The factors of i which usually appear have therefore got absorbed in the spatial γ matrices and in the definition of imaginary time. The γ_5 matrix also is hermitean: $\gamma_5^\dagger = (\gamma_1\gamma_2\gamma_3\gamma_4)^\dagger = \gamma_5$.

Introduce now an L^4 lattice and place spinors $\psi(n)$ and $\bar{\psi}(n)$ on each of its site $n = (n_1, n_2, n_3, n_4)$. A derivative can then naturally be approximated by $\partial_\mu \psi = [\psi(n + a\hat{\mu}) - \psi(n - a\hat{\mu})]/2a$. This is not the only way to discretize it. Indeed, one could take asymmetric forward or backward derivatives, which are, however, not (anti)hermitean. One can add further terms to it which have higher powers of a. These are usually used to improve the behaviour of lattice fermions to make them mimic continuum fermions. As a consequence of this discretization the Dirac action becomes:

$$S_F = \sum_{n,l} \bar{\psi}(n) M_{n,l} \psi(l),$$ (5.66)

where

$$M_{n,l} = \frac{1}{2} a^3 \sum \gamma_\mu (\delta_{l,n+a\hat{\mu}} - \delta_{l,n-a\hat{\mu}}) + a^4 m \, \delta_{n,l}.$$ (5.67)

The Grassmann variables can be integrated out to yield the free fermion partition function,

$$\mathcal{Z} = \det M.$$ (5.68)

Absorbing the a^3 factor in the equation above by rescaling the fermion fields to have no dimensions, and defining a Fourier transform of ψ one can solve the free Dirac equation on the lattice: $\tilde{\psi}(k) = \sum_n \psi(n) \exp(2\pi i k \cdot n / L)$, with its inverse defined by $\psi(n) = \sum_k \tilde{\psi}(k) \exp(-2\pi i k.n/L)/L^4$. Then the lattice Dirac equation,

$$\sum_\mu \gamma_\mu \left(\bar{\psi}(n + a\hat{\mu}) - \psi(n - a\hat{\mu}) \right) + 2am \, \psi(n) = 0,$$ (5.69)

simplifies to

$$i \sum_\mu \gamma_\mu \sin k_\mu a + am = 0.$$ (5.70)

. This clearly has the correct continuum limit. As $a \to 0$, $\sin k_\mu a \simeq k_\mu a$, and one obtains $i \not{k} + m = 0$, as expected.

Note, however, that there are other allowed solutions as well. To see this, one rewrites the momentum integral (shown below for a specific dimension μ) as follows. Let $r_\mu = k_\mu - \pi/a$ for a given μ, giving $dr_\mu = dk_\mu$. Then

$$\int_{-\pi/a}^{\pi/a} dk_\mu = \int_{-\pi/a}^{-\pi/2a} dk_\mu + \int_{-\pi/2a}^{\pi/2a} dk_\mu + \int_{\pi/2a}^{\pi/a} dk_\mu = \int_{-\pi/2a}^{\pi/2a} (dk_\mu + dr_\mu),$$ (5.71)

where we have substituted r_μ in the first and third integrals and rearranged terms. Thus as $a \to 0$, k_μ and r_μ both will contribute. This is the famous *fermion doubling problem* on the lattice. In fact, due to the four dimensions, we have $2^4 = 16$ fermions. Under the $k_\mu \to r_\mu$ transformation, the sin term also changes sign, leading effectively to a $\gamma_\mu \to -\gamma_\mu$ transformation. Each of the doubled fermion has thus a different set of γ-matrices. (with different signs). This leads to alternating sign for their γ_5, and thus chiralities in pairs. The lattice keeps chirality exact at all stages of cut-off when the bare mass $m = 0$.

Many solutions have been proposed for the fermion doubling problem of lattice QCD. They all share the common feature of having some undesirable property. Indeed, thanks to a no-go theorem by Nielsen and Ninomiya, we know that the solution will involve either non-locality (discretization of the derivative with many, possibly infinite, terms) or loss of chirality or loss of reality (non-hermitean Hamiltonian). In spite of the strong enthusiasm for a solution, called the overlap fermions, we will discuss here the two most widely used options.

1. *Wilson Fermions:* Wilson in his seminal paper on lattice gauge theories proposed to modify the Dirac matrix by adding a second derivative term which acts like an extra mass for the undesired (15) doublers. His modified Dirac matrix (for the dimensionless fermion fields) is:

$$M_{pn} = \frac{1}{2} \sum_\mu [(r + \gamma_\mu)\delta_{p+\hat{\mu}a,n} + (r - \gamma_\mu)\delta_{p-\hat{\mu}a,n}]$$
$$+ (ma + 4r)\delta_{p,n}, \tag{5.72}$$

   where r is a parameter usually taken to be unity. In that case, $\frac{1}{2}(1 \pm \gamma_\mu)$ is a projection operator. Thus only part of the spinor propagates. Note also that even for $m = 0$ the action has no chiral symmetry for any nonzero r.

   The Wilson fermions break all chiral symmetries completely. They are recovered only in the continuum limit. This is specially bad for any finite temperature studies where chiral symmetry restoring transition may be crucial. There have been such investigations but the results may still not be reliable due to the smallness of temporal lattices or equivalently coarse lattice spacings.

2. *Staggered fermions:* Also referred to as Kogut-Susskind fermions, these are most often used in finite temperature studies as they do have some chiral symmetry for $m = 0$. However, for nonzero lattice spacing, they mix the flavour components with the spin and thus do not have a clear definition for quark flavour. One uses single component χ in place of the four component ψ to define their Dirac matrix,

$$M_{pn} = \frac{1}{2} \sum_\mu (-1)^{x_1+\ldots+x_{\mu-1}} \left(\delta_{p+\hat{\mu}a,n} - \delta_{p-\hat{\mu}a,n}\right)$$
$$+ ma\delta_{p,n}. \tag{5.73}$$

Taking the continuum limit for these fermions needs a little care in interpretation. One can show that 16χ on a 2^4 hypercube give rise to quarks of four flavours, each having the usual four component spinors. Thus in the continuum, these fermions describe $N_f = 4$ flavour always. How to get the $2 + 1$ flavours, observed in nature? One takes appropriate root of the determinant (*e.g.*, 1/4th root for one flavour), which is justified from a perturbation theory viewpoint.

Problem 13: Show that in momentum space,

$$M_k = ma + i\gamma_\mu \sum \sin k_\mu a + r \sum_\mu (1 - \cos k_\mu a),$$

and the r-term kills 15 zeroes of M_k but gives the usual relation in the continuum limit.

5.4.2   Gauge fields

Once quarks on the space-time lattice have been introduced, the gluon (gauge) fields can be introduced by simply demanding invariance under rotation by $SU(N)$ to be a local symmetry. Let

$\psi(x) \rightarrow \psi'(x) = V_x \psi(x)$ be such a local gauge rotation with $V_x \in SU(N)$ as usual. Then the quark action $\bar{\psi}'(m) M' \psi'(n)$ becomes $\bar{\psi} V_m^\dagger M' V_n \psi$ under this rotation. Introduce $U_x^\mu \in SU(N)$ as gauge field associated with a directed link from x to $x + \hat{\mu}$, as shown in Fig. 5.9. Let $U_{x-\mu}^{\mu\,\dagger}$ denotes the link in opposite direction, as shown in the same figure. Defining further gauge transformation of the link fields by $U_x^{\mu'} = V_x U_x^\mu V_{x+\mu}^\dagger$, one can ensure that the fermion action is invariant under it. One needs a gauge invariant gluon action for the fields to be dynamical. The above transformation of the link (gauge) variables implies that a gauge invariant gluon action will result only from closed loops of U, as shown in Fig. 5.9. Defining the smallest such square loop on a lattice to be a *plaquette*, one sees that the product along the directed path gives rise to a matrix

$$U_P = U_x^\mu \, U_{x+\mu}^\nu \, U_{x+\nu}^{\mu\,\dagger} U_{x+\mu}^{\nu\,\dagger}.$$

It is easy to check that $\operatorname{Tr} U_P$ is gauge invariant, leading to an action for the gluons,

$$S_G = \beta \sum_{\text{all plaquettes } P} \left(1 - \frac{1}{N} \operatorname{Re} \operatorname{Tr} U_p \right). \qquad (5.74)$$

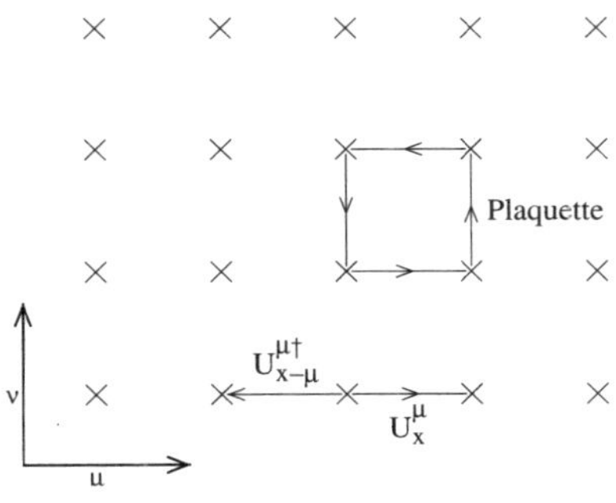

Figure 5.9: Gluon fields and action on a space-time lattice.

Problem 14: Does Eq. (5.74) reduce to the usual $F^{\mu\nu}F_{\mu\nu}$ action in the continuum limit?

(Hint: define a vector potential by $U_x^\mu = \exp\left(iga\, A_\mu^a(x)t^a\right)$ such that $x = (i+j)a/2$ and show that it does so, provided $\beta = 2N/g^2$.)

Combining all the ingradients so far from Eqs. (5.66) and (5.74) along with either Eq. (5.72) or (5.73), one obtains the full expression for the partition function of QCD at finite temperature on a space-time lattice :

$$\mathcal{Z} = \int_{bc} \prod dU_x^\mu \prod d\psi(x)\, d\bar\psi(x)\, e^{-S_F-S_G}, \qquad (5.75)$$

where bc denote the boundary conditions on gluon and quark fields,

$$U_{\vec{x},0}^\mu = U_{\vec{x},1/T}^\mu,$$
$$\psi(\vec{x},0) = -\psi\left(\vec{x},\frac{1}{T}\right), \qquad \bar\psi(\vec{x},0) = -\bar\psi\left(\vec{x},\frac{1}{T}\right).$$

Since the functional integral over fermions involve non-commuting Grassmann variables, one usually integrates the fermions out, using

$$\int d\psi\, d\bar\psi\, \exp\left(\bar\psi_m M_{mn}\psi_n\right) = \det M. \qquad (5.76)$$

Assuming the fermionic determinant to be positive, which it is for staggered fermions, and using the identity $\det M = \exp(\mathrm{Tr}\ln M)$, the QCD partition function (5.75) for staggered fermions can be re-written as

$$\mathcal{Z} = \int_{bc} \prod dU_x^\mu\, \exp\left(-S_G + \frac{1}{8}N_f \mathrm{Tr}\ln M^\dagger M\right). \qquad (5.77)$$

The factor of $N_f/8$ allows one to use the staggered fermions for an arbitrary number of flavours. As mentioned above, $N_f = 4$ in the continuum limit for these fermions for which the factor is half to account for the presence of both M and $M^\dagger$ in Eq. (5.77). The fermion determinant can be shown to be real but not positive for the Wilson fermions. Thus a factor of $N_f/2$ and M from Eq. (5.72) in Eq. (5.77) gives the QCD partition function for Wilson fermions.

5.4.3 Calculational techniques

From the partion function defined in the previous subsection, all physical quantities of interest can be obtained as thermal expectation values of appropriate operators. One needs to choose a suitable calculational technique to proceed further. There are various options available for the lattice QCD partion function (5.77) in view of its similarity with the conventional statistical mechanics systems such as the Ising model. We discuss some of these below, pointing out their domain of applications.

1. The strong coupling expansion is valid for $g^2 \to \infty$ or equaivalently $\beta \to 0$. The gluonic Boltzmann factor $\exp\left(-\beta \sum (1 - \frac{1}{n} Re\,\mathrm{Tr}\, U_p)\right)$ can be expanded in this limit. At $\beta = 0$, all the gauge variables U_x^μ are totally unconstrained and random and the partition function is dominated by quark contributions. I will refer you to the standard text books, cited at the end, for details. In this limit, one can

   (a) show that $V_{Q\bar{Q}} = \sigma r$ for $m_q \to \infty$ such that $\sigma a^2 = -\ln(\beta/2N)$, signalling confinement of quarks,

   (b) show for staggered fermions that the chiral condensate, $\langle \bar{\chi}\chi \rangle \neq 0$ and the pion mass, $m_\pi^2 \propto m_q$, signalling dynamical breaking of the chiral symmetry by the QCD vacuum,

   (c) calculate the hadron spectrum analytically to find it in qualitative agreement with the experiments,

   (d) show that $SU(3)$ gauge theory has a first order deconfinement phase transition and $SU(2)$ a second order one.

   One cannot show confinement in this limit in a theory with dynamical quarks. Since mesons can pop out of the vacuum for sufficiently large energies, the expected potential is anyway $V_{Q\bar{Q}} \neq \sigma r$ for very large r. One thus has no suitable order parameter to probe confinement.

It may be instructive to elaborate a little on (d). In the limit of large quark masses, the so-called quenched limit, QCD becomes a pure gauge theory (or Yang-Mills theory). Thus strong coupling limit of the quenched QCD provides a physical picture of the deconfinement transition which is useful to understand. One proceeds by constructing an effective action for the order parameter by integrating out other fields (which can be done in this limit). The definition of the order parameter in subsection 5.3.1 can be transcribed to the lattice to introduce $\Theta(\vec{x}) = \prod_{t=1}^{N_\beta} U_\mu^4(\vec{x}, t)$. Using this the effective action is defined by introducing a functional δ-function as below.

$$e^{-S_{\text{eff}}(L)} = \int \mathcal{D}U_\mu e^{-S_G} \prod_x \delta(L(\vec{x}) - \Theta(\vec{x})). \qquad (5.78)$$

It is easy to see that in general the effective action will be,

$$S_{\text{eff}}(L) = \int d^3x\, V(L(\vec{x})) + \int d^3x\, d^3y\, L^\dagger(x)\, S_2(\vec{x}-\vec{y})L(y) + \cdots. \qquad (5.79)$$

One uses strong coupling expansion and mean field theory to show that

$$\begin{aligned}
V(L) &= aL^\dagger L + bL^3 + c(L^\dagger L)^2, & \text{for} SU(3) \\
V(L) &= a'L^\dagger L + b'(L^\dagger L)^2 + c'(L^\dagger L)^3, & \text{for} SU(2)
\end{aligned} \qquad (5.80)$$

Further it is possible to show that in the strong coupling $a(T) > 0$, $b(T) > 0$ for high temperatures and $a(T)$ increases as T decreases yielding a first order phase transition for $SU(3)$ while for $SU(2)$, $b'(T) > 0$, $c'(T) > 0$ but $a'(T) < 0$ (respectively > 0) for low (high) temperatures, giving a second order phase transition when it passes through zero.

All this has been obtained with the lattice cut-off in place. One has to remove it. This can be done by successively increasing the order of the expansion but the radius of its convergence is finite, making us look for other methods for continuum limit. Nevertheless, many qualitative features can already be learnt from the strong coupling expansion.

2. The weak coupling is valid for small g and can be identified with the usual perturbation theory. Indeed, lattice is just yet another regulator for doing this expansion and all the usual results can be obtained. There are, of course, simpler and better methods to do this expansion. However, it serves as a bridge to relate many physical quantities to the continuum and also serves as a useful check.

3. Numercial methods are most commonly used which we describe in the next subsection after giving the reasons for their popularity.

5.4.4 Continuum limit and renormalization

As has been emphasised all through, one has to remove the lattice scaffolding eventually. It is this continuum limit of the lattice spacing $a \to 0$ which decides which calculational technique is the best to extract non-perturbative predictions of QCD. A dimensional quantity, such as mass of a particle, can be written as a dimensionless quantity on the lattice, ma. It can be obtained from a suitably chosen correlation function, *e.g.* consider $\psi(i)\gamma_5\psi(i) \equiv P(i)$. Then the correlation of $P(i)$ with $P(j)$ is the correlation function $\langle P(i)P(j)\rangle$. It decays exponentially with distance, $Ae^{-m_\pi a(i-j)}$ as $(i-j) \to \infty$, allowing the determination of the mass am_π.

Setting all quark masses to zero for simplicity, one sees that $ma \equiv 1/\xi = f(g^2)$. As $a \to 0$, $f(g^2)$ must also do so in order to obtain a nontrivial limit. Thus g must be $g(a)$. This implies that the lattice correlation length $\xi \to \infty$. A continuum limit thus corresponds to a critical point or a second order phase transition on the lattice. One needs to locate all such points where $\xi \to \infty$ to decide the fate of the lattice theory in the continuum limit and there are many such continuum theories. This divergence of correlation length does not occur in strong coupling expansion. A consequence of this specific infinite ξ due to vanishing a is that *all* dimensional quantities have a specific behaviour which is related to each other. *E.g.*, if $m_\pi a = \xi_\pi^{-1} \to \infty$ and $\sigma a^2 \sim \xi_\sigma^{-2} \to \infty$, m_π^2/σ^2 is a unique function independent of the lattice spacing. This is called scaling and is not valid in strong coupling, as can be explicitly checked by computing these quantities is the strong coupling expansion.

One can compute the rate change of g w.r.t. a for small enough g. This defines the usual β-function of a field theory: $\beta(g^2) = a\dfrac{d}{da}g(a) = -\mu\dfrac{d}{d\mu}g(\mu)$, where μ has dimensions of momentum. One knows this for QCD:

$$a\frac{dg}{da} = \beta_0\, g^3 + \beta_1\, g^5 + O(g^7), \tag{5.81}$$

where,

$$\beta_0 = \frac{33 - 2N_f}{48\pi^2}, \qquad \beta_1 = \frac{153 - 19N_f}{384\pi^2}. \tag{5.82}$$

The coefficients β_0 and β_1 are universal. They do not change with scheme and thus are the same for the lattice theory as well. For number of flavours N_f less than 16, QCD is asymptotically free, leading to $\beta_0 > 0$ above. One can integrate the equation to obtain,

$$a\Lambda_L = (\beta_0 g^2)^{-\beta_0/2\beta_1^2} \, \exp\left(-\frac{1}{2\beta_0 g^2}\right) \equiv f(g^2). \tag{5.83}$$

It is easily seen that $f(g^2) \to 0$, as $g \to 0$. Furthermore $f(g^2)$ has an essential singularity at $g = 0$; no expansion of $f(g^2)$ in g is possible. It is thus truely non-perturbative. One can also check that

$$m_\pi a = \frac{m_\pi}{\Lambda_L}\, a\Lambda_L, \qquad \sigma a^2 = \frac{\sigma}{\Lambda^2}\,(a\Lambda_L)^2,$$

which implies

$$\frac{m_\pi^2}{\sigma^2} = \text{constant}.$$

Note that in a theory with no dimensional parameters (all $m_q = 0$), we have obtained a variation of g^2 which depends on the dimansional lattice spacing a. In the above renormalization group equation, the effect of g^2 and a cancel each other and all physical quantities are related to an integration constant Λ_L which can be set by experiment. This is known as *dimensional transmutation*.

In a realistic world $(m_u + m_d)/2$, and m_s are two extra parameters in addition to Λ. However, these are small (compared to Λ_L which sets the scale of QCD) and thanks to this approximate chiral symmetry, $m_\pi^2 \propto (m_u + m_d)$ and $m_K^2 \propto (m_s + m_\mu)$. This then lays down a specific path:

1. One can check if lattice QCD does yield this chiral behaviour.

2. If yes, fix the quark masses using the two experiment determinations as input.

3. Use the experminatal information on m_ρ^2 to fix the unknown scale Λ_L.

Once this is done, all other physical quantities are predicted by the theory. If any of them disagrees strongly with experiments, there are no parameters to adjust for reconciliation and the theory, *i.e.*, QCD, has to be simply *wrong*.

Since most hadronic properties were well known before the advent of lattice QCD, a true prediction of the theory is at $T \neq 0$: the quark-hadron transition. If it is observed experimentally as predicted, then it would constiute a real triumph for QCD in even its non-perturbative regime. Indeed, it will also be the first experimental proof of confinement by 'deconfining': lattice QCD predicts confinement and also a deconfinement at high temperaures by using the same theoretical tools. If it is, on the other hand, definitely not seen and the prediction falsified, we have clearly to look for physics beyond standard model in a rather unexpected place.

The search for a method which allows evaluations at such values of g that the RGE is valid leads to the numerical Monte Carlo simulation technique as the choice for the best calculational tool for lattice QCD. Recall that for a physical observable A,

$$\langle A \rangle = \frac{1}{\int dU e^{-S}} \int dU \, e^{-S} \, A$$

If one could generate an ensemble of configurations of links $\{U\}$ such that the probability of finding a configuration C, $P(C) \propto e^{-S(C)}$ for it then $\langle A \rangle = \frac{1}{N} \sum_{i=1}^{N} A_i$ with A_i being value for configuration C and N being number of such configurations. The Monte Carlo simulation alogrithm due to Metropolis *et al.* is designed to generate such ensembles. In general, an algorithm is specified by a probability distrubution $P(C, C')$ for taking a configuration C to C'. Equilibrium is ensured by the detailed balance condition,

$$P(C, C')e^{-S(C')} = P(C, C')e^{-S(C)}$$

For gauge theory simulations, we begin with an arbitary starting configuration with all U_x^μ equal to identity (cold start) or random (hot start). Take a set of possible replacements U'. Usually this is done by writing $U_x^{\mu\prime} = \delta U \cdot U_x^\mu$, where $\delta U \in SU(3)$, is close to identity and is chosen randomly from a table which itself is made of random matrices but with a constraint that δU^{-1} is in the table. For new U', evaluate $S(U')$ and accept the change if

$$\exp\left(-\left[S(U') - S(U)\right]\right) = \exp\left(-\left[S_{new} - S_{old}\right]\right) > r,$$

where r is a random number between 0 and 1: $0 < r < 1$. Otherwise the change is rejected, retaining the old U. The above choice of $P(C, C')$ constitutes the Metropolis algorithm.

One may try this process for a given link more than once. One speaks then of many Metropolis hits. Usually 3–8 hits may be chosen. Depending on how close are the elements in the set to unity and how many hits are made, one may have 50% chance that $U_x^\mu \to U_x^{\mu\prime}$. One speaks of a 50% acceptance for the alogrithm in that case. One does this for each link step by step, leading to an iteration/sweep of the entire lattice. Typically, few hundred such sweeps may be needed for equilibriation. This is determined by monitoring the values of, *e.g.*, the plaquette $\mathrm{Tr}\,U_p$. For a cold (hot) start, it begins from unity (respectively zero) and eventually reaches a value reflecting the coupling β. Averaging then the physical observables over many sweeps, one can obtain their expectation values. Of course, as can be expected from the statistical nature of the technique, depending upon the number N of configurations over which one averages, the determination will have statistical errors, $\sigma^2 = (\langle A^2 \rangle - \langle A \rangle^2)/(N-1)$. This formula, however, is true only when all individual A_i are independent which they are not (see the definition of the algorithm above). They have correlations in Monte Carlo time, called autocorrelations which have to be eliminated. One uses methods like Jackknife, binning of data etc., details of which will not be covered here. See the textbooks cited at the end.

In case of full QCD, *i.e.*, QCD with dynamical quarks the action S contains $Det\,M$, where the fermionic matrix M covers the space of spin, colour, flavour and $(\vec{x}, t)$. Calculating e^{-S} everytime a change is proposed in a single link, is too hard and time consuming. One needs intelligent but exact algorithms which will be efficient. One such algorithm is Hybrid Monte Carlo. Once again details can be found in references cited at the end.

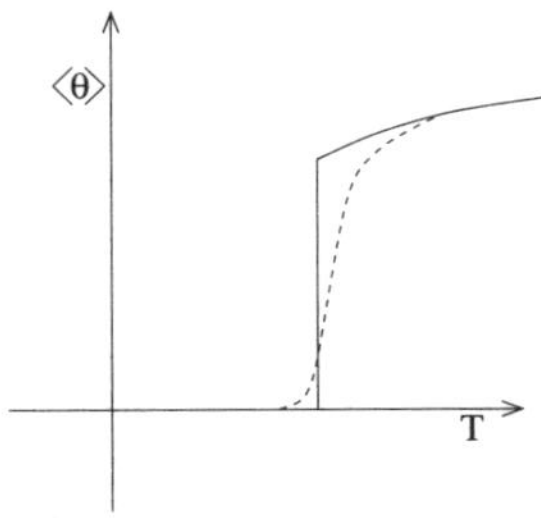

Figure 5.10: Rounding of the order parameter at first order transition.

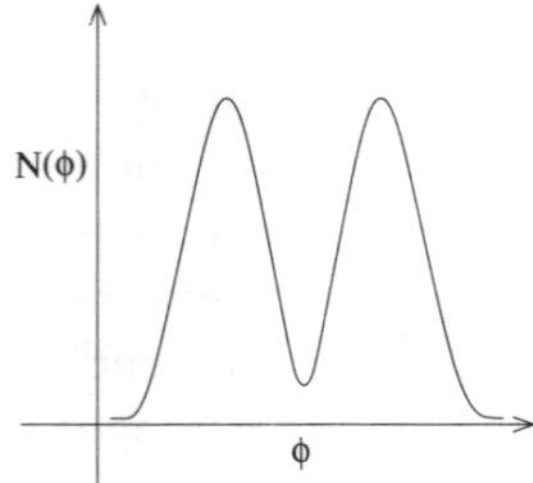

Figure 5.11: Histogram the order parameter at first order transition.

5.4.5　Physical observables

Having discussed the calculational techniques, let us now list the commonly studied physical obervables and discuss their important properties. As may have been clear by now, the two order parameters of interest for investigations at finite temperature are the chiral condensate:

$$\langle \bar{\psi}\psi \rangle \;=\; \frac{1}{N_\sigma^3 N_\beta N} \left\langle \sum_{m,i} \bar{\chi}_i(m)\chi_i(m) \right\rangle$$

$$=\; \frac{1}{N_\sigma^3 N_\beta N} \left\langle \frac{N_f}{4}\, Re\, \mathrm{Tr}\, M^{-1} \right\rangle$$

and the Polyakov loop

$$\langle L \rangle = \frac{1}{N_\sigma^3} \left\langle \sum_{\vec{x}} L(\vec{x}) \right\rangle . \tag{5.84}$$

One aims to look for phases with nonvanishing values for these to characterize these as the chiral symmetry phase and deconfined phase respectively. Note, however, that there is no spontaneous breaking of a symmetry (SSB) in a finite system. Due to possible tunnelling effects, the ground state always respects the symmetry. One usually adds a small symmetry breaking field. Taking first the limit of infinite volume, the symmetry breaking field is then sent to zero. Thus, the chiral condensate is defined as $\langle \bar{\psi}\psi \rangle = \lim_{m \to 0} \lim_{N_\sigma^3 \to \infty} \langle \bar{\psi}\psi \rangle$. For the Polyakov loop, $\langle |\sum L| \rangle$ is usually used.

Finite size of the lattice also renders the determination of the order of the phase transition, if any, and T_c difficult due to the round-off effects. A distinction between a discontinuous (first order) or a continuous (second order) transition is lost, as depicted in Fig.5.10. Nevertheless a first order phase transition is characterized by the coexistence of the system in both phases, leading to a frequency distribution of the order parameter, as shown in Fig.5.11 — away from the transition point or at a second order phase transition the distribution is a single Gaussian. One therefore uses the same finite volume effects to obtain information on T_c and the order of the phase transition. *E.g.*, as the volume V becomes larger, one should have lesser tunnellings, leading to much sharper peaks with deeper valley in Fig.5.11.

One can make the above criterion quantitative in form of finite size scaling which has been studied extensively in statistical mechanics. One knows from there that,

$$T_{c,N_\sigma} - T_{c,\infty} \propto N_\sigma^{-y_T} \qquad \text{and} \qquad \Delta T_{c,N_\sigma} \propto N_\sigma^{-y_T} \qquad (5.85)$$

permit an estimate of T_c. The exponent y_T distinguishes between a first order ($y_T = 3$) and second order ($y_T < 3$). The transition temperature on an N_σ^3 lattice, T_{c,N_σ}, can itself be defined in many ways, *e.g.*, from the location of the peaks of the susceptibility,

$$\chi_L^{max} \equiv N_\sigma^3 \left(\langle L^2 \rangle - \langle L \rangle^2 \right) \sim N_\sigma^{\rho_s}, \qquad (5.86)$$

or the specific heat,

$$C_v^{max} \equiv N_\sigma^3 \left(\langle \epsilon^2 \rangle - \langle \epsilon \rangle^2 \right) \sim N_\sigma^{\rho_{sp}}. \qquad (5.87)$$

The peak heights themselves also change with volume, as indicated above. One can deduce the nature of the transition from the corresponding exponents. For a first order transition, $y_T = 3 = \rho_s = \rho_{sp}$, while for a second order Ising like (or $SU(2)$ deconfinement transition), $y_T = 1/\nu = 0.63$, $\rho_s = \gamma/\nu = 1.93$, and $\rho_{sp} = \alpha/\nu = 0.18$.

Varying N_σ, one obtains the exponents which can be used to determine $T_{c,\infty}$ and the order of the phase transition. The above procedure yields a critical coupling β_c for a given N_β. It can be converted into physical units using the RGE equation:

$$T_c = \frac{1}{N_\beta \, a(\beta_c)} = \frac{\Lambda_L}{N_\beta} \frac{1}{a(\beta_c)\Lambda_L}.$$

This turns out to be not accurate enough since the two-loop RGE is only approximately obeyed at the values of β of interest. Moreover the quark mass has to be fixed as well in physical units. One therefore determines $m_\pi^2 = m_{ps}^2$ and $m_\rho^2 = m_v^2$ at $\beta = \beta_c$ but on a symmetric (zero temperature) lattice. Using m_ρ to fix the scale Λ_L and $m_{ps}^2/m_v^2 \propto m_q$ to determine the quark mass, one gets

$$\frac{T_c \, a}{m_\rho \, a} = \frac{\xi_\rho^{lattice}}{N_\beta}.$$

One also needs to obtain from the QCD lattice simulations various thermodynamical quantities, such as the energy density or the pressure. To obtain them, one introduces different lattice spacings, a_σ, a_β in the spatial and temporal directions. This changes the gluonic action to

$$S_G = \frac{2N}{g_\sigma^2}\xi^{-1} \sum_{x\,\mu<\nu<4} P_x^{\mu\nu} + \xi \frac{2N}{g_\beta^2} \sum_{x\,\mu<4} P_x^{\mu 4} , \qquad (5.88)$$

where $\xi = a_\sigma/a_\beta$ with $P_x^{\mu\nu} = 1 - Re\,Tr\,U_p/N$. In the fermionic action S_F, one has to introduce a $\xi^{-1}\gamma_F$ factor multiplying the M_{mn}^4 term. The couplings will now be function of both spacings but can be conveniently choosen to be $g_\sigma^2(a,\xi)$, and $g_\beta^2(a,\xi)$. The energy density is then given by

$$\epsilon = -\frac{1}{V}\frac{\partial}{\partial(1/T)}\ln \mathcal{Z}\bigg|_V = +\frac{\xi^2}{N_\sigma^3 N_\beta \, a^4}\frac{\partial}{\partial \xi}\ln \mathcal{Z}\bigg|_a, \qquad (5.89)$$

and the pressure is

$$P = \frac{\partial}{\partial V}(T\ln Z)_T = \frac{\xi}{3N_\sigma^3 N_\beta \, a^3}\frac{\partial}{\partial a}\ln Z\bigg|_{a_\beta}$$

$$= \frac{\xi}{3N_\sigma^3 N_\beta a^3} \left[\left(\frac{\partial \ln Z}{\partial a} \right)_\xi + \left(\frac{\partial \ln Z}{\partial \xi} \right)_a \frac{\xi}{a} \right].$$

Therefore,

$$\epsilon - 3P = \frac{\xi}{N_\sigma^3 N_\beta a^3} \left(\frac{\partial \ln Z}{\partial a} \right)_\xi. \qquad (5.90)$$

After taking the derivatives, one sets $\xi = 1$. Two methods have been used in the literature to obtain ϵ and P.

1. Calculate ϵ and P directly as above. One then needs to evaluate $\partial g_i^2 / \partial \xi^2$ etc., called the Karsch coefficients, since he first evaluated them in lowest order perturbation theory.

2. Noticing that the pressure is also $\partial \Omega / \partial \beta = 3(\bar{P}_s + \bar{P}_\beta)$ in quenched QCD (some additional derivatives with respect to quark mass for the fermionic terms), one integrates $\int_{\beta_0}^{\beta} 3(\bar{P}_s + \bar{P}_\beta) = \Omega(\beta) = -P(\beta)$ to obtain the pressure. Subtracting Ω_{vac} by evaluating the same quantity on a symmetric lattice, and using the relation $\epsilon - 3P$ (with non-perturbative β-function), one obtains a fully non-perturbative determination of these quantities.

Another set of interesting thermodynamical quantity of interest is that of various quark number susceptibilities (QNS) at $\mu = 0$. One can take two partial derivatives with respect to the chemical potential to obtain this. It can be argued that if $q\bar{q}$ mode dominate QGP, then singlet QNS (baryonic) ought to be large in QGP phase but since even the lowest baryon, proton, is too heavy, it will be small in hadronic phase. It is a measure of fluctuations which could be seen if QGP is formed.

Many interesting results exist from lattice QCD simulations, and more are coming out. The key features to emerge are: (i) a sudden jump in thermodynamic quantities at a temperature where the order parameters suggest a chiral symmetry breaking, confined phase to change into chirally symmetric, deconfined quark gluon-plasma phase; (ii) $T_c \sim 150$–180 MeV, depending on the number of flavours; (iii) an energy density of 1–2 GeV/fm^3 above the transition,; (iv) strongly non-perturbative plasma with the thermodynamical quantities differing from their perturbative limits sizeably; (v) obstruction to dimensional reduction by fermions, ie., fermionic bound state screening lengths tend to be smaller than those of gluonic and agreement of the experimentally determined Wroblewski parameter from the heavy ion collisions in RHIC and CERN with the lattice determinations.

5.5   Heavy ion collisions

The lattice QCD simulations, discussed in the previous chapter, indicate strongly the existence of a quark-gluon plasma phase at sufficiently high energy densities. Not surprisingly, therefore, one can expect to produce QGP in laboratory, provided

- the produced energy density is large, *i.e.*, in excess of 1–3 GeV/fm^2,

- the system is size big enough: $L \gg \Lambda_{QCD}^{-1} \sim 1$ fm,

- many particles are in thermal equilibrium.

These restriction makes one turn to heavy ion collisions at high energies. Since for a symmetric collisions of AA nuclei, $L \sim 2R_A \sim 2.4A^{1/3}$ fm, it is $\sim 14 fm$ for heavy nuclei like lead or gold with $A \sim 200$. Clearly, a heavy ion collision also produces many particles. So the key questions are whether they reach a thermal equilibrium, and if yes, how do we ensure that the desired energy density is attained. A crucial issue is the magnitude of thermalization time. It should be such that the evolution of the system proceeds as a QGP or as a medium for long enough time to have observable consequences.

Bjorken in his seminal paper in 1983 provided a beautiful physical framework to address many of these questions and also provided a formula to estimate the initial energy density. Although recent theoretical efforts focus on improving the framework, it continues to be the backbone of the discussion of all the experimantal results obtained so far.

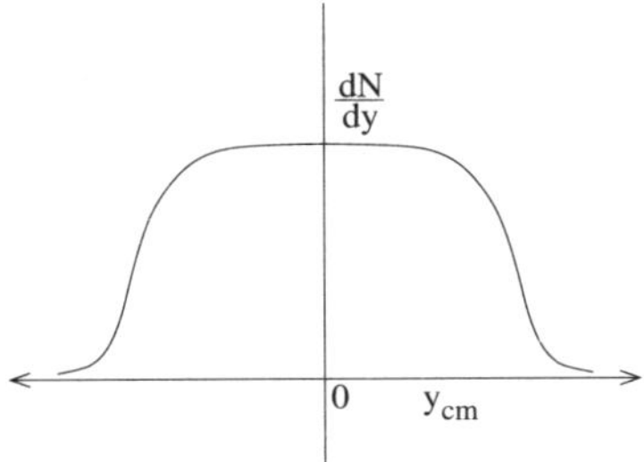

Figure 5.12: Plateau of the rapidity distribution.

From the known data on the distributions of the hadrons produced in hadron-hadron and hadron-nucleus collisions, he extracted three hypotheses which he assumed were true for nucleus-nucleus collisions. Let us first define a kinematic variable, called rapidity of a particle: $y = \frac{1}{2}\ln[(E + P_L)/(E - P_L)]$ for a particle with momentum P_L in the direction of the beam and $P_L^2 + P_T^2 + m^2 = E^2$

> **Problem 15:** Find out what happens to y under a Lorentz boost in the longitudinal (beam) direction. Show it becomes $y + y_1$.

Now we can state the three hypotheses of Bjorken.

1. There exists a central plateau in the (single) particle (inclusive) distributions in *h-h* collisions as a function of y. This is schematically shown in Fig. 5.12. As seen there, near $y \simeq 0$ in the center of mass frame, the spectrum is invariant under Lorentz boost (see problem above). Physically this implies that a 250 GeV $p\bar{p}$ collision at 90° (*i.e.*, $P_L = 0$) looks the same as 10 GeV + 6.25 TeV collision at 90°. The rapidity plateau is a central theme in all discussions on heavy ion collisions and QGP.

2. A rapidity plateau is assumed to exist for A-A collisions as well. When Bjorken's paper was written, this assumption

did not have as much support from data as in Fig. 5.12. Only recently data has become available and at high collision energies, it seems to be borne out.

3. There exists a 'leading-baryon' effect. The net baryon number of a projectile or target resides within 1–2 units of its rapidity. Again this was empirically supported by the known data for $pp, p\alpha$, and $\alpha\alpha$ at large $\sqrt{s}/A$.

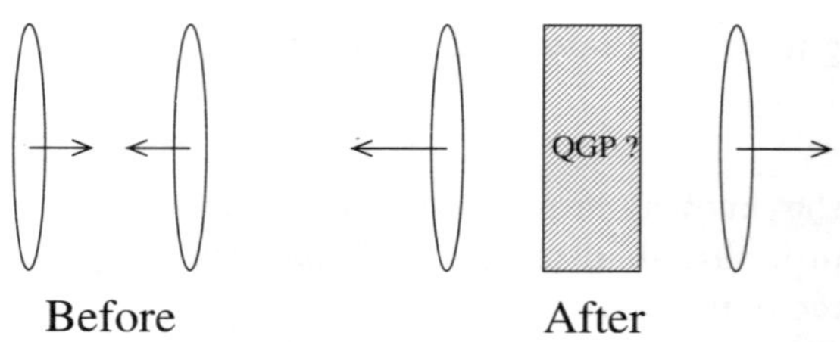

Figure 5.13: A typical central heavy ion collision.

In order to compute the energy density attained in a central collision in Bjorken's picture, let us consider a head-on nucleus-nucleus (AA) collision with the same mass number A. As a result of the above hypotheses, the collision will look as depicted in Fig.5.13. Most of the baryon number will fly away as the target and projectile recede from each other after the collision, leaving behind an almost baryon-free region with many particles in it. Let us concentrate on a thin slab of thickness d at the center. The energy contained within this slab is given by,

$$E = Ny\frac{d\langle E\rangle}{dy} = N\frac{d}{t}\frac{d\langle E\rangle}{dy},$$

where N is the incident number of nucleons, the average energy of produced particles per unit rapidity is $d\langle E\rangle/dy$ at y, which is the rapidity of the slab. This yields an estimate for the energy density $\epsilon = E/\mathcal{A}.d$, where $\mathcal{A}$ is effective area of incidence, leading to $\epsilon = \dfrac{N}{2t\mathcal{A}}\dfrac{d\langle E\rangle}{dy}$. Here t is the time since the first contact of the colliding nuclei. If quark-gluon plasma is formed at time τ_i then substituting it for t above, one will obtain the initial energy density. It may be useful to get an idea of the typical numbers. Writing $N/\mathcal{A} = \mathcal{A}/\pi(1.2.\mathcal{A}^{1/3})^2 = \mathcal{A}^{1/3}/(4.5\,\text{fm}^2) \equiv d_0^{-2}$, one has for $A \sim 200$, $d_0 \simeq 0.7$ fm for lead-lead collisions. To proceed further, we need to choose a collision energy. We summarise, in the table 5.1, the details of the past, current and future heavy ion collisions programs.

It may be noted that Indian laboratories/universities have significant participation in many of these. The WA98 experiment in CERN has a group from the Universities of Jammu, Panjab and Rajasthan and from VECC and SINP, Kolkata and IOP, Bhubaneswar. The BARC, Mumbai, BHU, Varanasi and VECC, Kolkata have collaborations in Phenix and STAR while a collaboration from most of these groups is active on the Alice experiment at LHC, CERN.

From the experiments already done, we have an idea of the particle density produced at the respective energies. Extrapolating to higher energies, we can make a reasonable guess for the future. Similarly the average energy of the particles produced is also known. A crucial key parameter in the formula for initial energy density above is the formation time for quark-gluon plasma. Taking it to be 1 fm, and using,

$$\epsilon = \frac{1}{2td_0^2}\frac{d\langle E\rangle}{dy} = \langle E_T\rangle\frac{d_0^2}{2t}\frac{dn}{dy},$$

one finds that $\epsilon^{SPS} = 3.2\,\text{GeV/fm}^3$, $\epsilon^{RHIC} = 5.5\,\text{GeV/fm}^3$. These are, of course, way above the minimum required on the basis of estimates from lattice QCD. Several authors have argued for lower thermalization time as $\sqrt{s}$ increases. For example, Shuryak uses perturbative $gg \to gg$ cross sections to argue that gluons will thermalize faster by a factor of 3.5. It would have been nice to have some reliable estimate for thermalization time and its $\sqrt{s}$-dependence. Naively, whatever quanta may be contained in the slab will collide with each other sufficiently many times and then may produce thermal equilibrium. Many diffrent types of ideas

Machine	$\sqrt{s}_{NN} = \frac{\sqrt{s}}{A}$	A-B	Remarks
Alte. Grad. Sync. BNL, USA	5 GeV	Au-Au	
SpS CERN Geneva 6–7 major expts	$\sim$ 17 GeV $\sim$ 19 GeV	Pb-Pb S-U S-S	Most complete and interesting results
RHIC, BNL 4 major expts Phenix, Star, Phobos & Brahms	55 GeV 130–200 GeV	Au-Au Au-Au	July 2000 2000-02
LHC, CERN Major detector: Alice	5500 GeV	Pb-Pb	2006–07?

Table 5.1: Relativisitc Heavy Ion Experiments

and models for production of thermalization period have been discussed in the literature:

1. Colour flux tube model: non-perturbative particle production.

2. Perturbative QCD (pQCD) inspired model: McLerran-Venugopalan model, colour glass condenstate.

3. Minijets $\rightarrow$ string fragmentation: Hijing, Venus event generators.

4. pQCD plus kinetic theory: parton cascade model.

Most address the issue of particle production dn/dy but have very little to offer on whether and when thermalization must take place. Recently, there have been attempts to study thermalization in QCD using a perturbative approach by Mueller, Son and others. For details of these, interested reader may look up the proceedings of Quark Matter conferences and the books edited by Hwa, cited at the end.

Once thermalization is established, hydrodyanamic equations can be used to model the expansion of the QGP fluid. The above estimate of initial energy density can be used as a boundary condition for the hydrodynamic expansion. If one makes a longitudinal Lorentz boost, then these initial conditions remain unchanged due to assumption (2) of Bjorken. Throughout the central region, initial conditions imposed at a thermalization proper time τ after collision are invariant with respect to Lorentz boosts. This implies that subsequent evolution of the QGP will also be similarly boost invariant.

In the Landau hydrodynamical model of expansion, one obtains the evolution of energy density $\epsilon(x)$, pressure $P(x)$ and the temperature $T(x)$. Defining a unit vector $u_\mu(x)$ such that $u_\mu u^\mu = 1$ and assuming the expanding fluid to be perfect, the energy momentum tensor is $T_{\mu\nu} = (\epsilon + p)u_\mu u_\nu - g_{\mu\nu}P$. Energy-momentum conservation tells us that $\partial^\mu T_{\mu\nu} = 0$. Contracting with u_ν or $g_{\nu\lambda}u_\nu u_\lambda$ various hydrodynamic equations can be obtained. For details, see the review by Cleymans *et al* cited at the end.

Ignoring the expansion in transverse direction for simplicity (it can be incorporated in a conceptually straightforward manner), the proper-time evolution of the system can be worked out as follows: $\tau = \sqrt{t^2 - z^2}$ and at $\tau = \tau_0$, $\epsilon = \epsilon_0$ is initial energy density. Defining a space-time rapidity as $y = \frac{1}{2}\ln[(t+z)/(t-z)]$, one has various quantities as function of τ and y: $\epsilon(\tau,y), P(T,y)$ etc.

One can simplify by using boost invariance. *E.g.*, the initial condition is $\epsilon(\tau_0, y) = \epsilon(\tau_0) = \epsilon_0$. Solutions of the hydrodynamical equations with such boost invariance will have $\epsilon(\tau), P(\tau), T(\tau)$ etc. One can show that the hydrodynamical equations simplify and one has

$$\frac{d\epsilon}{d\tau} = -\frac{\epsilon + P}{\tau}, \qquad \frac{ds}{d\tau} = -\frac{s}{\tau}, \tag{5.91}$$

implying that $s\tau$ is a constant. This, in turn, means that the entropy remains constant during the expansion of QGP.

It is possible to proceed still farther without using a specific equation of state: $\epsilon = f(T)$, $P = f_1(T)$. Writing $\epsilon = \epsilon(P)$ and $P = P(T)$, Eq.(5.91) can be re-written as,

$$\frac{d\epsilon}{d\tau} = \frac{d\epsilon}{dP}\frac{dP}{dT}\frac{dT}{d\tau} = -\frac{\epsilon + P}{\tau} = -\frac{Ts}{\tau}. \tag{5.92}$$

Using the canonical relations, $P = \frac{T}{V}\ln Z$ and $s = \frac{1}{V}\frac{\partial}{\partial T}(T\ln Z)$, one has $dP/dT = s$. Since, $v_s^{-2} = \partial\epsilon/\partial P$, where v_s is the velocity of sound in the medium (QGP here), Eq. (5.92) implies

$$\frac{1}{T}\frac{dT}{d\tau} = -\frac{v_s^2}{\tau}.$$

If QGP were to be an ideal relativistic gas of quarks and gluons, we would have $v_s^2 = 1/3$. Substituting it above and integrating, one obtains, $T \propto \tau^{-1/3}$. Thus the hydrodynamical expansion proceeds with cooling of the plasma as proper time increases. Since $\epsilon \propto T^4$, the energy density drops proportional to $\tau^{-4/3}$. Lattice

QCD does have some results for v_s^2 which can be used to obtain evolution of T, ϵ for the interacting QCD plasma. Generically, the equations of hydrodynamics imply hydrodynamical flow and cooling by expansion. One expects therefore that after some time the temperature would fall sufficiently and the plasma will condense in the hadrons. The hadronic gas would have its own value for the velocity of sound but the expansion and cooling would continue as long as particles in the fluid are interacting. With increasing volume, the mean free paths of particles increase. As they become equal to size of the system, particles start to decouple, which is refered to as *free streaming* or *freeze-out*. It is usually assumed that this happens suddenly or fast enough in τ, as soon as $T(\tau)$ falls below $T_{\text{freezeout}}$.

A common scenario, discussed in the literature dealing with QGP signals, is the mixed phase of QGP and hadron gas. If the transition from QGP to hadrons is of first order, as is the case in simple bag models (albeit by construction) or in quenched approximation to QCD (weak first order), then the the system will exist in this mixed phase and Eq. (5.91) can be used to find out how long it will stay in the mixed phase. Recall that at T_c, $P_{QGP} = P_{had}$. Therefore the discontinuity in the entropy at T_c is the same as the latent heat: $\Delta s = \Delta \epsilon$. The latent heat in the bag model (in full QCD) is $L/T_c^4 = 15$ ($\simeq 0$ or at most ≤ 2). Note the almost an order of magnitude decrease in going from the naive bag model to (lattice) QCD. This can have severe consequences for the analyses of plasma signals in heavy ion collisions. Let $f(\tau)$ be the fraction in QGP phase at time τ. Then

$$S(\tau) = f(\tau)S_{QGP}(T_C) + (1 - f(\tau))\, S_{had}(T_c).$$

During the transition from QGP to mixed phase to hadrons, the temperature remains pegged at T_c. Clearly the entropy is fully in the QGP phase at the start of the condensation to hadrons at time τ_1. Hence, using Eq. (5.91),

$$\tau f(\tau)S_{QGP} + \tau(1 - f(\tau))S_{had} = \tau_1 S_{QGP}.$$

Dividing out by S_{QGP} and denoting the ratio $S_{had}(T_c)/S_{QGP}(T_c)$ by r, the equation can be rearranged as

$$f(\tau) = \frac{\tau_1 - r\tau}{\tau(1 - r)}. \tag{5.93}$$

At τ_2, where the mixed phase is converted completely into hadrons, $f(\tau_2) = 0$, yielding $\tau_2 = \tau_1/r$. The duration of mixed phase is thus $\Delta\tau = \tau_2 - \tau_1 = (1 - r^{-1})\tau_1$. For a strong first order phase transition, such as that for the bag model, one has a small r, about $\sim 1/16$. This gives rise to a very long lived mixed phase. On the other hand, lattice QCD estimates of r are much smaller, leading to a very short lived mixed phase.

5.6　Signals of QGP

As discussed in the previous chapter, heavy ion collisions offer the possibility of producing quark-gluon plasma by meeting many of the necessary conditions. Although, it is not clear that this can be made from a theoretically plausible scenario into a compelling one, we saw that QGP condenses into hadrons by expanding and cooling. This makes the detection of QGP experimentally both a tricky and challenging task. Irrespective of whether it was produced, the final products are always the stable hadrons, photons or leptons. One therefore needs signals of QGP which exploit the high temperature and the unique phase that may have been created to distinguish between the end-products or the patterns of their production. Many such signals have been proposed and we attempt to give a glimpse of the principles behind these signals by considering some specific ones below.

Due to their association with the shorter time scales at which QGP phase may have possibly existed, the so-called hard probes, have potentially a good chance of signalling the presence of QGP. Prominent such hard probes considered so far are: J/ψ suppression, jet quenching, dileptons and photons. They all are characterised by the common property of being processes calculable in perturbative QCD. Thus, one expects them to be reasonably well understood in usual (anticipated) non-QGP situations like hadron-hadron collisions. Consequently, one hopes to be able to get reasonable estimates of the background although there are pQCD limitations as well. By 'background' here we mean contributions from known sources that can be definitely identified as conventional.

5.6.1 J/ψ suppression

The most widely discussed hard probe is that of suppression of J/ψ production cross section. Let us first describe what is J/ψ. It is a bound state of heavy quark, charm and its antiquark, *i.e.*, the $c\bar{c}$ bound state. Its mass is 3.1 GeV. Experimentally, it is observed using its $J/\psi \to \mu^+\mu^-$ decay channel. Plotting the mass distribution of the produced dilepton ($\mu^+\mu^-$ or e^+e^-) spectrum, *i.e.*, $dN/dM_{\mu^+\mu^-}$ vs. $M_{\mu^+\mu^-} = (p_{\mu^+} + p_{\mu^-})^2$, J/ψ shows up as a narrow peak over the continuum background of $\mu^+\mu^-$. The background itself can be shown to come from what is known as a Drell-Yan process (see the lectures on perturbative QCD in this volume), and is calculable in pQCD (at least for sufficiently large M). The experimentally observed dilepton spectrum with the narrow J/ψ peak is shown schematically in Fig.5.14.

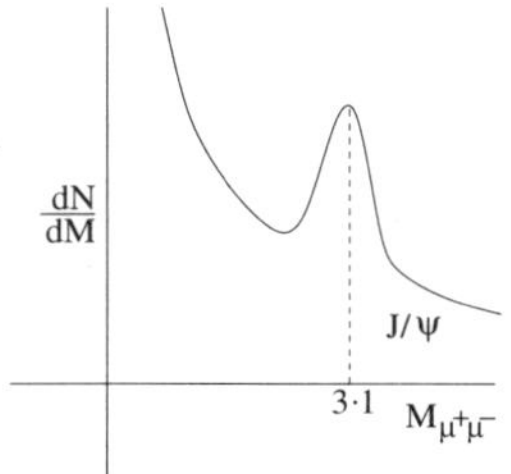

Figure 5.14: Sketch of the J/ψ peak on $\mu^+\mu^-$ background.

Figs. 5.15 and 5.16 display the lowest order pQCD diagrams contributing to the continuum and the peak. Writing p_1 and p_2 for the 4-momenta of quark q and antiquark $\bar{q}$ in the nucleon-

nucleon center-of mass frame, one can relate them to the momenta of the parent nucleons by momentum fractions x_1 and x_2 in the relativistic limit: $p_1 = x_1(E, 0, 0, E)$, and $p_2 = x_2(E, 0, 0, -E)$. Clearly, the 4-momentum q of the photon in Fig.5.15 is $p_1 + p_2$.

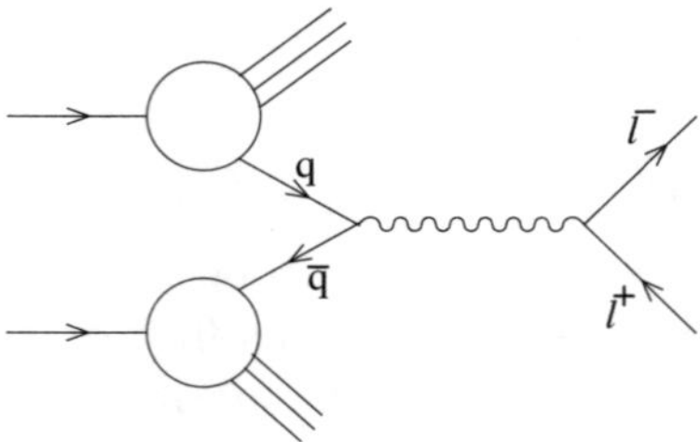

Figure 5.15: Lowest order diagram for the DY process.

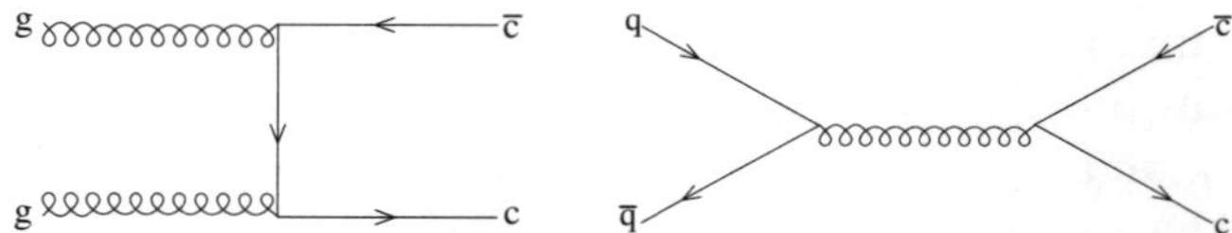

Figure 5.16: Lowest order diagrams for J/ψ hadroproduction.

Let us write $q = (M \cosh y, 0, 0, M \sinh y)$. Using the above equations defining p_1 and p_2, one can easily show that $x_1 = Me^y/\sqrt{s}$, $x_2 = Me^{-y}/\sqrt{s}$ and $M^2 = x_1 x_2 s$ with $s = 4E^2$. One also has $y = \frac{1}{2}\log(x_1/x_2)$. Note that the dilepton mass M and the rapidity y are the kinematic variables, varying which the experimentally observable distribution $d\sigma/dM^2 dy$ can be investigated. Using the parton model, one has

$$\frac{d\sigma}{dM^2 dy} = \frac{4\pi\alpha_{em}^2}{9M^2 s} \int \frac{dx_1 dx_2}{x_1 x_2} \sum_i e_i^2 \left(q_i(x_1)\bar{q}_i(x_2) + (1 \leftrightarrow 2)\right)$$
$$\times \; \delta\left(1 - \frac{M^2}{x_1 x_2 s}\right) \delta\left(y - \frac{1}{2}\ln\frac{x_1}{x_2}\right).$$

This, of course, is the lowest order contribution. At the next order, the quarks can emit a gluon or a gluon and quark can fuse to yield a photon and a gluon in the final state. Such higher order contributions turn out to be almost as large and have to be included. Phenomenologically, this has been traditionally done by fixing the normalization from the experimental data by a constant K. The higher order contributions do provide an effective K-factor of about two, as required by the data. Having thus fixed the continuum background for J/ψ, the cross-section for J/ψ-production can be obtained from the data as the area under the peak. Again pQCD can be used to estimate the $\sigma^{J/\psi}$ reasonably. The corresponding lowest order diagrams of Fig.5.16 can be evaluated similarly to obtain the cross section for charm-anticharm $(c\bar{c})$ pair. One needs an additional, non-perturbative, input to convert it to the cross section for J/ψ. There are currently two popular models to do this. In the colour evaporation model, the simpler of the two, it is assumed that the $c\bar{c}$ pair hadronizes to produce J/ψ if the invariant mass of the pair lies between $4m_c^2$ and $4m_D^2$, where m_D is the mass of the lightest meson with one charm quark. Further the probability to convert in this interval is a free parameter $f_{J/\psi}$, giving rise to,

$$\sigma^{J/\psi} \;=\; f_{J/\psi} \int_{4m_c^2}^{4m_D^2} d\hat{s} \int dx_1 dx_2 \Big[g(x_1)g(x_2)\sigma_{gg}(\hat{s}) + $$
$$\left(q(x_1)\bar{q}(x_2) + 1 \leftrightarrow 2 \right) \sigma_{q\bar{q}}(\hat{s}) \Big] \, \delta(\hat{s} - x_1 x_2 S),$$

where the partonic cross section σ_{gg} and $\sigma_{q\bar{q}}$ can be computed using Feynman diagrams as usual. The other, color octet model, uses the J/ψ wavefunction and other non-perturbative matrix elements for the hadronization part. Both the models can explain the pp data for the total cross section, $\sigma^{J/\psi}$ or the differential cross section $d\sigma^{J/\psi}/dp_T$ Does it work equally well for the J/ψ production in A-A? We will comment on this later.

5.6.2   Probe of deconfinement

Let us now review in brief why J/ψ-suppression may be a probe of deconfinement. The salient features of the original Matsui-Satz argument, given in 1986, are:

1. The J/ψ is a $c\bar{c}$ bound state and thus has too high a mass to be produced by fusion of thermal gluons. Thus production of the $c\bar{c}$ pair is itself unlikely unless T is very high (not even at RHIC/LHC, where the expected T is less than 1 GeV and average momenta of gluons will be $\sim T$ or less).

2. Due to its large mass, $c\bar{c}$ pair is created at early times. Soft interactions convert $c\bar{c}$ to J/ψ. The larger time scale for hadronization means that the pair "sees" the hot thermalized medium during hadronization.

3. If a deconfined QGP is produced, the potential experienced by the pair is not a linear confining one, $V_{Q\bar{Q}} \neq \sigma r$, but screened. The J/ψ may, therefore, not be able to form if the (Debye) screening $r_D(T)$ is smaller than the radius $r_{J/\psi}(T)$ of the $c\bar{c}$ bound state in QGP.

4. As QGP cools and hadronizes, the unbound $c\bar{c}$ pair is randomized in the medium and cannot bind anymore. This leads to the suppression of J/ψ if QGP is formed.

Of course, the plasma formation should then also imply an enhanced D-production, where D mesons contain only one charm quark or antiquark. It turns out to be experimentally difficult to see this enhancement as its underlying $\sigma^{c\bar{c}}$ (open) is already quite large. This argument is clearly valid for other heavy resonances like Υ or the other bound states of $c\bar{c}$ such as χ, η etc. One needs to estimate the correponding Debye screening length and determine the range of temperatures where the suppression occurs. One can treat the heavy quark-antiquark system in a non-relativistic potential model to obtain r_D. These have been obtained by solving the Schrödinger equation for $V(r) = V_0(r)e^{-r/r_D(T)}$, with $V_0(r) = (\sigma r - \alpha/r)$ having the usual linear plus coulomb term.

One increases r_D to find if the known bound state with appropriate quantum numbers at zero temperature still exists. For $r_D < r_D^c$, where there is no bound state, one can say that it has dissolved. The resulting pattern suggests $T_\chi \sim T_c$, $T_\Upsilon \gg T_{J/\psi} \sim 1.25\, T_c$. While the treatment using the nonrelativistic Schrödinger equation may be alright, one still needs to include additional effects, such as collisions with the particles in the thermal heat bath.

The Debye screening length $r_D(T)$ can also be obtained by studying appropriate correlation functions in finite temperature QCD. Using perturbation theory, one finds,

$$r_D^{-1}(T) = \sqrt{1 + \frac{N}{3}g(T)T}.$$

In order to obtain it non-perturbatively from lattice QCD, one uses the $\langle L(x)L^\dagger(x)\rangle$ correlation function. It is expected to decay as $\sim e^{-r/r_D}$ and the decay length is thus $r_D(T)$. It has been found in simulations that (a) $r_D \simeq 0.2\, T^{-1}$ at $T/T_c \simeq 2$, which corresponds to about 0.1 fm and (b) perturbation theory does not work at such low temperatures. Typically r_D^{pert} is 3 times larger.

An interesting open problem is to find out whether lattice QCD can tell us anything directly about J/ψ at such temperatures. The formalism which we discussed in Section 4 is being applied with new innovations in techniques to extract this information. The main point is to obtain the spectral function at finite temperature.

Let us summarise the expectations again. If QGP does form in heavy ion collisions, then ideally one should not get any J/ψ's but it background the continuum DY is affected very little (unless the temperatures attained are very high). Due to the finite size of the produced fireball as well as its finite lifetime, some J/ψ escape. Therefore, the J/ψ peak is not completely gone and one has a (partial) suppression of J/ψ.

5.6.3 The NA50 experiment at CERN

Ideally to establish suppression one needs to compare a known non-plasma source with the expected plasma source. This suggests comparing AA, pA and pp collisions under the same experimental conditions. However, due to the low production rates in pp, it may not be feasible experimentally. One may assume that the peripheral collisions have no plasma (since the expeected ϵ is small) and compare the central events in AA collisions with them.

The NA50 experiment at CERN used this idea to establish J/ψ suppression in heavy ion collisions. The experiment has (1) a dimuon spectrometer, with an acceptance region (*i.e.*, the region where they can detect reliably) defined by $0 \leq y^{cm}_{\mu^+\mu^-} \leq 1$ and $|\cos\theta_{cs}| < 1/2$, (2) an electromagnetic caloriemeter to measure neutral transverse energy E_T, and (3) a zero degree caloriemeter to measure the energy E_{ZDC}. The two caloriemeters together can provide a reliable information on how central or peripheral the collision is (*i.e.*, how small or large the impact parameter is) and one can thus figure out the corresponding cross-section.

Investigating the continuum background, they noted that (1) for the AA collisions, the DY describes the shape of their data well for $M > 3$

$$\sigma_{DY}^{Pb-Pb} \propto (A_{pb})^2 \quad \text{or,} \quad \sigma_{DY}^{A\cdot B} \propto A \cdot B$$

and (2) dimuons come below that mass scale from open charm as well:

$$gg \to c\bar{c} \to D\bar{D} \to \mu^-\mu^+ + K\bar{K} + \nu\bar{\nu}.$$

An enhanced contribution from charm seems necessary. Their observation on DY is in agreement with other experiments as well. In the case of the peak J/ψ, it was observed that the total production cross section falls as $(A \cdot B)^\alpha$, with $\alpha = 0.91$. Again this has been observed by other experiments as well. Thus J/ψ seems to be always suppressed in heavy ion collisions. It seems natural to ask whether pQCD can explain it. For both the models discussed above in subsection 5.6.1 the most likely answer is no. Indeed, as in case of DY, one should naively expect it to be also $\sigma_{J/\psi}^{AB} \alpha A \cdot B$. One may expect some modifications due to the so-called EMC effect. which refers to the observed modification of the structure function of a quark inside a nucleus compared to a free proton. Depending on the average momentum fraction x for J/ψ production, one expects a suppression. Unfortunately, the gluon density in A is known very poorly. Nevertheless, the $\sigma_{AB}^{J/\psi}$ seems unlikely to decrease as rapidly in pQCD as observed. This failure could be due to a problem with pQCD description at $M = M_{J/\psi}$ scale which is too low. It may be that the nuclear size comes into play and breaks the simple hadronization models used earlier.

A currently accepted picture for explaining the suppression is as follows. The $c\bar{c}$ pair formed in nucleus collides with nucleons inside and breaks. One thus talks of a *pre-resonant J/ψ* absorption by nucleons in the nucleus. In a Glauber-like picture the decrease in the J/ψ cross section is then accounted for by a free parameter, σ_{abs}, which is the J/ψ-absorption cross section for the nucleon.

$$
\begin{aligned}
\sigma_{hA \to J/\psi} &= \int d^2b\, dz\, \sigma_{hp \to J/\psi} \exp\left(-\int_z^\infty dz'\sigma_{abs}\rho(b, z')\right) \\
&\simeq \sigma_{hp \to J/\psi} A e^{-L(A)/\lambda}, \qquad \text{if} \quad \lambda = \frac{1}{\rho\sigma_{abs}} \gg R \\
&\simeq \sigma_{hp \to J/\psi} \cdot A^\alpha, \qquad \alpha < 1.
\end{aligned}
$$

This is generalized to AB by estimating the path length L for those collisions. Note that $\sigma_{abs}^{J/\psi N}$ is *not* an experimentally measured or even a measurable quantity. Nor can it be computed. It appears as a free parameter in the fitting procedure.

It turns out that the above picture does account for the suppression observed in most hadron-nucleus or nucleus-nucleus collisions but *overestimates* the J/ψ cross section in Pb-Pb collisions compared to the NA50 results. This anamolous suppression of J/ψ in Pb-Pb collision has been interpreted by many as a signal that QGP formed in those collisions. However, some authors have had alternative pictures to explain the data while some have emphasised the suppression itself to be not so significant statistically. Indeed, for the total cross section it seems to be acceptable at a 95% confidence level.

5.6.4   Dileptons and photons

Some of the other widely regarded signals for QGP are direct photons and dileptons. Since the strongest interaction they exhibit are electromagnetic, one expects them to escape the quark-gluon plasma immediately thanks to their small cross sections. A primary difficulty with them is, however, that all stages of collisions contribute. Thus information of various stages appears in an integrated form in their distributions. Moreover, their spectrum turns out to be merely an indicator of the temperature of the plasma. They act as a good thermomemter but not as a phase identifier.

Technically, one needs the informations on details of space-time evolution of QGP and therefore its equation of state and the dynamics of phase transition. There are theoretical problems in calculating the distributions as well since perturbation theory is used while its domain of reliability may be restricted to very high T region. Infra-red divergencies make it even more unreliable for low mass photons and/or small p_T. Deeply virtual photons with far off-shell momenta seem necessary to control the infra-red problems.

Another practical problem is the presence of many conventional sources which yield dileptons (or photons) and their dominant range, *e.g.*, (a) DY, (b) $D\bar{D}$ and (c) hadronic sources. It turns out that the thermal rate is basically $\propto \exp(-M_{\ell\ell}/T)$. Thus it dies out quickly with the mass of the pair. On the other hand, in the low mass region it may be swamped by usual hadronic decays. It is not clear that any window in M exists for thermal effects to dominate with certainty.

5.7   Summary

Let us summarize the salient points.

1. Many simple models of hadrons, like the Bag model, suggest a phase transition to a new state of matter, the quark-gluon plasma (QGP).

2. Properties of QGP can be derived from first principles since we know QCD is the theory of strong interaction.

3. The grand-canonical partition function

$$\mathcal{Z} = \text{Tr} \ \exp\left(-\left[\hat{H} - \mu\hat{N}\right]/T\right)$$

can be cast as a functional integral

$$\mathcal{Z} = \int_{b.c.} \mathcal{D}A_\mu \mathcal{D}\psi \mathcal{D}\bar{\psi} e^{-S_{QCD}}.$$

This is true for any field theory, *e.g.*, the electroweak theory. Therefore the formalism we developed can be used for any field theory at finite T/μ.

4. The best calculational tool for QGP from QCD is the formulation of QCD on a discrete space-time lattice. It *predicts* QGP phase transition at $T \sim 150$–170 MeV. The necessary energy density is predicted to be ~ 1–3 GeV/fm^3. Furthermore, the QGP phase is predicted to be non-perturbative for $T/Tc \sim 1$–10.

5. The distributions and yields of soft hadrons are consistent with the freezeout idea in Bjorken scenario μ-T phase diagram from experiments.

6. The production of J/ψ is suppressed in AA collisions and even, perhaps anomalously, in the Pb-Pb collision.

The big question is whether there is sufficient evidence to the claim that QGP has been seen to be created in the laboratory and whether we need the next generation RHIC or LHC experiments. It seems that one has at best only a circumstantial evidence. Cleaner and stronger evidence is needed to make unequivocal claims. This needs higher collision energy $\sqrt{s}$, *i.e.*, an experiment like RHIC or even LHC.

A better proof may also come by investigating the functional dependence of the signals and, in particular, the hard probes on $\sqrt{s}$.

Let us end by a list of things that need to be done:

- Examine alternate explanations for observations and see if they can be ruled out. *E.g.*, $e^{-m_T/T}$ may be possible in string fragmentation or J/ψ may be suppressed by co-movers. As long as these such alternatives remain viable, a strong case for the existence of quark-gluon plasma is not tenable.

- One needs to establish (both theoretically and experimentally) that equilibriation does occur. Moreover, the estimates for ϵ, τ_{eq} etc., which go into further analysis of signals, are well understood and cross-checked.

- There is as yet no 'smoking gun' for the quark-gluon plasma. Almost all signals, even when they have been experimentally observed, seem to have alternative conventional (*i.e.*, not involving QGP) explanations. This raises the question: Can it not be devised at all? It will be nice to answer this question one way or the other.

Bibliography

[1] **Quark Gluon Plasma:**

[1a] J. Cleymans, R.V. Gavai, and E. Suhonen, Phys. Rept. **130**, 217 (1986).

[1b] B. Muller, Physics of the Quark-Gluon Plasma, Springer-Verlag (1985).

[1c] R.C. Hwa, Quark Gluon Plasma, World Scientific (1990).

[1d] R.C. Hwa, Quark Gluon Plasma 2, World Scientific (1995).

[2] **Finite Temperature Field Theory:**

[2a] M. Le Bellac, Thermal Field Theory, Cambridge University Press (1996).

[2b] J.I. Kapusta, Finite-Temperature Field Theory, Cambridge University Press (1989).

[3] **Lattice Field Theory:**

[3a] M. Creutz, Quarks, Gluons and Lattices, Cambridge University Press (1985).

[3b] H.J. Rothe, Lattice Gauge Theories: An Introduction, World Scientific (1998).

[3c] I. Montvay and G. Munster, Quantum Fields on a Lattice, Cambridge University Press (1997).

[3d] M. Creutz, Quantum Fields on the Computer, World Scientific (1992).

[4] **Grassmann Calculus:**

[4a] F.A. Berezin, The Method of Second Quantization, Academic Press (1966).

[5] **Perturbation Theory at High Temperature:**

[5a] P. Arnold and C.-x. Zhai, Phys. Rev. **D51**, 1906 (1995); e-Print Archive: hep-ph/9410360.

[5b] C.-x. Zhai and B. Kastening, Phys. Rev. **D52**, 7232 (1995); e-Print Archive: hep-ph/9507380.

[6] **Free Energy and Perturbation Theory:**

[6a] E. Braaten and A. Nieto, Phys. Rev. **D53**, 3421 (1996); e-Print Archive: hep-ph/9510408.

[7] **Advanced Reading Material:**
Proceedings of International Symposium on Lattice Field Theory,

[7a] Lattice 2000: Nucl. Phys. Proc. Suppl. **94** (2001).

[7b] Lattice 2001: Nucl. Phys. Proc. Suppl. **106** (2002).

[7c] Lattice 2002: Nucl. Phys. Proc. Suppl. **119** (2003).

[7d] Lattice 2003: To appear in Nucl. Phys. Proc. Suppl. (2004).

Proceedings of the International Conferences on Ultra-relativistic Nucleus-Nucleus Collisions,

[7e] Quark Matter 1999: Nucl. Phys. **A661** (1999).

[7f] Quark Matter 2001: Nucl. Phys. **A698** (2002).

[7g] Quark Matter 2002: Nucl. Phys. **A715** (2003).

[7h] Quark Matter 2004: To appear in J. Phys. **G**.

Chapter 6

$N = 1$ Supersymmetric Gauge Theories

Debashis Ghoshal

6.1 Introduction & the supersymmetry algebra

At least at a non-technical level, all of us have heard that super-symmetry (called *susy* for short), is a symmetry that mixes bosonic and fermionic degrees of freedom in a dynamical system. We will make this notion more precise during the course of these lectures; however before getting into that let us spend a few minutes to recall the motivation for the exercise we are going to undertake.

Of course the most obvious reason is that there have been some progress in understanding the dynamics of supersymmetric gauge theories. Any symmetry gives us a useful handle in analysing the behaviour of a physical system, and supersymmetry is no exception. Indeed it turns out that supersymmetry imposes powerful constraints in the way physical system can behave and consequently makes such system accessible beyond the domain of perturbation theory.

A more physical motivation —although admittedly biased by the presently held paradigm in high energyb physics—is that the interactions of elementary particles well below the Planck scale is described by a local quantum field theory that is approximately supersymmetric. Also it is a non-abelian gauge theory. This fact has been established for the electroweak sector of the standard model for a long time; while a more recent evidence for the role of gauge theory in strong interactions is the announcement of evidence in favour of quark-gluon plasma. Special behaviour of supersymmetric gauge theories could hopefully therefore be directly reflected in the physics of elementary particles.

Finally, by the end of the decade, we shall know whether supersymmetry operates in nature, at least in the most expected way.

We shall not say anything more about applications of supersymmetric gauge theories in elementary particle physics. Some aspects are discussed in the other lectures of this volume.

A word about references. For the formalism of supersymmetry I have drawn heavily from the review article by Lykken[1] and the classic text by Wess and Bagger[2]. In the later parts dealing with applications and the more modern developments I have used the reviews by Peskin[3] and Seiberg[4]. All these contain much more than what I shall be able to present in nine lectures and could be used for further reading. Throughout these lectures we shall use Weyl spinors. It is also possible to formulate supersymmetry using Majorana spinors, as is done *e.g.* in Refs.[5, 6]. Lastly I have not attempted to be cite the original references to the literature in many cases. Some of the reviews in the bibliography may be consulted for this purpose.

These notes are based on lectures given at the XV SERC School on Theoretical High Energy Physics held in Kolkata and at the VIII Winter School in High Energy Physics held in Seoul, Korea. I would like to thank the organisers for the kind hospitality and the participants for their enthusiastic involvement. I am grateful to Dileep Jatkar, Sunil Mukhi, Probir Roy, Tapobrata Sarkar and Ashoke Sen for many useful discussions.

Let us begin with supersymmetry.

The idea was first proposed by Golfand and Likhtman[7] in 1971. Somehow it did not gain popularity, nor was it widely known perhaps, until the 1974 paper of Wess and Zumino[8]. These authors constructed a field theory action that has a remarkable new kind of symmetry: it is invariant under infinitesimal variation of bosons (respectively fermions) that is proportional to fermions (bosons). Schematically

$$\delta_\xi \phi_B \sim \xi \psi_F,$$
$$\delta_\xi \psi_F \sim \xi \partial \phi_B,$$

where, ξ is an infinitesimal fermionic parameter. That ξ is fermionic follows from matching spin and statistics of two sides of the above equations. Also we see from the first equation that ξ has mass dimension $-1/2$, which brings in the derivative of the bosonic field ϕ_B in the second. Thus already from dimensional analysis we see that such a symmetry must mix with spacetime symmetries — translation in the above. So supersymmetry is a 'spacetime symmetry' as opposed to internal symmetries that do not mix with spacetime transformations. Indeed Gol'fand and Likhtman begin by asking whether it is possible to extend the algebra of spacetime symmetries such that the Poincaré algebra (consisting of spacetime translations, rotations and boosts), is a proper subalgebra of the extended symmetry.

Before we go ahead and start writing equations, let us remind ourselves about transformation properties of spinors, *i.e.* how they are defined, since they will play a crucial role in our consideration. This will also help set up our notation and convention.

The world we live in has three space and one time directions. It is (locally) flat[1] and isotropic. Symmetries of this spacetime are

- Translations in space and time directions — infinitesimal translations generated by P_μ, $\mu = 0, 1, 2, 3$.

- Lorentz transformations (rotations and boosts) — infinitesimal Lorentz transformations generated by $M_{\mu\nu}$. ($M_{\mu\nu} = -M_{\nu\mu}$ are antisymmetric.)

---

[1] The effect of gravity is negligible at energies far below the Planck scale — in the usual domain of elementary particle physics.

They satisfy the following algebra (Poincaré algebra)

$$
\begin{aligned}
[P_\mu, P_\nu] &= 0, \\
[M_{\mu\nu}, P_\lambda] &= i\left(\eta_{\nu\lambda} P_\mu - \eta_{\mu\lambda} P_\nu\right), \\
[M_{\mu\nu}, M_{\lambda\sigma}] &= i\left(\eta_{\nu\lambda} M_{\mu\sigma} - \eta_{\nu\sigma} M_{\mu\lambda} + \eta_{\mu\sigma} M_{\nu\lambda} - \eta_{\mu\lambda} M_{\nu\sigma}\right).
\end{aligned}
\tag{6.1}
$$

Our convention for the metric is the standard one in particle physics: $\|\eta_{\mu\nu}\| = \mathrm{diag}(+1, -1, -1, -1)$.

Quantum fields (and elementary particles described by the excitation of these fields) transform covariantly under Lorentz transformations. A trivial example is a scalar field ϕ

$$
\phi(x) \to \phi'(x') = \phi(x)
$$

More precisely, as

$$
\begin{aligned}
x^\mu &\to x'^\mu = \Lambda^\mu{}_\nu x^\nu, \\
\phi(x) &\to \phi'(x) = \phi(\Lambda^{-1} x).
\end{aligned}
$$

In the above, Λ is a finite Lorentz transformation, and we have adopted the so called *active* point of view.

This is clearly the simplest possible behaviour of a field which has just one 'component'. For a multi-component field, there could be mixing between its various components. For example, for a vector V_μ

$$
V^\mu(x) \to V'^\mu(x) = \Lambda^\mu{}_\nu V^\nu(\Lambda^{-1} x).
$$

This is the defining relation of a vector. In its infinitesimal form, $\Lambda^\mu{}_\nu = \delta^\mu_\nu - i\omega^{\rho\sigma}(M_{\rho\sigma})^\mu{}_\nu$, where

$$
(M_{\rho\sigma})^\mu{}_\nu = i\delta^\mu_\rho \eta_{\sigma\nu} - i\delta^\mu_\sigma \eta_{\rho\nu}
\tag{6.2}
$$

is the matrix representation of the Lorentz generators on vectors[2], and ω's are infinitesimal angles and velocities parametrising rotations and boosts. The commutation relations (6.1) are equivalent to the statements that P_μ is a vector operator, and that $M_{\mu\nu}$ defines a rank 2 anti-symmetric tensor.

In order to define a spinor, let us do the following.

---

[2] On scalars $M_{\mu\nu} = 0$, *i.e.* $\Lambda = \mathbf{1}$.

Exercise: *Define the infinitesimal generators of*

$$\text{rotations} \quad L_i \;=\; \frac{1}{2}\epsilon_{ijk}M_{jk}, \quad \text{and}$$
$$\text{boosts} \quad K_i \;=\; M_{0i}; \quad i,j = 1,2,3. \tag{6.3}$$

Express the commutators between $M_{\mu\nu}$ in terms of the generators L_i's and K_i's. Show that the combinations

$$\mathbf{J}^{\pm} = \frac{1}{2}\left(\mathbf{L} \pm i\mathbf{K}\right) \tag{6.4}$$

commute with each other and separately satisfy the angular momentum algebra:

$$\begin{aligned}
[J_i^+, J_j^+] &= i\epsilon_{ijk}J_k^+, \\
[J_i^-, J_j^-] &= i\epsilon_{ijk}J_k^-, \\
[J_i^+, J_j^-] &= 0.
\end{aligned} \tag{6.5}$$

This exercise shows that the Lorentz algebra is (almost) a product of two independent angular momentum algebras. Transformation properties of fields which transform covariantly[3] under Lorentz algebra are determined by their behaviour under the two angular momentum algebras.

Recall that the transformation properties of fields/states under rotation in three dimensions, (that is representations of angular momentum), are labelled by *spin j*, where $j = 0, \frac{1}{2}, 1, \frac{3}{2}, \cdots$ can take any half-integer value. Lorentz transformation properties of fields are therefore labelled by a pair of half-integers (j_+, j_-), where $j_{\pm} = 0, \frac{1}{2}, 1, \frac{3}{2}, \cdots$ independently. For example, $(0,0)$ is the scalar representation. It corresponds to the choice $\mathbf{J}^{\pm} = \mathbf{0}$.

If we take $J_i^+ = \frac{1}{2}\sigma_i$, ($\sigma_i, i = 1,2,3$ are the three Pauli matrices); and $\mathbf{J}^- = \mathbf{0}$, the corresponding field transforms in the $(\frac{1}{2},0)$ representation. This two-component field is called a *left chirality*

---

[3]Fields that transform covariantly are said to be in a *representation* of the algebra.

spinor. We shall label this as $\psi_\alpha = \begin{pmatrix} \psi_1 \\ \psi_2 \end{pmatrix}$. Under a Lorentz transformation

$$\psi_\alpha \rightarrow \psi'_\alpha = \left(\delta_\alpha^\beta - i\omega^{\rho\sigma}(M_{\rho\sigma})_\alpha{}^\beta \right) \psi_\beta. \tag{6.6}$$

On the other hand, if we choose $\mathbf{J}^+ = \mathbf{0}$ and $J_i^- = \frac{1}{2}\sigma_i$, the field transforms as $(0, \frac{1}{2})$ representation. Again we have a two-component field called a *right chirality spinor*. We shall adopt a convention in which this field is denoted by $\bar{\chi}^{\dot\alpha} = \begin{pmatrix} \bar{\chi}^{\dot 1} \\ \bar{\chi}^{\dot 2} \end{pmatrix}$. Under a Lorentz transformation

$$\bar{\chi}^{\dot\alpha} \rightarrow \bar{\chi}'^{\dot\alpha} = \left(\delta_{\dot\beta}^{\dot\alpha} - i\omega^{\rho\sigma}(M_{\rho\sigma})^{\dot\alpha}{}_{\dot\beta} \right) \bar{\chi}^{\dot\beta}. \tag{6.7}$$

The left and right chirality spinor representations are complex as is evident from the definition (6.4) of the generators.

One may now work out, with the help of the above exercise, that for a left chirality spinor ψ_α (respectively right chirality spinor $\bar{\chi}^{\dot\alpha}$), the Lorentz generators are given by the following matrices

$$(M_{\mu\nu})_\alpha{}^\beta \equiv (\sigma_{\mu\nu})_\alpha{}^\beta = \frac{i}{4} \left[(\sigma_\mu)_{\alpha\dot\gamma}(\bar{\sigma}_\nu)^{\dot\gamma\beta} - (\sigma_\nu)_{\alpha\dot\gamma}(\bar{\sigma}_\mu)^{\dot\gamma\beta} \right],$$

$$(M_{\mu\nu})^{\dot\alpha}{}_{\dot\beta} \equiv (\bar{\sigma}_{\mu\nu})^{\dot\alpha}{}_{\dot\beta} = \frac{i}{4} \left[(\bar{\sigma}_\mu)^{\dot\alpha\gamma}(\sigma_\nu)_{\gamma\dot\beta} - (\bar{\sigma}_\nu)^{\dot\alpha\gamma}(\sigma_\mu)_{\gamma\dot\beta} \right]. \tag{6.8}$$

In the above we have introduced the notation

$$\begin{aligned} \sigma^\mu = \bar{\sigma}_\mu &= (\mathbf{1}, \, \vec{\sigma}) \\ \bar{\sigma}^\mu = \sigma_\mu &= (\mathbf{1}, -\vec{\sigma}), \end{aligned}$$

where $\mathbf{1}$ is the 2×2 identity matrix.

The more familiar Dirac spinor is made up of one left and one right chirality spinor

$$\Psi_D = \begin{pmatrix} \psi_\alpha \\ \bar{\chi}^{\dot\beta} \end{pmatrix}.$$

On a Dirac spinor the Lorentz generators take the form $M_{\mu\nu} = \frac{1}{4}[\gamma_\mu, \gamma_\nu]$, where

$$\gamma^\mu = \begin{pmatrix} 0 & \sigma^\mu \\ \bar{\sigma}^\mu & 0 \end{pmatrix}.$$

The nomenclature left- and right-chirality spinors used above may now be related to the familiar notion of chirality.

Notice the index structure of the matrices σ_μ and $\bar{\sigma}_\mu$. The former has undotted-dotted indices while the latter has dotted-undotted ones.

Exercise: *Write the Clifford algebra*

$$\{\gamma^\mu, \gamma^\nu\} = 2\eta^{\mu\nu}$$

in terms of the σ^μ and $\bar{\sigma}^\mu$ matrices.

Exercise: *Show that*

$$(\sigma_{\mu\nu})^\dagger = \bar{\sigma}_{\mu\nu}$$

and

$$\begin{aligned}
\sigma_{\mu\nu} &= \frac{i}{2}\epsilon_{\mu\nu}{}^{\lambda\rho}\sigma_{\lambda\rho}, \\
\bar{\sigma}_{\mu\nu} &= -\frac{i}{2}\epsilon_{\mu\nu}{}^{\lambda\rho}\bar{\sigma}_{\lambda\rho},
\end{aligned} \tag{6.9}$$

where, $\epsilon_{0123} = +1$ in our convention. The Eqs.(6.9) mean that the rank 2 antisymmetric tensor $\sigma_{\mu\nu}$ ($\bar{\sigma}_{\mu\nu}$ respectively) is (anti-)self-dual.

A Lorentz vector transforms in the $(\frac{1}{2}, \frac{1}{2})$ representation. Therefore one can make a vector by combining left and right chirality spinors. In other words, a vector may be thought to have two spinor indices, one undotted (left type) and one dotted (right type). The transition to this description from the more familiar one is done with the help of the σ^μ matrices, (which may be thought of as Clebsch-Gordon coefficients):

$$V_\mu \rightarrow V_{\alpha\dot{\beta}} = V_\mu(\sigma^\mu)_{\alpha\dot{\beta}}.$$

Exercise: *Since a vector index can be traded with a pair of spinor indices, for a second rank tensor $T_{\mu\nu}$ we may define*

$$T_{\mu\nu} \rightarrow T_{\alpha\beta\dot\alpha\dot\beta} = T_{\mu\nu}(\sigma^\mu)_{\alpha\dot\alpha}(\sigma^\nu)_{\beta\dot\beta}.$$

Breaking the above in symmetric and antisymmetric pieces one has

$$\begin{aligned}
T_{\alpha\beta\dot\alpha\dot\beta} &= T_{(\alpha\beta)[\dot\alpha\dot\beta]} + T_{[\alpha\beta](\dot\alpha\dot\beta)} + T_{[\alpha\beta][\dot\alpha\dot\beta]} + T_{(\alpha\beta)(\dot\alpha\dot\beta)} \\
&\equiv \epsilon_{\dot\alpha\dot\beta}T_{(\alpha\beta)} + \epsilon_{\alpha\beta}T_{(\dot\alpha\dot\beta)} + \epsilon_{\alpha\beta}\epsilon_{\dot\alpha\dot\beta}T + T_{(\alpha\beta)(\dot\alpha\dot\beta)},
\end{aligned}$$

where, $T_{(\dot\alpha\dot\beta)} = -\frac{1}{2}\epsilon^{\alpha\beta}T_{\alpha\beta\dot\alpha\dot\beta}$, etc. Show that the first (respectively second) term on the RHS above correspond to antisymmetric (anti-)self-dual part of the the tensor $T_{\mu\nu}$, while last term is the traceless symmetric part and the third term is the trace.

Show that from the anti-symmetric Lorentz generators $M_{\mu\nu}$, we get two sets of tensors $M_{\alpha\beta}$ and $\bar{M}_{\dot\alpha\dot\beta}$ in terms of which the commutation relations read as follows:

$$[M_{\alpha\beta}, M_{\gamma\delta}] = \frac{1}{2}\left(\epsilon_{\alpha\gamma}M_{\beta\delta} + \epsilon_{\alpha\delta}M_{\beta\gamma} + \epsilon_{\beta\gamma}M_{\alpha\delta} + \epsilon_{\beta\delta}M_{\alpha\gamma}\right),$$

and similarly for $[\bar{M}_{\dot\alpha,\dot\beta}, \bar{M}_{\dot\gamma,\dot\delta}]$ while $[M, \bar{M}] = 0$. Also show that

$$M_{\alpha\beta}\psi_\gamma = \frac{1}{2}\left(\epsilon_{\gamma\alpha}\psi_\beta + \epsilon_{\gamma\beta}\psi_\alpha\right),$$

and a similar relation for $\bar{M}\bar\psi$.

There is one last thing we need to do before we get back to supersymmetry. Recall that one can define tensors by taking products of vectors. These have multiple indices. Similarly, one can define 'spinor-tensors' that have multiple spinor indices and transform like products of spinors. Indeed, the above 'redefinition' of vector is such an example. Now consider the tensor $\epsilon_{\alpha\beta}$,

$$||\epsilon_{\alpha\beta}|| = -\,||\epsilon^{\alpha\beta}|| = \begin{pmatrix} 0 & -1 \\ 1 & 0 \end{pmatrix}, \tag{6.10}$$

in our convention.

> **Exercise:** *Show that $\epsilon_{\alpha\beta}$ is a (numerically) invariant tensor under Lorentz transformation.*

With the help of this ϵ-tensor, we can write the transpose of the spinor ψ_α

$$\left(\psi^T\right)_\alpha \equiv \psi^\alpha,$$

by 'raising the index' as:

$$\psi^\alpha = \epsilon^{\alpha\beta}\psi_\beta \qquad \Rightarrow \qquad \psi_\alpha = \epsilon_{\alpha\beta}\psi^\beta = -\psi^\beta\epsilon_{\beta\alpha}. \tag{6.11}$$

Notice that the combination $\psi^T\chi = \psi^\beta\chi_\beta = \epsilon^{\beta\alpha}\psi_\alpha\chi_\beta$ is Lorentz invariant. herefore $\epsilon^{\alpha\beta}$ behaves like the 'metric' for the left-chirality spinors.

Similarly one can define the 'metric' $\epsilon_{\dot\alpha\dot\beta}$ on the dotted spinors[4], and lower/raise indices by

$$\bar\psi_{\dot\alpha} = \epsilon_{\dot\alpha\dot\beta}\bar\psi^{\dot\beta} \qquad \Rightarrow \qquad \bar\psi^{\dot\alpha} = \epsilon^{\dot\alpha\dot\beta}\bar\psi_{\dot\beta} = -\bar\psi_{\dot\beta}\epsilon^{\dot\beta\dot\alpha}. \tag{6.13}$$

Notice that due to the antisymmetry of the spinors under exchange, one has to be careful in ordering them while contracting indices. In the convention we shall adopt, undotted indices are contracted from NW to SE, *i.e.* in the $\searrow$ direction,

$$\psi\chi = \psi^\alpha\chi_\alpha = -\chi_\alpha\psi^\alpha;$$

and dotted indices from SW to NE, *i.e.* in the $\nearrow$ direction

$$\bar\psi\bar\chi = \bar\psi_{\dot\alpha}\bar\chi^{\dot\alpha} = -\bar\chi^{\dot\alpha}\bar\psi_{\dot\alpha}$$

respectively.

---

[4]In our convention,

$$||\epsilon_{\dot\alpha\dot\beta}|| = -\,||\epsilon^{\dot\alpha\dot\beta}|| = \begin{pmatrix} 0 & -1 \\ 1 & 0 \end{pmatrix}, \tag{6.12}$$

In defining supersymmetry, we augment the infinitesimal generators of the Poincaré algebra (P_μ and $M_{\mu\nu}$), by the fermionic generators

$$Q_\alpha, \qquad \alpha = 1, 2,$$
$$\bar{Q}^{\dot\alpha}, \qquad \dot\alpha = 1, 2;$$

that is by a left-chirality spinor Q_α and its hermitian conjugate

$$(Q_\alpha)^\dagger = \bar{Q}_{\dot\alpha}.$$

These generators obey the following (anti-)commutation relations.

$$
\begin{aligned}
[Q_\alpha, P_\mu] = [\bar{Q}_{\dot\alpha}, P_\mu] &= 0, \\
[Q_\alpha, M_{\mu\nu}] &= \frac{1}{2}(\sigma_{\mu\nu})_\alpha{}^\beta Q_\beta, \\
[\bar{Q}_{\dot\alpha}, M_{\mu\nu}] &= \frac{1}{2}\bar{Q}_{\dot\beta}(\bar{\sigma}_{\mu\nu})^{\dot\beta}{}_{\dot\alpha}, \qquad (6.14)\\
\{Q_\alpha, Q_\beta\} = \{\bar{Q}_{\dot\alpha}, \bar{Q}_{\dot\beta}\} &= 0, \\
\{Q_\alpha, \bar{Q}_{\dot\beta}\} &= 2\sigma^\mu{}_{\alpha\dot\beta} P_\mu.
\end{aligned}
$$

In the above, the second and third line simply define left- and right-chirality spinors. So the last three equations are really the new relations. The last of these states that the effect of two successive supersymmetry transformations is the same as that of a spacetime translation. Recall that we had a glimpse of this fact earlier from our dimensional consideration.

Notice that, in conformity with the spin-statistics theorem, the supersymmetry generators anticommute.

There are a few things that a standard course on supersymmetry would have discussed in more detail. We shall just gloss over them.

- The first of this is the fact there is not much option in extending the Poincaré algebra. In 1967 Coleman and Mandula proved a theorem that may roughly be stated as follows[9]:

If the infinitesimal generators of the symmetries of a quantum field theory are bosonic, *i.e.* they obey commutation relations, then under some reasonable physical assumption, the corresponding symmetry algebra is a product of the Poincaré algebra and an internal symmetry algebra.

Haag, Łopuszanski and Sohnius[10] showed that if in addition, anticommuting spinor generators are allowed, (*extended*) supersymmetry is the only possible generalisation of the Poincaré algebra.

- In extended supersymmetry there are N sets of spinor generators Q_α^A, $\bar{Q}_{A\dot\alpha}$, $A = 1, 2, \cdots, N$; which satisfy the following modified relations

$$
\begin{aligned}
\{Q_\alpha^A, \bar{Q}_{B\dot\beta}\} &= 2\delta_B^A \sigma_{\alpha\dot\beta}^\mu P_\mu \\
\{Q_\alpha^A, Q_\beta^B\} &= \epsilon_{\alpha\beta} Z^{AB} \\
\{\bar{Q}_{A\dot\alpha}, \bar{Q}_{B\dot\beta}\} &= -\epsilon_{\dot\alpha\dot\beta} Z_{AB}^*.
\end{aligned}
\tag{6.15}
$$

The generators Z^{AB} commute with all the other generators of the extended supersymmetry algebra.

- In addition, the spinors Q^A transform in the N dimensional *i.e.* defining representation of the group of N dimensional unitary matrices U(N) (SU(4) for $N = 4$). For our case $N = 1$, there is a U(1) symmetry. The generators Q_α and $\bar{Q}_{\dot\alpha}$ may be assigned charges $+1$ and -1 respectively under this symmetry. If R is the generator of this U(1), we have

$$
\begin{aligned}
[R, Q_\alpha] &= Q_\alpha \\
[R, \bar{Q}_{\dot\alpha}] &= -\bar{Q}_{\dot\alpha},
\end{aligned}
\tag{6.16}
$$

and $[R, P_\mu] = 0$, $[R, M_{\mu\nu}] = 0$. This is a chiral symmetry, and is in general anomalous. We shall make use of this symmetry in our discussion of effective field theories.

- The supersymmetry algebra is a generalisation of a Lie algebra, and hence is constrained by (generalised) *Jacobi identities*. The structure of these identities are as follows.

$$(-1)^{\varepsilon_A \varepsilon_C} \left[\{A, B\}, C\right\} \;+\; (-1)^{\varepsilon_B \varepsilon_A} \left[\{B, C\}, A\right\}$$
$$+\; (-1)^{\varepsilon_C \varepsilon_B} \left[\{C, A\}, B\right\} = 0,$$

where, ε is 0 (respectively 1) for bosonic (fermionic) operators, and the mixed bracket notation $[\cdot, \cdot\}$ stands for an anticommutator when both the operators are fermionic and a commutator otherwise.

6.2   Representations of supersymmetry on states

We shall now discuss the irreducible representations of the supersymmetry algebra, *i.e.* a collection of bosonic and fermionic states/fields that transform covariantly under supersymmetry. First we shall discuss particle states as supersymmetry representations, and come back to the representation on (quantum) fields in the next lecture.

To begin with let us recall that particle representations of the Poincaré algebra are labelled by the eigenvalues of the following Casimir operators

- $P^2 = P_\mu P^\mu$ with eigenvalue m^2 (mass square),

- $W^2 = W_\mu W^\mu$ where $W_\mu = \frac{1}{2}\epsilon_{\mu\nu\rho\sigma}P^\nu M^{\rho\sigma}$ is the Pauli-Lubanskí vector.

Exercise: *Show that*

1. *for massive particles,* i.e. $m^2 \neq 0$, $W^2 = -m^2 j(j+1)$, *where* $j = j_+ + j_-$;

2. *for massless particles,* i.e. $m^2 = 0$, $W_\mu = \lambda P_\mu$, *where* λ *is the helicity.*

Therefore mass and spin/helicity are the quantum numbers that label particle states. In order to describe which quantum numbers label representations of supersymmetry, notice that

$$[P_\mu, Q_\alpha] = [P_\mu, \bar{Q}_{\dot\alpha}] = 0.$$

Hence P^2 commute with the new generators, and continue to be a Casimir of the enlarged symmetry. However,

$$[M_{\mu\nu}, Q_\alpha] \neq 0 \qquad\qquad [M_{\mu\nu}, \bar{Q}_{\dot\alpha}] \neq 0,$$

and W^2 is no longer a Casimir of the supersymmetry algebra. We need to make the following modification. Define

$$\begin{aligned}
B_\mu &= W_\mu - \frac{1}{4}\,\bar{Q}_{\dot\alpha}\bar{\sigma}_\mu^{\dot\alpha\beta}Q_\beta, \\
C_{\mu\nu} &= B_\mu P_\nu - B_\nu P_\mu.
\end{aligned} \qquad (6.17)$$

Exercise: *Show that* $[C_{\mu\nu}, Q_\alpha] = 0.$

Therefore the second Casimir of the supersymmetry algebra is $C^2 = C_{\mu\nu}C^{\mu\nu}$. Supersymmetry multiplets are labelled by mass and eigenvalue of the operator C^2.

We are now ready to construct the representations of the supersymmetry algebra on particle states, *i.e.* on asymptotic on-shell physical states.

First, let us consider massive states, *i.e.* $m^2 \neq 0$. In this case, one can go to the rest frame and make the choice $P_\mu = (m, \mathbf{0})$. With this choice, we find that

$$\begin{aligned}
\{Q_\alpha, \bar{Q}_{\dot\beta}\} &= 2\sigma^\mu_{\alpha\dot\beta}P_\mu \\[4pt]
&= 2m \begin{pmatrix} 1 & 0 \\ 0 & 1 \end{pmatrix},
\end{aligned}$$

or, explicitly in terms of the components

$$\begin{aligned}
\{Q_1, \bar{Q}_{\dot1}\} &= 2m \\
\{Q_2, \bar{Q}_{\dot2}\} &= 2m \\
\{Q_1, \bar{Q}_{\dot2}\} &= 0 = \{Q_2, \bar{Q}_{\dot1}\}.
\end{aligned}$$

We have here two pairs of fermionic creation/annihilation operators.

We shall (arbitrarily) choose the dotted spinorial generators $\bar{Q}_{\dot{\alpha}}$, ($\dot{\alpha} = 1, 2$), to be the creation operators and the undotted ones annihilation operators. Further we may rescale Q's by $1/\sqrt{2m}$ to define conventionally normalised creation/annihilation operators

$$
\begin{aligned}
a_\alpha &= \frac{1}{\sqrt{2m}}\, Q_\alpha, & \alpha &= 1, 2; \\
a_\alpha^\dagger &= \frac{1}{\sqrt{2m}}\, \bar{Q}_{\dot{\alpha}}, & \dot{\alpha} &= 1, 2.
\end{aligned}
\tag{6.18}
$$

Let us define a state $|\Omega\rangle$ such that

$$
a_\alpha|\Omega\rangle = 0, \quad \text{for } \alpha = 1, 2.
$$

$|\Omega\rangle$ is a (Clifford) vacuum state with respect to the fermionic creation/annihilation operators.

What are the quantum numbers that label this state? Thanks to our discussion on Casimirs, we know the answer to this question. One quantum number is of course the mass m. To find the other one:

> **Exercise:** *Show that on massive states, in the rest frame,*
>
> $$
> B_i = - m \left(L_i - \frac{1}{4m}\bar{Q}\sigma_i Q \right) \equiv - m\tilde{L}_i
> $$
>
> *and hence,*
>
> $$
> \begin{aligned}
> C_{0i} &= - mB_i = m^2\tilde{L}_i \\
> C_{ij} &= 0.
> \end{aligned}
> $$
>
> *Therefore,* $C^2 = 2C_{0i}C^{0i} = 2m^4\tilde{L}_i\tilde{L}^i.$

It is easy to check that the generators $\tilde{L}_i$ obey the angular momentum algebra. So the eigenvalues of $\tilde{L}^2$ may be labelled by

$\tilde{j}(\tilde{j}+1)$, where $\tilde{j}$ can take any half-integral value. Actually, with our choice of $|\Omega\rangle$,

$$\tilde{L}_i|\Omega\rangle = L_i|\Omega\rangle.$$

Hence, acting on $|\Omega\rangle$, $\tilde{j} = j$ label the spin $j = j_+ + j_-$. Being an eigenstate of spin, $|\Omega\rangle$ is labelled by

$$|\Omega\rangle = |m; j, j_3\rangle, \qquad j_3 = -j, -j+1, \cdots, j-1, j.$$

The (Clifford) vacuum $|\Omega\rangle$ is $(2j+1)$-fold degenerate.

The excitations over the vacuum $|\Omega\rangle$ are defined by the fermionic creation operators $a_1^\dagger$ and $a_2^\dagger$. We have

$$\begin{array}{ccc} & |\Omega\rangle & \\ a_1^\dagger|\Omega\rangle & & a_2^\dagger|\Omega\rangle \\ & a_1^\dagger a_2^\dagger|\Omega\rangle & \end{array} \qquad (6.19)$$

i.e. a total of $4(2j+1)$ states in the supersymmetry mutiplet.

Recall that $a_\alpha^\dagger \sim \bar{Q}_{\dot\alpha}$ transforms as a $\left(0, \frac{1}{2}\right)$ spinor. In particular, (by a choice of convention), $a_1^\dagger$ (respectively $a_2^\dagger$) has L_3 eigenvalue $+\frac{1}{2}$ $(-\frac{1}{2})$. The spins of the different states in a massive supermultiplet are

$$\begin{array}{ccccc} \text{state} & |\Omega\rangle & a_1^\dagger|\Omega\rangle & a_2^\dagger|\Omega\rangle & a_1^\dagger a_2^\dagger|\Omega\rangle \\ \text{spin} & j_3 & j_3 + \frac{1}{2} & j_3 - \frac{1}{2} & j_3 \end{array} \qquad (6.20)$$

If j is an integer, the first and the last states are bosonic, and the second and third ones are fermionic. The statistics is opposite when j is a half odd integer. As an example, consider $j = 0$. There are then four states in the supersymmetry multiplet, two of these have spin $j_3 = 0$ and the other two have $j_3 = \pm\frac{1}{2}$. The spin zero states may be combined into a scalar and a pseudo-scalar while the spin half states describe the degrees of freedom of a Weyl fermion.

This matching of bosonic and fermionic degrees of freedom in a multiplet is a remarkable property of supersymmetry. We can easily prove the following

Theoerem: *Every representation of supersymmetry algebra contains an equal number of bosonic and fermionic states.*

Proof: Let us define the operator $(-1)^{N_F}$ whose eigenvalues are $+1$ on bosonic and -1 on fermionic states. By definition

$$(-1)^{N_F} Q_\alpha = - Q_\alpha (-1)^{N_F}.$$

Now taking a trace of the representation, (which we assume to be finite dimensional for it to be well defined), we have

$$\mathrm{tr}\left[(-1)^{N_F}\{Q_\alpha, \bar{Q}_{\dot\beta}\}\right] = \mathrm{tr}[(-1)^{N_F} Q_\alpha \bar{Q}_{\dot\beta}$$
$$+ (-1)^{N_F} \bar{Q}_{\dot\beta} Q_\alpha\}]$$

i.e.
$$\mathrm{tr}\left[(-1)^{N_F} P_\mu\right] = 0,$$

where we have used the supersymmetry algebra in the LHS and the cyclic property of trace and the identity involving $(-1)^{N_F}$ and Q_α in the RHS. For a fixed value of P_μ in a given multiplet, we then have, $\mathrm{tr}[(-1)^{N_F}] = 0$, which proves the assertion.

Now let us discuss the supersymmetry representation on massless states. We can choose a reference frame such that

$$P_\mu = (E, 0, 0, E)$$

Exercise: *Show that on massless states $W_0 = \lambda E$, $W_3 = \lambda E$, and hence*

$$B_0 = W_0 - \frac{1}{4}\bar{Q}Q$$

$$B_3 = W_3 + \frac{1}{4}\bar{Q}\sigma_3 Q.$$

Also that the only non-vanishing component of $C_{\mu\nu}$ is

$$C_{03} = E(B_0 - B_3) = -\frac{1}{2}E\bar{Q}_{\dot{2}}Q_2,$$

whence, $C^2 = 0$.

In our chosen basis the supersymmetry algebra takes the following form

$$\{Q_\alpha, \bar{Q}_{\dot{\beta}}\} = 4E \begin{pmatrix} 1 & 0 \\ 0 & 0 \end{pmatrix},$$

or, explicitly in components

$$\begin{aligned}
\{Q_1, \bar{Q}_{\dot{1}}\} &= 4E \\
\{Q_2, \bar{Q}_{\dot{2}}\} &= 0 \\
\{Q_1, \bar{Q}_{\dot{2}}\} &= 0 = \{Q_2, \bar{Q}_{\dot{1}}\}.
\end{aligned}$$

As in the massive case, let us define $|\Omega\rangle$ annhilated by the annihilation operators $a_1 \sim Q_1$ and $a_2 \sim Q_2$. In addition, since $\{Q_2, \bar{Q}_{\dot{2}}\} = 0$, we have

$$\langle\Omega|Q_2\bar{Q}_{\dot{2}}|\Omega\rangle = 0,$$

i.e, the excitation $\bar{Q}_{\dot{2}}|\Omega\rangle$ is a null state, or $\bar{Q}_{\dot{2}}$ is zero in the operator sense.

This leaves us with only one pair of creation/annihilation operators $a = \frac{1}{2\sqrt{E}}Q_1$ and $a^\dagger = \frac{1}{2\sqrt{E}}\bar{Q}_{\dot{1}}$, which satisfy $\{a, a^\dagger\} = 1$. The massless supersymmetry consists of the states

$$\begin{aligned}
|\Omega\rangle &\quad : \text{a non-degenerate state of helicity } \lambda \\
a^\dagger|\Omega\rangle &\quad : \text{a non-degenerate state of helicity } \lambda + \tfrac{1}{2}.
\end{aligned} \tag{6.21}$$

Notice that the massless supersymmetry multiplet is not a CPT eigenstate. One needs two pairs of irreducible massless multiplets, *i.e.* four states of helicity $\lambda, \lambda + \frac{1}{2}$ and $-\lambda, -\frac{1}{2} - \lambda$ to complete a CPT eigenstate.

We just finished discussing how certain bosonic and fermionic particle states form representations of supersymmetry. In other words, we have a set of states which transforms covariantly under supersymmetry variation. These are asymptotic on-shell states. In order to construct a quantum field theory, however, we need to know how general off-shell states form representations of supersymmetry. This can be done by considering bosonic and fermionic fields and studying their behaviour under supersymmetry variation. A far more economic and elegant approach is in terms of what will be called *superfields*. Different bosonic and fermionic fields that mix under supersymmetry transformations can be thought of as components of this single superfield. This approach is also advantageous from a practical point of view, as many properties of supersymmetry are manifest when expressed in terms of superfields. These concepts were introduced by Salam and Strathdee[11].

To do this, however, we need to make a digression to discuss the algebra and calculus of Grassmann variables. To this end, let us introduce spinor parameters θ^α, $\bar{\theta}_{\dot{\alpha}}$ $(\alpha, \dot{\alpha} = 1, 2)$ — (notice the index assignment) — which satisfy the relations

$$\begin{aligned}
\{\theta^\alpha, \theta^\beta\} &= 0, \\
\{\bar{\theta}_{\dot{\alpha}}, \bar{\theta}_{\dot{\beta}}\} &= 0, \\
\{\theta^\alpha, \bar{\theta}_{\dot{\beta}}\} &= 0,
\end{aligned} \tag{6.22}$$

as also $[x^\mu, \theta^\alpha] = 0$, $[x^\mu, \bar{\theta}_{\dot{\alpha}}] = 0$. These are anticommuting analogues of a complex variable z, and are called Grassmann numbers or variables. The pair $(\theta, \bar{\theta})$ can be taken to parametrise infinitesimal supersymmetry variation

$$\delta_{\text{susy}} = (\theta^\alpha Q^\alpha + \bar{\theta}_{\dot{\alpha}} \bar{Q}^{\dot{\alpha}}) \equiv (\theta Q + \bar{\theta} \bar{Q}). \tag{6.23}$$

Due to the anticommuting nature of the parameters $(\theta, \bar{\theta})$, the combinations $\theta Q \equiv \theta^\alpha Q_\alpha$ and $\bar{\theta} \bar{Q} \equiv \bar{\theta}_{\dot{\alpha}} \bar{Q}^{\dot{\alpha}}$ satisfy the *commutation* relations

$$\begin{aligned}
[\theta Q, \bar{\theta} \bar{Q}] &= 2(\theta \sigma^\mu \bar{\theta}) P_\mu, \\
[\theta Q, \theta Q] &= 0 = [\bar{\theta} \bar{Q}, \bar{\theta} \bar{Q}].
\end{aligned} \tag{6.24}$$

All the algebraic relations of supersymmetry are now expressed in terms of commutators. (We shall also consider the replacement $P_\mu \rightarrow - y^\mu P_\mu$, where y^μ is an infinitesimal parameter for translation.)

The infinitesimal variations may now be exponentiated to define a finite transformation

$$G(y, \theta, \bar\theta) = \exp\left\{i\left(-y^\mu P_\mu + \theta^\alpha Q_\alpha + \bar\theta_{\dot\alpha}\bar Q^{\dot\alpha}\right)\right\}.$$

Notice the dimensions of the parameters $[y] = M^{-1}$ and $[\theta] = [\bar\theta] = M^{-1/2}$.

Parenthetical comments: Actually, we should have written

$$G(y, \theta, \bar\theta) = \exp\left\{i\left(-y{\cdot}P + \theta Q + \bar\theta\bar Q\right)\right\}\exp\left\{-\frac{i}{2}\omega^{\mu\nu}M_{\mu\nu}\right\},$$

but we left out the Lorentz transformation part. It is consistent to set that part to identity. In other words, we are parametrising a coset space defined by the quotient of the super-Poincare group by its Lorentz subgroup.

Notice also that since the anti-commutators of Q, $\bar Q$ are non-zero, the following forms

$$\exp\left\{i\left(-y{\cdot}P + \theta Q\right)\right\}\exp\left\{i\bar\theta\bar Q\right\}$$
$$\exp\left\{i\left(-y{\cdot}P + \bar\theta\bar Q\right)\right\}\exp\left\{i\theta Q\right\}$$

and the one we gave earlier for G are not all equivalent. It is also consistent to work with either of the above forms and get the same results. However the explicit differential operator form for the generators Q and $\bar Q$ will be different in each case.

Notice that in the above the Grassmann parameters $(\theta^\alpha, \bar\theta_{\dot\alpha})$ appear in the same footing as the coordinates y^μ. So the full parameter space is labelled by

$$(y^\mu, \theta^\alpha, \bar\theta_{\dot\alpha}); \quad \mu = 0, \cdots, 3; \quad \alpha, \dot\alpha = 1, 2.$$

We should think of this space as 4 normal (*i.e.* bosonic) plus 4 anti-commuting (*i.e.* fermionic) extension of our familiar space-time. (The dimension of the extended space is sometimes written as (4|4).) This is called the $N = 1$ *rigid superspace*.

Just as it is advantageous to construct relativistic quantum field theory in a manifestly Lorentz covariant formalism, it is of great advantage to formulate supersymmetric theories in superspace.

6.3   Superspace & superfields

We can define functions in superspace — these are going to be the superfields — and differentiate and integrate them with respect to the coordinates $(x^\mu, \theta^\alpha, \bar\theta_{\dot\alpha})$. To explain the rules on differentiation and integration, let us consider the simpler example of a $(1|1)$ dimensional superspace. This has only two coordinates (x, θ) and $\theta^2 = 0$. Due to the nilpotence of the coordinate θ, Taylor expansion of a function $f(x, \theta)$ in superspace in terms of θ terminate after the linear term:

$$f(x, \theta) = f_0(x) + \theta f_1(x), \tag{6.25}$$

where $f_0(x)$ and $f_1(x)$ are functions of of the commuting coordinate x. Using

$$\frac{d}{d\theta}\,(\theta) = 1 \qquad\qquad \frac{d}{d\theta}\,(1) = 0, \tag{6.26}$$

it follows that

$$\frac{d}{d\theta}\,(f(x, \theta)) = f_1(x). \tag{6.27}$$

(Notice that the dimension $[d/d\theta]$ is $M^{1/2}$.)

Now since we want the integral of a total derivative to vanish, we define the following rules of integration

$$\int d\theta = 0, \qquad\qquad \int d\theta\, \theta = 1, \tag{6.28}$$

which gives

$$\int d\theta \frac{d}{d\theta} f(x, \theta) = \int d\theta\, f_1(x) = 0.$$

The integral so defined is invariant under translation of θ by an arbitrary constant ξ:

$$\int d(\theta + \xi)\, f(x, \theta + \xi) = \int d\theta\, [f_0(x) + (\theta + \xi)f_1(x)]$$

$$= \int d\theta\, \theta f_1(x)$$

$$= \int d\theta\, f(x, \theta).$$

It is a curious fact that integration and differentiation in θ are equivalent!

- $\dfrac{d}{d\theta} f(x, \theta) = f_1(x)$

- $\displaystyle\int d\theta\, f(x, \theta) = f_1(x)$

And consistent with these rules $[d\theta] = M^{1/2}$, unlike in ordinary space. Finally, we can define a delta function by

$$\delta(\theta) = \theta$$

leading to the expected result $\displaystyle\int d\theta\, \delta(\theta) = 1$.

Coming back to our $(4|4)$ dimensional superspace, we have the following rules ($\partial_\alpha \equiv \partial/\partial\theta^\alpha$ and $\bar{\partial}^{\dot\alpha} \equiv \partial/\partial\bar{\theta}_{\dot\alpha}$):

$$\begin{aligned}
\partial_\alpha \theta^\beta &= \delta_\alpha^\beta, \\
\partial_\alpha \theta_\beta &= \partial_\alpha(\epsilon_{\beta\gamma}\theta^\gamma) = -\epsilon_{\alpha\beta}, \\
\bar{\partial}^{\dot\alpha}\bar{\theta}_{\dot\beta} &= \delta_{\dot\beta}^{\dot\alpha}, \\
\bar{\partial}^{\dot\alpha}\bar{\theta}^{\dot\beta} &= -\epsilon^{\dot\alpha\dot\beta}.
\end{aligned} \tag{6.29}$$

One can also 'raise' index of ∂_α by chain rule of differentiation

$$\partial^\alpha \equiv \frac{\partial}{\partial\theta_\alpha} = \frac{\partial\theta^\beta}{\partial\theta_\alpha}\frac{\partial}{\partial\theta^\beta} = \partial^\alpha(\epsilon^{\beta\gamma}\theta_\gamma)\partial_\beta = -\epsilon^{\alpha\beta}\partial_\beta.$$

The following is a compilation of useful results that will come in handy in our subsequent calculations:

$$
\begin{aligned}
\partial_\alpha \left(\theta^\beta \theta^\gamma\right) &= \delta_\alpha^\beta \theta^\gamma - \delta_\alpha^\gamma \theta^\beta \\
\partial_\alpha (\theta\theta) &= 2\theta_\alpha \\
\bar\partial^{\dot\alpha} (\bar\theta\bar\theta) &= 2\bar\theta^{\dot\alpha} \\
\partial^2 (\theta\theta) &= 4 \\
\bar\partial^2 (\bar\theta\bar\theta) &= 4
\end{aligned}
\tag{6.30}
$$

It is straightforward to derive the above.

As for integration, we shall define the following convention

$$
\begin{aligned}
d^2\theta &= -\frac{1}{4} d\theta^\alpha d\theta^\beta \epsilon_{\alpha\beta} = \frac{1}{2} d\theta^1 d\theta^2, \\
d^2\bar\theta &= -\frac{1}{4} d\bar\theta_{\dot\alpha} d\bar\theta_{\dot\beta} \epsilon^{\dot\alpha\dot\beta} = -\frac{1}{2} d\bar\theta_{\dot 1} d\bar\theta_{\dot 2} \\
d^4\theta &\equiv d^2\theta \, d^2\bar\theta;
\end{aligned}
\tag{6.31}
$$

so that,

$$
\begin{aligned}
\int d^2\theta \, \theta\theta &= +1, \\
\int d^2\bar\theta \, \bar\theta\bar\theta &= +1.
\end{aligned}
\tag{6.32}
$$

After this long detour we get back to the transformation generated by

$$
G(x^\mu, \theta^\alpha, \bar\theta_{\dot\alpha}) = \exp\left[i(-x^\mu P_\mu + \theta^\alpha Q_\alpha + \bar\theta_{\dot\alpha} \bar Q^{\dot\alpha}) \right].
$$

This generator is unitary since $(\theta^\alpha Q_\alpha)^\dagger = \bar Q_{\dot\alpha} \bar\theta^{\dot\alpha} = \bar\theta_{\dot\alpha} \bar Q^{\dot\alpha}$. If we consider two such successive transformations, the result is

$$
\begin{aligned}
G(x,\theta,\bar\theta)\, G(y,\xi,\bar\xi) \;=\; \exp\Big[&- i(x^\mu + y^\mu - i\theta^\alpha \sigma^\mu_{\alpha\dot\beta} \bar\xi^{\dot\beta} \\
&i\xi^\alpha \sigma^\mu_{\alpha\dot\beta} \bar\theta^{\dot\beta}) P_\mu \;+\; i\,(\theta^\alpha + \xi^\alpha)\, Q_\alpha + i\,(\bar\theta_{\dot\alpha} + \bar\xi_{\dot\alpha})\, \bar Q^{\dot\alpha}\Big].
\end{aligned}
$$

Exercise: *Show the above. Hint: You will need the Baker-Campbell-Hausdorf formula*

$$e^A\, e^B \;=\; \exp\Big(A + B + \frac{1}{2!}[A,B] + \frac{1}{3!}\Big(\frac{1}{2}[[A,B],B] + \frac{1}{2}[A,[A,B]]\Big) + \cdots\Big).$$

The successive applications of superspace transformations generate the following motion is terms of the superspace coordinates

$$(x,\theta,\bar\theta) \xrightarrow{\;G(y,\xi\bar\xi)\;} (x + y + i\xi\sigma\bar\theta - \theta\sigma\bar\xi, \theta + \xi, \bar\theta + \bar\xi),$$

which is given by the following differential operators

$$y^\mu P_\mu \;=\; i\,y^\mu \frac{\partial}{\partial x^\mu},$$
$$\xi^\alpha Q_\alpha \;=\; \xi^\alpha\Big(\partial_\alpha - i\sigma^\mu_{\alpha\dot\beta}\bar\theta^{\dot\beta}\partial_\mu\Big),$$
$$\bar\xi_{\dot\alpha}\bar Q^{\dot\alpha} \;=\; \bar\xi_{\dot\alpha}\Big(-\bar\partial^{\dot\alpha} + i\bar\sigma^{\mu\,\dot\alpha\beta}\theta_\beta\partial_\mu\Big).$$

Alternatively, we may write[5],

$$P_\mu \;=\; i\,\frac{\partial}{\partial x^\mu},$$
$$Q_\alpha \;=\; \partial_\alpha - i\sigma^\mu_{\alpha\dot\beta}\bar\theta^{\dot\beta}\partial_\mu, \qquad (6.33)$$
$$\bar Q_{\dot\alpha} \;=\; \bar\partial_{\dot\alpha} - i\theta^\beta\sigma^\mu_{\beta\dot\alpha}\partial_\mu.$$

The above differential operator representation leads to the anti-commutator

$$\{Q_\alpha, \bar Q_{\dot\beta}\} = -2\sigma^\mu_{\alpha\dot\beta}P_\mu,$$

in apparent disagreement with (6.14), due to the extra minus sign. There is, however, no contradiction, as what we witness here is the difference in the active and passive points of view of symmetry transformations. The differential operator representation is in terms of superspace coordinates.

---

[5] Since Q and $\bar Q$ are not hermitian operators, we have rescaled them by factors of i, without anything going wrong.

We shall now define a general scalar superfield $\Phi(x,\theta,\bar{\theta})$ in our $(4|4)$ dimensional $N = 1$ rigid superspace. A scalar function satisfies

$$\Phi(x',\theta',\bar{\theta}') = \Phi(x,\theta,\bar{\theta}),$$

and hence

$$\delta_\xi\Phi = \left(\xi Q + \bar{\xi}\bar{Q}\right)\Phi. \tag{6.34}$$

Let us now consider the Taylor expansion of Φ in powers of θ and $\bar{\theta}$.

$$
\begin{aligned}
\Phi(x,\theta,\bar{\theta}) \;=\;& \phi(x) + \theta\psi(x) + \bar{\theta}\bar{\chi}(x) + (\theta\theta)m(x) \\
&+ (\bar{\theta}\bar{\theta})n(x) + (\theta\sigma^\mu\bar{\theta})v_\mu(x) + (\theta\theta)\bar{\theta}\bar{\lambda}(x) \\
&+ (\bar{\theta}\bar{\theta})\theta\eta(x) + (\theta\theta)(\bar{\theta}\bar{\theta})d(x).
\end{aligned}
\tag{6.35}
$$

Each term in the above expansion is a field in physical (Minkowski) spacetime. In particular

	Fields	*Type*	Bose dof	Fermi dof
•	$\phi(x), m(x)$ $n(x), d(x)$	scalars	4×2	0
•	$\psi_\alpha(x), \eta_\alpha(x)$	L-spinors	0	2×4
•	$\bar{\chi}^{\dot\alpha}(x), \bar{\lambda}^{\dot\alpha}(x)$	R-spinors	0	2×4
•	$v_\mu(x)$	vector	4×2	0

$$\tag{6.36}$$

The fields $\{\phi(x), \psi_\alpha(x), \bar{\chi}^{\dot\alpha}(x), \cdots\}$ are called the components of the superfield Φ.

Let us remark parenthetically that the above is the most general possible expansion for a scalar superfield, since *e.g.* $\theta\bar{\sigma}^\mu\theta = -\theta\sigma^\mu\bar{\theta}$ and $(\sigma^\mu\bar{\theta})_\alpha(\theta\sigma_\mu\bar{\theta}) = 2\theta_\alpha(\bar{\theta}\bar{\theta})$, etc. (See Appendix 6.11 for this type of manipulations.)

Now consider the supersymmetry variation of the scalar superfield (6.34). In terms of the component fields this leads to the following relations

$$
\begin{aligned}
\delta_\xi\phi \;&=\; \xi\psi + \bar{\xi}\bar{\psi}, \\
\delta_\xi\psi \;&=\; 2\xi m + (\sigma^\mu\bar{\xi})(v_\mu + i\partial_\mu\phi), \\
\delta_\xi\bar{\chi} \;&=\; 2\bar{\xi}n + (\bar{\sigma}^\mu\xi)(-v_\mu + i\partial_\mu\phi),
\end{aligned}
$$

$$\delta_\xi m = \bar{\xi}\bar{\lambda} - \frac{i}{2}(\partial_\mu\psi)\sigma^\mu\bar{\xi},$$

$$\delta_\xi n = \xi\eta + \frac{i}{2}\xi\sigma^\mu(\partial_\mu\bar{\psi}), \qquad (6.37)$$

$$\delta_\xi v_\mu = \xi\sigma_\mu\bar{\lambda} + \eta\sigma_\mu\bar{\xi} + \frac{i}{2}(\partial_\mu\psi)\xi - \frac{i}{2}(\partial_\mu\bar{\chi})\bar{\xi},$$

$$\delta_\xi\bar{\lambda} = 2\bar{\xi}d + i(\bar{\sigma}^\mu\xi)\partial_\mu m + \frac{i}{2}\bar{\xi}(\partial^\mu v_\mu),$$

$$\delta_\xi\eta = 2\xi d + i(\sigma^\mu\bar{\xi})\partial_\mu n - \frac{i}{2}\xi(\partial^\mu v_\mu),$$

$$\delta_\xi d = \frac{i}{2}\xi\sigma^\mu(\partial_\mu\bar{\lambda}) - \frac{i}{2}(\partial_\mu\eta)\sigma^\mu\bar{\xi}.$$

Exercise: *(i) Derive the above relations. (You will need to use the Fierz identities given in Appendix 6.11.)*
(ii) Work out

$$\left(\delta_{\xi_1}\delta_{\xi_2} - \delta_{\xi_2}\delta_{\xi_1}\right)\phi = -2i\left(\xi_1\sigma^\mu\xi_2 - \xi_2\sigma^\mu\xi_1\right)\partial_\mu\phi.$$

6.4 Chiral & vector superfields

The result obtained at the end of the last lecture shows that the general scalar superfield forms a basis for an off-shell linear representation of supersymmetry:

- Supersymmetry variations of component fields are proportional to each other (also involving derivatives).

- Therefore the supersymmetry algebra closes, *i.e.* supersymmetry variations involve only those fields present in the mutiplet and no others.

However, there are a large number of component fields. It turns out that the set is not the minimal one. In other words, the scalar superfield representation is *reducible*[6].

---

[6]It is not fully reducible though. That is, the scalar superfield cannot be written as a direct sum of irreducible superfields.

In an effort to reduce the number of component fields in (6.35), let us try to set one of the spinor fields, say $\bar{\chi}$ to zero. To make this consistent with supersymmetry, we should also require that its supersymmetry variation vanishes, and so on. In the end, we have to impose the following set of constraints

$$\bar{\chi}(x) = 0$$
$$v_\mu(x) = i\partial_\mu\phi(x)$$
$$n(x) = 0$$
$$\eta(x) = 0$$
$$\bar{\lambda}(x) = -\tfrac{i}{2}(\partial_\mu\psi)\sigma^\mu$$
$$d(x) = -\tfrac{1}{4}\Box\phi(x) \tag{6.38}$$

leading to a *reduced* scalar superfield

$$\Phi_R = \phi + \theta\psi + (\theta\theta)m + i(\theta\sigma^\mu\bar\theta)\partial_\mu\phi$$
$$+ \frac{i}{2}(\theta\theta)\left((\partial_\mu\psi)\sigma^\mu\bar\theta\right) - \frac{1}{4}(\theta\theta)(\bar\theta\bar\theta)\Box\phi.$$

Exercise: *Check the mutual consistency of the constraints imposed in (6.38). This demonstrates that the reduced scalar superfield Φ_R, (with fewer number of components than the general scalar superfield Φ), defines an off-shell linear representation of the supersymmetry algebra.*

Let us now define

$$y^\mu = x^\mu + i\theta^\alpha\sigma^\mu_{\alpha\dot\beta}\bar\theta^{\dot\beta}. \tag{6.39}$$

Using Taylor expansion, and some spinor identities, one can rewrite the restricted superfield Φ_R as

$$\Phi_R(y,\theta) = \phi(y) + \theta\psi(y) + (\theta\theta)m(y), \tag{6.40}$$

which shows that the restricted superfield is a function of y and θ, but has no explicit dependence on $\bar\theta$. Had it not been for the $\bar\theta$ in the definition of y in (6.39), we could have concluded that Φ_R is independent of $\bar\theta$, that is

$$\bar\partial_{\dot\alpha}\,\Phi_R(y,\theta) \overset{!}{=} 0.$$

This is of course not the case. Another problem is that

$$[\bar{\partial}_{\dot\alpha}, \xi Q] = i\xi^\beta \sigma^\mu_{\beta\dot\alpha} \partial_\mu,$$

that is, $\bar{\partial}_{\dot\alpha}$ does not commute with supersymmetry variation. Hence imposing a constraint like $\bar{\partial}_{\dot\alpha}\Phi_R = 0$ is not consistent with supersymmetry. In other words, $\bar{\partial}_{\dot\alpha}\Phi_R$ is not a superfield. Thankfully from our experience with tensor analysis, we know what to do in such a situation: define an appropriate covariant derivative to impose the constraint consistently. The covariant derivative defined as

$$\bar{D}_{\dot\alpha} = -\bar{\partial}_{\dot\alpha} - i\theta^\beta \sigma^\mu_{\beta\dot\alpha} \partial_\mu \tag{6.41}$$

leads to a consistent way to impose the constraint

$$\bar{D}_{\dot\alpha}\Phi = 0 \tag{6.42}$$

on the general scalar superfield (6.35) to restrict it to Φ_R.

> **Exercise:** *Show that $[\bar{D}_{\dot\alpha}, \xi Q] = 0$, or equivalently $\{\bar{D}_{\dot\alpha}, Q_\beta\} = 0$. Also show that $\{\bar{D}_{\dot\alpha}, \bar{Q}_{\dot\beta}\} = 0$, and $\{\bar{D}_{\dot\alpha}, \bar{D}_{\dot\beta}\} = 0$.*

> **Exercise:** *Show that $\bar{D}_{\dot\alpha} y^\mu = 0$ and $\bar{D}_{\dot\alpha}\theta^\beta = 0$.*

The last exercise shows that a superfield constrained by (6.42) is a function of y and θ only, and has no explicit dependence on $\bar{\theta}$.

Definition: A scalar superfield Φ constrained by the (spinorial) chirality condition $\bar{D}_{\dot\alpha}\Phi = 0$ is called a *chiral superfield*.

We have already found an example of a chiral superfield in (6.40):

$$\Phi(y, \theta) = \phi(y) + \sqrt{2}\theta\psi(y) + \theta\theta F(y). \tag{6.43}$$

(In the above, a scaling $\psi \to \sqrt{2}\psi$ and a change in notation $m \to F$ has been done to conform with standard notation in the literature.) A chiral superfield has the following

- complex scalar ϕ dof $= 2$ bose $[\phi] = M$,
- complex L-spinor ψ dof $= 4$ fermi $[\psi] = M^{3/2}$,
- complex scalar F dof $= 2$ bose $[F] = M^2$.

Under infinitesimal supersymmetry transformation

$$\begin{aligned}
\delta_\xi \phi &= \sqrt{2}\xi\psi \\
\delta_\xi \psi &= \sqrt{2}\xi F + \sqrt{2}i\sigma^\mu\bar{\xi}\,\partial_\mu\phi \\
\delta_\xi F &= -\sqrt{2}i\partial_\mu\psi\sigma^\mu\bar{\xi}.
\end{aligned} \qquad (6.44)$$

An important property of a chiral superfield is that the product of two chiral superfields is again a chiral superfield. This follows from the chain rule of covariant differentiation.

The notion of an anti-chiral superfield is immediate:

Definition: A scalar superfield that satisfies the condition $D_\alpha\Phi = 0$, where

$$D_\alpha = \partial_\alpha + i\sigma^\mu_{\alpha\dot\beta}\bar\theta^{\dot\beta}\partial_\mu, \qquad (6.45)$$

is an anti-chiral superfield.

> **Exercise:** *Show that $\{D_\alpha, Q_\beta\} = 0$. Also show that $\{D_\alpha, \bar{Q}_{\dot\beta}\} = 0$, and $\{D_\alpha, D_\beta\} = 0$, and finally*
>
> $$\{D_\alpha, \bar{D}_{\dot\beta}\} = -2i\sigma^\mu_{\alpha\dot\beta}\partial_\mu.$$

> **Exercise:** *Let $y^\dagger = x - i\theta\sigma\bar\theta$. Show that $D_\alpha y^{\mu\dagger} = 0$ and $D_\alpha\bar\theta^{\dot\beta} = 0$.*

An anti-chiral superfield is therefore a function of $y^\dagger$ and $\bar\theta$ only. In particular, if $\Phi(y,\theta)$ is a chiral superfield (6.43), $\Phi^\dagger(y^\dagger,\bar\theta)$ is an anti-chiral superfield

$$\Phi^\dagger(y^\dagger,\bar\theta) = \phi^*(y^\dagger) + \sqrt{2}\bar\theta\bar\psi(y^\dagger) + \bar\theta\bar\theta F^*(y^\dagger).$$

As before, the product of two anti-chiral superfields is again an anti-chiral superfield. However the product $\Phi^\dagger\Phi$ is neither a chiral nor an anti-chiral superfield. Same applies to the sum $(\Phi + \Phi^\dagger)$.

> **Exercise:** *Show that the imposing the conditions $D_\alpha\Phi = 0$ and $\bar{D}_{\dot\alpha}\Phi = 0$ simultaneously on a scalar superfield reduces it to a constant.*

Exercise: *Define*

$$\varphi = \Phi\big|_{\theta=\bar{\theta}=0}$$
$$\Psi_\alpha = D_\alpha\Phi\big|_{\theta=\bar{\theta}=0}$$
$$\mathcal{F} = D^\alpha D_\alpha\Phi\big|_{\theta=\bar{\theta}=0}.$$

Express φ, Ψ and $\mathcal{F}$ in terms of the component fields ϕ, ψ and F. Compute the transformation laws for φ, Ψ and $\mathcal{F}$ using the differential operator representation for $Q, \bar{Q}$.

Exercise: *Show that in terms of the variables $(y, \theta, \bar{\theta})$, the covariant derivatives may be written as*

$$D_\alpha = \partial_\alpha + 2i\sigma^\mu_{\alpha\dot{\beta}}\bar{\theta}^{\dot{\beta}}\frac{\partial}{\partial y^\mu},$$

$$\bar{D}^\alpha = -\partial^\alpha - 2i\bar{\theta}_{\dot{\beta}}\bar{\sigma}^{\mu\dot{\beta}\alpha}\frac{\partial}{\partial y^\mu},$$

$$\bar{D}_{\dot{\alpha}} = -\partial_{\dot{\alpha}},$$
$$\bar{D}^{\dot{\alpha}} = \partial^{\dot{\alpha}}. \tag{6.46}$$

Exercise: *Compute the expressions for Φ^2, Φ^3 and $\Phi^\dagger\Phi$ in the component field expansion.*

There is another way to reduce the number of fields in a general scalar superfield. For a scalar superfield $V(x, \theta, \bar{\theta})$, let us impose the covariant reality condition:

$$V(x, \theta, \bar{\theta}) = V^\dagger(x, \theta, \bar{\theta}). \tag{6.47}$$

In components, this leads to the following set of constraints

$$\phi = \phi^*$$
$$(\psi_\alpha)^* = \bar{\chi}^{\dot{\alpha}}$$
$$m = n^*$$
$$v_\mu = (v_\mu)^*$$
$$\eta_\alpha = (\bar{\lambda}^{\dot{\alpha}})^*$$
$$d = d^* \tag{6.48}$$

Therefore, a scalar superfield restricted by the reality condition has four real scalars, (which may be combined into two complex scalars), one real vector and two complex Weyl spinors, (or equivalently, one real Majorana spinor). There are altogether eight bosonic and eight fermionic degrees of freedom.

We can already construct an example of a real superfield trivially from a chiral superfield. If Λ is a chiral superfield, the sum $(\Lambda + \Lambda^\dagger)$ is a real superfield. In fact, this means that a real superfield is not uniquely determined. Given a real superfield $V(x, \theta, \bar{\theta})$, it is possible to construct another one

$$V(x, \theta, \bar{\theta}) \to V(x, \theta, \bar{\theta}) + \Lambda(y, \theta) + \Lambda^\dagger(y^\dagger, \bar{\theta}). \qquad (6.49)$$

This is a 'gauge freedom' in defining a real superfield. If we expand a real superfield in components

$$\begin{aligned}
V(x, \theta, \bar{\theta}) \;=\;& \rho(x) + \theta\chi(x) + \bar{\theta}\bar{\chi}(x) + \theta\theta M(x) + \bar{\theta}\bar{\theta} M^*(x) \\
&-(\theta\sigma^\mu\bar{\theta})A_\mu(x) \;+\; i\theta\theta\,\bar{\theta}\bar{\lambda}(x) - i\bar{\theta}\bar{\theta}\,\theta\lambda(x) + \frac{1}{2}\theta\theta\,\bar{\theta}\bar{\theta}\,D,
\end{aligned}$$

and use the freedom in

$$\begin{aligned}
\left(\Lambda + \Lambda^\dagger\right)(x, \theta, \bar{\theta}) \;=\;& (\phi + \phi^*) + \sqrt{2}(\theta\psi + \bar{\theta}\bar{\psi}) + \theta\theta F + \bar{\theta}\bar{\theta} F^* \\
&+ i\theta\sigma^\mu\bar{\theta}(\partial_\mu\phi - \partial_\mu\phi^*) + \frac{i}{\sqrt{2}}\theta\theta(\bar{\theta}\bar{\sigma}^\mu\partial_\mu\psi) \\
&- \frac{i}{\sqrt{2}}\bar{\theta}\bar{\theta}(\theta\sigma^\mu\partial_\mu\bar{\psi}) - \frac{1}{4}\theta\theta\,\bar{\theta}\bar{\theta}(\Box\phi + \Box\phi^*),
\end{aligned}$$

we find that

$$\begin{aligned}
\delta\rho &= \phi + \phi^* = 2\,\mathrm{Re}\,\phi \\
\delta\chi &= \sqrt{2}\psi \\
\delta M &= F \\
\delta A_\mu &= -i\partial_\mu(\phi - \phi^*) = \partial_\mu(2\,\mathrm{Im}\,\phi) \qquad (6.50) \\
\delta\lambda &= \frac{1}{\sqrt{2}}\sigma^\mu\partial_\mu\bar{\psi} \\
\delta D &= -\frac{1}{2}\Box(\phi + \phi^*) = 2\,\mathrm{Re}\,\phi.
\end{aligned}$$

Let us first point out the most interesting aspect of the 'gauge transformation' in (6.50) above:

$$\delta A_\mu(x) = \partial_\mu(2\,\mathrm{Im}\ \phi(x)) = \partial_\mu \varepsilon(x),$$

where, $\varepsilon(x)$ is a real scalar field. This is therefore the usual abelian gauge transformation for the vector field $A_\mu(x)$ in the components of the real superfield V. For this reason, a real superfield is also known as a *vector superfield*.

The more general superfield transformation (6.49) means that any superfield action invariant under the above abelian gauge transformation is also independent of several component fields of V. In particular, we may choose $\mathrm{Re}\,\phi$, F and ψ to set ρ, m and χ to zero respectively. This partially fixes the gauge freedom in $V \to V + \Lambda + \Lambda^\dagger$, and is called the *Wess-Zumino gauge*. After this gauge fixing, the vector superfield takes the form

$$V_{WZ} = -\,(\theta\sigma^\mu\bar\theta)A_\mu + i\theta\theta\,\bar\theta\bar\lambda - i\bar\theta\bar\theta\,\theta\lambda + \frac{1}{2}\theta\theta\,\bar\theta\bar\theta\,D. \qquad (6.51)$$

It should be stressed that the Wess-Zumino gauge imposes no restriction on $\mathrm{Im}\,\phi$, therefore it does *not* fix the abelian gauge freedom of the component vector field A_μ.

There are four real bosonic degrees of freedom in the component fields of V_{WZ}: $(4-1) = 3$ from the vector A_μ, and one from the real scalar D. Also there are four real fermionic ones from λ.

6.5 More on vector superfields

Another way to reformulate Wess-Zumino gauge fixing is to observe that, without any loss of generality, a vector superfield may be decomposed as

$$V = V_{WZ} + \Lambda + \Lambda^\dagger, \qquad (6.52)$$

where Λ is a chiral superfield.

Exercise: *Work out the component expansion for V_{WZ}^2 and V_{WZ}^3.*

While the Wess-Zumino condition allows one to get rid of the superfluous fields in a vector superfield and retain only the relevant ones, it is *not* invariant under supersymmetry variation. In other words, the WZ gauge condition is not covariant. It is, however, possible to give a covariant description in which only the necessary fields in a vector superfield are retained. To this end let us define the following covariant derivatives on a vector superfield.

$$
\begin{aligned}
W_\alpha &= -\frac{1}{4}\left(\bar{D}\bar{D}\right) D_\alpha V(x,\theta,\bar{\theta}) \\
&= -\frac{1}{8}\left(\bar{D}\bar{D}\right)\left(e^{-2V} D_\alpha e^{+2V}\right), \\
\bar{W}_{\dot{\alpha}} &= -\frac{1}{4}\left(DD\right)\bar{D}_{\dot{\alpha}} V(x,\theta,\bar{\theta}) \\
&= \frac{1}{8}\left(DD\right)\left(e^{+2V}\bar{D}_{\dot{\alpha}} e^{-2V}\right).
\end{aligned}
\tag{6.53}
$$

Notice that

1. The construction/definition of the W_α $(\bar{W}_{\dot{\alpha}})$ ensures that[7]

$$
\bar{D}_{\dot{\alpha}} W_\alpha = 0 \qquad\qquad \left(D_\alpha \bar{W}_{\dot{\alpha}} = 0\right), \tag{6.54}
$$

   that is, W_α (resp. $\bar{W}_{\dot{\alpha}}$) is a (anti-)chiral superfield. Since this field also carries a spinor index it is a chiral spinor superfield.

2. However, W_α is not a general chiral superfield, since it satisfies

$$
DW = \bar{D}\bar{W}. \tag{6.55}
$$

3. The fields W_α and $\bar{W}_{\dot{\alpha}}$ are both invariant under the gauge transformation (6.49).

Exercise: *Prove the two statements mentioned above.*

It is important to note that the two superfields W_α and $\bar{W}_{\dot{\alpha}}$ are invariant under the full superfield gauge transformation. Therefore even the abelian gauge invariance of the component vector

---

[7]Since the $\bar{D}$'s anticommute and has only two independent components, $\bar{D}^3 = 0$.

field A_μ (that the Wess-Zumino gauge did not fix), is no longer available. Consequently, the components of these (anti-)chiral superfields can be calculated in the Wess-Zumino gauge without any loss of generality. That is the exercise we shall do now. In order to simplify the computation, observe that being a chiral superfield, W_α is a function of y and θ only, (similarly $\bar{W}_{\dot\alpha}$ is a function of $y^\dagger$ and $\bar\theta$), with no explicit dependence on $\bar\theta$ (resp. θ).

Exercise: *Carry out this computation in detail to show that*

$$
\begin{aligned}
W_\alpha &= -i\lambda_\alpha(y) + \theta_\alpha D(y) + i\theta^\beta \sigma^{\mu\nu}_{\beta\alpha} F_{\mu\nu}(y) \\
&\quad + (\theta\theta)\sigma^\mu_{\alpha\dot\beta}\partial_\mu \bar\lambda^{\dot\beta}(y), \\
\bar{W}_{\dot\alpha} &= +i\bar\lambda_{\dot\alpha}(y^\dagger) + \bar\theta_{\dot\alpha} D(y^\dagger) - i\bar\theta_{\dot\beta}\bar\sigma^{\mu\nu\dot\beta}_{\dot\alpha} F_{\mu\nu}(y^\dagger) \\
&\quad + (\bar\theta\bar\theta)\bar\sigma^{\mu\beta}_{\dot\alpha}\partial_\mu\lambda_\beta(y^\dagger).
\end{aligned}
\tag{6.56}
$$

Hint: Start by writing the real superfield in Wess-Zumino gauge V_{WZ} as a function of $(y,\theta,\bar\theta)$ and use results from (6.46).

We see that the component fields in the spinor chiral superfield W are a left-handed spinor λ, a scalar D and the field strength $F_{\mu\nu}$. It is therefore also called a *field strength supermultiplet*. Moreover, since $\sigma^{\mu\nu}$ is self-dual (6.9), only the self-dual part of $F_{\mu\nu}$ contributes to the degrees of freedom in W.

One can generalise this result in which the vector field in the vector supermultiplet has abelian gauge invariance to one with non-abelian gauge invariance. In order to do that, however, we have to rewrite the abelian gauge condition in a way that is generalisable to the non-abelian case. Recall the Wess-Zumino gauge freedom $V \to V+\Lambda+\Lambda^\dagger$. With a scaling $\Lambda \to i\Lambda$, this is equivalent to

$$
e^V \to e^{-i\Lambda^\dagger}\, e^V\, e^{+i\Lambda}.
\tag{6.57}
$$

Let us now elevate the superfields to Lie algebra valued superfields

$$
\begin{aligned}
V &\to V_{ij} = t^a_{ij} V_a, \\
\Lambda &\to \Lambda_{ij} = t^a_{ij} \Lambda_a,
\end{aligned}
$$

where t^a are hermitian generators of some Lie algebra satisfying

$$\left[t^a, t^b \right] = i f^{abc} t^c, \tag{6.58}$$

and normalised so that

$$\mathrm{tr}\left(t^a t^b \right) = C(R)\, \delta^{ab},$$
$$C(R) = \frac{\dim R}{\dim \mathbf{g}}\, C_2(R). \tag{6.59}$$

In the above, $\dim R$ and $C_2(R)$ are the dimension and the quadratic Casimir in the representation R and $\dim \mathbf{g}$ is the dimension of the Lie algebra $\mathbf{g}$.

To first order in gauge parameter superfield Λ

$$\delta V = i(\Lambda - \Lambda^\dagger) + \frac{i}{2}\left[V, (\Lambda + \Lambda^\dagger) \right]$$
$$+ \frac{i}{12}\left[V, [V, (\Lambda - \Lambda^\dagger)] \right] + \cdots . \tag{6.60}$$

The linear term $\delta V = i(\Lambda - \Lambda^\dagger)$ still allows for a choice of WZ gauge which, as in the abelian case, does not fix the non-abelian gauge freedom of the component vector fields A_μ^a. Once we fix the WZ gauge, we have $V_{WZ}^a = (\theta \sigma \bar\theta) A_\mu^a + \cdots$, and also $(\Lambda + \Lambda^\dagger) = -2i\mathrm{Re}\,(\phi)$, $(\Lambda - \Lambda^\dagger) = 2i(\theta\sigma\bar\theta)\partial_\mu \mathrm{Im}\,(\phi)$. Therefore the third term and above in (6.60) vanish in the WZ gauge as they have too many θs or $\bar\theta$s. The relation

$$\delta V_{WZ} = i(\Lambda - \Lambda^\dagger) + \frac{i}{2}\left[V, (\Lambda + \Lambda^\dagger) \right] \tag{6.61}$$

implies the usual non-abelian gauge transformation for the component fields A_μ (non-abelian gauge field), λ (a spinor in the adjoint representation) and D (auxiliary scalar field in the adjoint representation).

As in the abelian case, one can define the constrained (anti-)chiral spinor superfield W_α (and $\bar W_{\dot\alpha}$) by

$$W_\alpha = -\frac{1}{8}\bar D^2 e^{-2V} D_\alpha e^{2V},$$
$$\bar W_{\dot\alpha} = +\frac{1}{8}D^2 e^{+2V} \bar D_{\dot\alpha} e^{-2V}. \tag{6.62}$$

These superfields transform homogeneously under gauge transformations:

$$W_\alpha \to e^{-2i\Lambda} W_\alpha e^{+2i\Lambda}, \qquad \bar{W}_{\dot\alpha} \to e^{-2i\Lambda^\dagger} W_\alpha e^{+2i\Lambda^\dagger}. \qquad (6.63)$$

Once again one can compute the component fields of W_α and $\bar{W}_{\dot\alpha}$ in the WZ gauge. Explicitly

$$W_\alpha = -\frac{1}{4}\bar{D}^2 D_\alpha V_{WZ} + \frac{1}{2}\bar{D}^2(V_{WZ} D_\alpha V_{WZ}) - \frac{1}{4}\bar{D}^2 D_\alpha V_{WZ}^2, \quad (6.64)$$

and similarly for $\bar{W}_{\dot\alpha}$. After some algebra, the result is the expected non-abelian generalisation

$$\begin{aligned}
W_\alpha &= -i\lambda_\alpha(y) + \theta_\alpha D(y) + i\theta^\beta \sigma^{\mu\nu}_{\beta\alpha} F_{\mu\nu}(y) \\
&\quad + (\theta\theta)\sigma^\mu_{\alpha\dot\beta}\nabla_\mu \bar{\lambda}^{\dot\beta}(y), \\
\bar{W}_{\dot\alpha} &= i\bar{\lambda}_{\dot\alpha}(y^\dagger) + \bar{\theta}_{\dot\alpha} D(y^\dagger) - i\bar{\theta}_{\dot\beta}\sigma^{\mu\nu\dot\beta}{}_{\dot\alpha} F_{\mu\nu}(y^\dagger) \\
&\quad - (\bar{\theta}\bar{\theta})\bar{\sigma}^{\mu\dot\beta}_{\dot\alpha}\nabla_\mu \lambda_\beta(y^\dagger),
\end{aligned} \qquad (6.65)$$

where, $F_{\mu\nu} = \partial_\mu A_\nu - \partial_\nu A_\mu + i[A_\mu, A_\nu]$ is the non-abelian field strength and ∇_μ is the gauge covariant derivative, *e.g.* $\nabla_\mu \bar{\lambda} = \partial_\mu \bar{\lambda} + i[A_\mu, \bar{\lambda}]$.

> **Exercise:** *Show the transformation property (6.63) and work out the component field expressions (6.65).*

6.6 Wess-Zumino model, supersymmetry breaking

There are some properties that we saw before, but did not take proper notice of, are the following. First, the supersymmetry variation of the $\theta\theta\bar{\theta}\bar{\theta}$, *i.e.* the highest, component of a scalar superfield

$$\delta_\xi d(x) = \frac{i}{2}\partial_\mu \left(\xi\sigma^\mu \bar{\lambda}(x) - \eta(x)\sigma^\mu \bar{\xi}\right) \qquad (6.66)$$

is a total (spacetime) derivative. Therefore

$$\int d^4x\, \delta_\xi d(x) = 0. \qquad (6.67)$$

In case of a vector superfield, the variation is

$$\delta_\xi D(x) = i\partial_\mu \left(\xi\sigma^\mu\bar\lambda(x) - \lambda(x)\sigma^\mu\bar\xi\right). \tag{6.68}$$

Therefore, for *any* vector superfield $V(x,\theta,\bar\theta)$

$$2\kappa \int d^4x \int d^4\theta\, V(x,\theta,\bar\theta), \tag{6.69}$$

where κ is a parameter, is invariant under $N = 1$ supersymmetry transformation. With this observation, we immediately have a way of constructing actions which are invariant under ($N = 1$ global) supersymmetry. As an example, let us consider the chiral superfield $\Phi(y,\theta)$, for which $\Phi^\dagger\Phi$ is a vector superfield. Therefore,

$$\int d^4x \int d^4\theta\, \Phi^\dagger\Phi \tag{6.70}$$

is a supersymmetric action.

Of course, (6.69) with V a vector superfield is also an example of a supersymmetric action. However, on dimensional ground we see that the parameter κ has mass dimension two, while the action (6.70) involving $\Phi^\dagger\Phi$ does not require a dimensionful parameter. Moreover (6.69) is linear in the component fields, while (6.70) is quadratic. Nevertheless, we shall have occassion to come back to (6.69) later.

Secondly, the supersymmetric variation of the $\theta\theta$, *i.e.* again the highest, component of a chiral superfield Φ

$$\delta_\xi F(x) = -\sqrt2 i\partial_\mu\psi(x)\sigma^\mu\bar\xi = \partial_\mu\left(-\sqrt2 i\psi(x)\sigma^\mu\bar\xi\right) \tag{6.71}$$

is a also total derivative. Therefore

$$\int d^4x \left(\int d^2\theta\Phi(y,\theta) + \int d^2\bar\theta\Phi^\dagger(y^\dagger,\bar\theta)\right) \tag{6.72}$$

is invariant under ($N = 1$ global) supersymmetry for *any* chiral superfield Φ.

Consider, for example the chiral superfield Φ^2 leading to

$$m \int d^4x \int d^2\theta\, \Phi^2(y,\theta) + \text{h.c.}, \tag{6.73}$$

where, the complex parameter m has mass dimension one. As before we can, and shall, consider the effect of the term

$$\lambda \int d^4x \int d^2\theta \, \Phi(y, \theta) + \text{h.c.}, \tag{6.74}$$

with $[\lambda] = M^2$, later in this lecture.

The advantage of the superfield method lies in the fact that an action written in terms of superfields in manifestly invariant under supersymmetry transformations. Let us consider the Wess-Zumino model defined by the lagrangian density[8]

$$\mathcal{L}_{WZ} = \int d^4\theta \, \Phi^\dagger \Phi - \int d^2\theta \left(\frac{1}{2}m\Phi^2 + \frac{1}{3}g\Phi^3 \right) + \text{h.c.} \tag{6.75}$$

When expanded in terms of the component fields, the first term above contains the canonical kinetic terms for a complex scalar field ϕ and a Weyl fermion ψ, while the others describe interactions including Yukawa interactions (see exercise in lecture 6.4):

$$\begin{aligned} \mathcal{L}_{WZ} &= \eta^{\mu\nu}(\partial_\mu\phi^*)(\partial_\nu\phi) + F^*F - \left(m\phi + g\phi^2\right)F \\ &\quad - \left(m^*\phi^* + g^*(\phi^*)^2\right)F^* - i\bar\psi\bar\sigma^\mu\partial_\mu\psi \\ &\quad + \frac{1}{2}(m\psi\psi + m^*\bar\psi\bar\psi) + g\phi\psi\psi + g^*\phi^*\bar\psi\bar\psi. \end{aligned} \tag{6.76}$$

Notice that there is no term involving the derivatives of F. It is therefore an auxiliary field with algebraic equation of motion

$$F^* - m\phi - g\phi^2 = 0. \tag{6.77}$$

This is readily solved and we use this to eliminate F from the action. After eliminating the auxiliary field, the bosonic part of the action S_{WZ} takes the form

$$\mathcal{L}_{WZ}^B = (\partial_\mu\phi^*)(\partial^\mu\phi) - \mathcal{V}(\phi), \tag{6.78}$$

---

[8]The Wess-Zumino action defines the most general unitary renormalisable supersymmetric theory for a single chiral superfield.

where the scalar potential

$$\mathcal{V}(\phi) = F^*F = |m|^2\phi^*\phi + (m^*g\phi + mg^*\phi^*)\phi^*\phi + |g|^2(\phi^*\phi)^2. \quad (6.79)$$

The potential is positive definite. Consequently, the hamiltonian and the total energy are also positive definite.

The last property is actually a general property of supersymmetric theories. We can prove this from the supersymmetry algebra (6.14). Explicitly, using

$$\begin{aligned} \{Q_1, \bar{Q}_1\} &= 2P_0 + 2P_3, \\ \{Q_2, \bar{Q}_2\} &= 2P_0 - 2P_3, \end{aligned}$$

the hamiltonian $\mathcal{H} \equiv P_0$ can be expressed as

$$\mathcal{H} = \frac{1}{4}\left(Q_1\bar{Q}_1 + \bar{Q}_1 Q_1 + Q_2\bar{Q}_2 + \bar{Q}_2 Q_2\right). \quad (6.80)$$

The expectation value for the energy density in any state $|\Psi\rangle$ is thus

$$\begin{aligned} \langle\Psi|\mathcal{H}|\Psi\rangle &= \frac{1}{4}\left(||\bar{Q}_1|\Psi\rangle||^2 + ||Q_1|\Psi\rangle||^2 + ||\bar{Q}_2|\Psi\rangle||^2 + ||Q_2|\Psi\rangle||^2\right) \\ &\geq 0, \end{aligned} \quad (6.81)$$

always positive definite.

This implies that states with vanishing energy density are supersymmetric ground states of the theory. They are ground states because with zero energy we reach the minimum possible value of energy, and they are supersymmetric because

$$\langle\Omega|\mathcal{H}|\Omega\rangle = 0 \quad \Leftrightarrow \quad \begin{matrix} Q_\alpha|\Omega\rangle = 0 \\ \bar{Q}_{\dot\alpha}|\Omega\rangle = 0 \end{matrix}, \quad \text{for all } \alpha, \dot\alpha = 1, 2. \quad (6.82)$$

In other words, supersymmetry is preserved in ground states with zero energy. Conversely, in ground states with non-zero (positive, as always with supersymmetry) energy, supersymmetry is *spontanously broken*.

Coming back to the Wess-Zumino model, let us assume that m and g are real for simplicity. In this case, the scalar potential is

$$\mathcal{V}(\phi) = |F|^2 = m^2|\phi|^2 + mg(\phi + \phi^*)|\phi|^2 + g^2|\phi|^4, \qquad (6.83)$$

while the auxiliary field is

$$F^* = m\phi + g\phi^2 = \left(mA + g(A^2 - B^2)\right) + i\left(mB + 2gAB\right), \quad (6.84)$$

where, the complex scalar $\phi = A + iB$ is written in terms of two real fields. We see that F can be made to vanish for

$$\langle B \rangle = 0 \quad \text{and} \quad \langle A \rangle = 0, -(m/g).$$

Since $F = 0$, (which guarantees that the potential $\mathcal{V}$ vanishes), supersymmetry is unbroken in the Wess-Zumino model. The supersymmetric vacuum states are parametrised by the above choices of vacuum expectation values for the scalar fields.

The superpotential of the Wess-Zumino model

$$W(\Phi) = \frac{1}{2}m\Phi^2 + \frac{1}{3}g\Phi^3 \qquad (6.85)$$

goes to $-W(\Phi)$ (modulo an unimportant constant shift) under the transformation

$$\Phi \to -\frac{m}{g} - \Phi$$

which interchanges the two ground states. However, the sign of $W(\Phi)$ can be rotated away by a $U_R(1)$ symmetry

$$\theta \to e^{-i\alpha}\theta \qquad \Rightarrow \qquad d^2\theta \to e^{+2i\alpha}d^2\theta. \qquad (6.86)$$

(Note that the above is unlike a commuting variable x for which x and dx transform in the same way. Recall that for anti-commuting variables, differentiation and intrgration are equivalent operations.) Suppose under this rotation

$$\Phi(y, \theta) \to \tilde{\Phi}(y, \theta) = e^{2in\alpha}\Phi(y, e^{-i\alpha}\theta), \qquad (6.87)$$

i.e. Φ has charge n. The condition that the action remains invariant, *i.e.*,

$$\int d^2\theta\, W\left(\Phi(y,\theta)\right) = \int d^2\theta\, W\left(\tilde{\Phi}(y,\theta)\right)$$

$$= \int d^2\theta\, W\left(e^{2in\alpha}\Phi(y, e^{-i\alpha\theta})\right),$$

implies that the superpotential must have charge $+2$ under $U_R(1)$ rotation

$$W(\Phi) \to e^{2i\alpha}W(\Phi). \tag{6.88}$$

For $\alpha = \pi/2$, this is a discrete $\mathbf{Z}_4$ symmetry, (under which the fermions are multiplied by a factor of i), relating the two ground states. However since a 2π rotation changes the sign of fermions, a $\mathbf{Z}_2 \subset \mathbf{Z}_4$ remains unbroken.

It is not possible to break supersymmetry spontaneously in the Wess-Zumino model involving a single chiral superfield. It turns out that for this one needs at least three chiral superfields. Before exhibiting this model, let us write the general form of a Wess-Zumino type action involving many chiral superfields

$$S = \int d^4x \left[\int d^4\theta\, \delta_{ij}\Phi_i^\dagger\Phi_j - \left(\int d^2\theta\, W\left(\{\Phi_i\}\right) + \text{h.c.}\right)\right]. \tag{6.89}$$

The superpotential

$$W\left(\{\Phi_i\}\right) = w_0 + \lambda_i\Phi_i + m_{ij}\Phi_i\Phi_j + g_{ijk}\Phi_i\Phi_j\Phi_k + \cdots \tag{6.90}$$

is a general functional of the chiral superfields only, *i.e.* it is a functional only of the Φ_i's and not the $\Phi_i^\dagger$'s. In other words, $W(\Phi)$ is an *analytic* function of Φ.

The bosonic potential in this case is

$$V\left(\{\phi_i, \phi_i^*\}\right) = \sum_i |F_i|^2 = \sum_i \left|\left[\frac{\delta W}{\delta\Phi_i}\right]_{\Phi_i=\phi_i}\right|^2. \tag{6.91}$$

In fact the name *superpotential* for $W(\Phi)$ derives from its close relation to the (bosonic) potential. Let us stress once again that the superpotential is an analytic function of the chiral superfields Φ_i. Analytic functions are also called *holomorphic*. Therefore the superpotential is a *holomorphic* function. This is not only true of the bare lagrangian, but also of the effective one, (if supersymmetry is to remain unbroken). There is, however, some subtlety in the interpretation of the last statement, and we shall return to this point later. At that time, we shall see the power of requirement of analyticity (or holomorphy), and how, for many theories, it is sufficient to determine the effective superpotential *exactly* and study its consequence on the dynamics of the theory.

Let us now study a model of spontaneous supersymmetry breaking. Consider the lagrangian (first proposed by O'Raifeartaigh[12] in 1975):

$$\mathcal{L} = \int d^4\theta \sum_{i=1}^{3} \Phi_i^\dagger \Phi_i - \int d^2\theta \left(\lambda\Phi_1 + m\Phi_2\Phi_3 + \frac{1}{2}g\Phi_1\Phi_2^2 \right) + \text{h.c.}$$

(6.92)

Notice that there is a term linear in Φ. This is of the form that was mentioned in the beginning of this lecture.

The algebraic equations of motion for the auxiliary fields are

$$\begin{aligned}
F_1^* &= \lambda + \frac{1}{2}g\phi_2^2, \\
F_2^* &= m\phi_3 + g\phi_1\phi_2, \\
F_3^* &= m\phi_2.
\end{aligned}$$

(6.93)

Clearly, F_1^* and F_3^* cannot be made to vanish simultaneously. The resulting scalar potential is

$$\begin{aligned}
\mathcal{V} &= |F_1|^2 + |F_2|^2 + |F_3|^2 \\
&= \lambda^2 + (m^2 + \lambda g)\,(\text{Re } \phi_2)^2 + (m^2 - \lambda g)\,(\text{Im } \phi_2)^2 \\
&\quad + m^2|\phi_3|^2 + 2mg(\phi_1\phi_2\phi_3^* + \phi_1^*\phi_2^*\phi_3) \\
&\quad + g^2|\phi_1|^2|\phi_2|^2 + \frac{1}{4}g^2\left(|\phi_2|^2\right)^2.
\end{aligned}$$

(6.94)

This is positive as long as $m^2 \geq \lambda g$. The potential (6.94) is minimised by choosing

$$\langle\phi_2\rangle = 0, \quad \langle\phi_3\rangle = 0, \quad \langle\phi_1\rangle = \text{unconstrained}.$$

The value of the potential at the minimum is

$$\mathcal{V}(\langle \phi_1 \rangle) = \lambda^2,$$

a positive definite quantity. Therefore supersymmetry is *spontaneously broken.*

> **Exercise:** *Compute the masses of the bosons and fermions in the O'Raifeartaigh model. There is at least one massless fermion when supersymmetry is spontaneously broken. This is analogous to the massless boson (Goldstone mode) that appears when a global symmetry is spontaneously broken, and is called a goldstino).*
>
> *Also show that the sum of the mass-square of the bosons equal to that of the fermions. This too is a generic feature of supersymmetric models, the property being manifest in vacua with unbroken supersymmetry.*

For $\lambda = 0$, there is no supersymmetry breaking, but this case illustrates another important feature of supersymmetric models. Namely, we have

$$\mathcal{V}(\langle \phi_1 \rangle, \langle \phi_2 \rangle = 0, \langle \phi_3 \rangle = 0) = 0 \qquad (6.95)$$

for arbitrary values of $\langle \phi_1 \rangle$. There are infinitely many supersymmetric vacua parametrised by the vacuum expectation value of the field ϕ_1. Thus the parameter space that labels supersymmetric vacua is the complex $\langle \phi_1 \rangle$-plane (see Fig.6.1(a)).

Let us further restrict to $m = 0$, *i.e.* we consider the simple model of two chiral superfields Φ_1 and Φ_2 with superpotential

$$W(\Phi_1, \Phi_2) = \frac{1}{2}\Phi_1\Phi_2^2. \qquad (6.96)$$

This corresponds to the bosonic potential

$$\mathcal{V}(\phi_1, \phi_2) = g^2 |\phi_1|^2 |\phi_2|^2 + \frac{1}{4}g^2 \left(|\phi_2|^2 \right)^2.$$

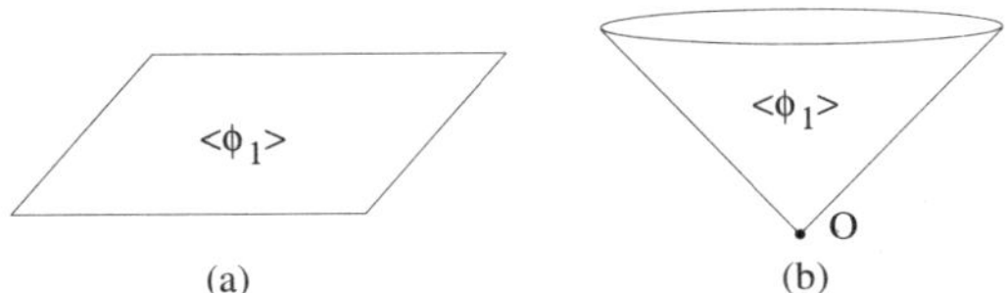

Figure 6.1: Moduli space of supersymmetric vacua (a) for potential (6.95) and (b) for superpotential (6.96) with singularity at the origin.

We see that $\mathcal{V}(\phi_1, \phi_2)$ vanishes for $\langle \phi_2 \rangle = 0$ with no condition on $\langle \phi_1 \rangle$. In the vacuum labelled by $\langle \phi_1 \rangle$ the field ϕ_2 gets an effective mass

$$m_2 = \sqrt{2}g|\langle \phi_1 \rangle|.$$

The origin of the parameter space is therefore a special point — the mass of the field ϕ_2 vanishes there. This is a *singular point* (or a *singularity*) in the sense that a heavy field becomes massless here.

The name for the parameter space that labels supersymmetric vacua is the *moduli space of vacua*. Fig.6.1(b) is the moduli space of vacua for the model described by (6.96).

> **Exercise:** *Analyse the behaviour of the moduli space of vacua for a theory of three chiral superfields* Φ_1, Φ_2 *and* Φ_3 *with superpotential*
>
> $$W(\Phi_1, \Phi_2, \Phi_3) = g\,\Phi_1\Phi_2\Phi_3.$$
>
> *The moduli space is shown in Fig.6.2. It consists of three branches meeting at a singular point.*

6.7 Lagrangians of supersymmetric gauge theories

In the last lecture, we saw how to write actions for chiral superfields leading to supersymmetric actions involving scalars and

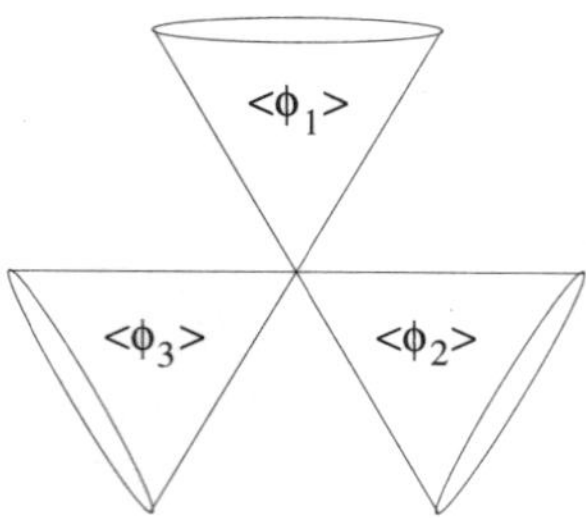

Figure 6.2: Moduli space of supersymmetric vacua described by the superpotential $W = g\,\Phi_1\Phi_2\Phi_3$. The origin is a singular point.

spin-half fermions. Constructing an action for a gauge theory is also along expected lines. We have the field strength supermultiplet which is a chiral spinor superfield W_α such that $\bar{D}_{\dot\alpha}W_\beta = 0$. Moreover, (in the non-abelian case[9]), it transforms homogeneously

$$W_\alpha \to e^{-2i\Lambda}W_\alpha e^{+2i\Lambda}.$$

Therefore,

$$\text{tr}\left(W^\alpha W_\alpha\right) \to \text{tr}\left(e^{-2i\Lambda}W^\alpha W_\alpha e^{+2i\Lambda}\right) = \text{tr}\left(W^\alpha W_\alpha\right)$$

is both gauge and Lorentz invariant. Further, since this is a chiral superfield, we can get a gauge invariant candidate term for the lagrangian by integrating over θ's: $\int d^2\theta\,\text{tr}\left(W^\alpha W_\alpha\right)$. In terms of the component fields

$$\int d^2\theta\,\text{tr}\left(W^\alpha W_\alpha\right) = \text{tr}\Big(D^2 - \frac{1}{2}F^{\mu\nu}F_{\mu\nu} - \frac{i}{4}F_{\mu\nu}F_{\rho\sigma}\epsilon^{\mu\nu\rho\sigma}$$
$$-2i\lambda^\alpha\sigma^\mu_{\alpha\dot\beta}\nabla_\mu\bar{\lambda}^{\dot\beta}\Big).$$

The problem with the above is that there is no dependence on the gauge coupling. Also the $F\tilde{F}$ term is imaginary. Both these

---

[9]In the abelian case, W_α is gauge invariant and no trace need to be taken.

problems are solved by defining the lagrangian

$$\mathcal{L}_{SYM} = \frac{1}{8\pi}\text{Im}\left[\tau \int d^2\theta \,\text{tr}\left(W^\alpha W_\alpha\right)\right], \qquad (6.97)$$

where,

$$\tau = \frac{\theta_{YM}}{2\pi} + i\,\frac{4\pi}{g^2}. \qquad (6.98)$$

Explicitly, in terms of component fields, the above lagrangian reads as follows

$$\begin{aligned}
\mathcal{L}_{SYM} &= \frac{1}{g^2}\,\text{tr}\left(D^2 - \frac{1}{2}F^{\mu\nu}F_{\mu\nu} - i\lambda^\alpha \sigma^\mu_{\alpha\dot\beta}\nabla_\mu\bar\lambda^{\dot\beta}\right)\\
&\quad - \frac{\theta_{YM}}{32\pi^2}\,tr\left(F_{\mu\nu}\tilde{F}^{\mu\nu}\right),
\end{aligned} \qquad (6.99)$$

where, $\tilde{F}^{\mu\nu} = \frac{1}{2}\epsilon^{\mu\nu\rho\sigma}F_{\rho\sigma}$ is the dual field strength.

Notice that in addition to the gauge fields A_μ, the supersymmetric Yang-Mills theory contains 'matter' fields: fermions λ and auxiliary fields D, both in the *adjoint* representation. The fermions λ^a are supersymmetric partners of the gauge fields, and are called *gluinos* or *gauginos*.

As we had mentioned in the last lecture, it is possible to add a term

$$2\kappa \int d^4x \int d^4\theta V(x,\theta,\bar\theta) = \kappa \int d^4x D(x).$$

However, this is gauge invariant only for an abelian vector superfield V. Therefore such a term may be added only to the lagrangian of the supersymmetric Maxwell theory. It turns out that this term leads to spontaneous breaking of supersymmetry[13]. The parameter κ is known as the Fayet-Iliopoulos parameter.

After having constructed actions for chiral superfields (matter) and vector superfields (gauge fields), we shall now discuss the coupling of matter to gauge fields. We shall begin with the coupling of matter to an abelian gauge field.

The first point to notice is that we cannot put the matter fields in the gauge supermultiplet. This is because all fields in the this multiplet must belong to the same representation as the gauge fields, *i.e.* in the abelian case they must all be charge neutral and belong to the adjoint in the non-abelian case.

Consider the chiral superfields Φ_i, $i = 1, 2, \cdots, n$. Under a global U(1) rotation

$$\Phi_i \to \Phi_i{}' = e^{-2iq_i\lambda}\Phi_i,$$

where, q_i and λ are real constants. (In particular, $\bar{D}_{\dot{\alpha}}q_i = 0$, $\bar{D}_{\dot{\alpha}}\lambda = 0$. Therefore, $\Phi_i{}'$ is also a chiral superfield.) The Wess-Zumino type lagrangian

$$\mathcal{L} = \int d^4\theta\, \Phi_i^\dagger\Phi_i - \int d^2\theta\,(m_{ij}\Phi_i\Phi_j + g_{ijk}\Phi_i\Phi_j\Phi_k + \text{h.c.})$$

is invariant under this rotation if

$$
\begin{aligned}
m_{ij} &= 0 \quad \text{whenever } q_i + q_j \neq 0, \\
g_{ijk} &= 0 \quad \text{whenever } q_i + q_j + q_k \neq 0.
\end{aligned}
$$

If we want to gauge this symmetry, *i.e.* make λ a function of x, since $\bar{D}_{\dot{\alpha}}\lambda(x) \neq 0$, we must promote the function $\lambda(x)$ to a chiral superfield $\Lambda(x,\theta)$ $(\bar{D}_{\dot{\alpha}}\Lambda = 0)$ such that $\Phi_i{}' = e^{-2iq_i\Lambda}\Phi_i$ is again a chiral superfield. This makes the superpotential gauge invariant. However, the kinetic term

$$\Phi_i^\dagger\Phi_i \to \Phi_i^\dagger e^{2iq_i(\Lambda^\dagger - \Lambda)}\Phi_i$$

is no longer invariant. This is familiar from gauging of non-supersymmetric theories. In order to restore invariance under local phase rotations, we need to introduce a vector superfield $V(x,\theta,\bar{\theta})$ such that

$$
\begin{aligned}
e^V \to e^{V'} &= e^{-i\Lambda^\dagger}e^V e^{i\Lambda} \\
\textit{i.e,} \quad V \to V' &= V + i(\Lambda - \Lambda^\dagger).
\end{aligned}
\tag{6.100}
$$

The gauge invariant lagrangian is then

$$
\begin{aligned}
\mathcal{L} =\ & \frac{1}{2}\int d^2\theta\, W^\alpha W_\alpha + \int d^4\theta\, \Phi_i^\dagger e^{2q_i V}\Phi_i \\
& - \int d^2\theta\,(m_{ij}\Phi_i\Phi_j + g_{ijk}\Phi_i\Phi_j\Phi_k) + \text{h.c.}
\end{aligned}
\tag{6.101}
$$

Due to the presence of the exponential in the lagrangian, it is not clear if the theory is renormalisable. We can, however, evaluate it in the Wess-Zumino gauge where

$$\Phi^\dagger e^{2qV_{WZ}} \Phi = \Phi^\dagger \left(1 + 2qV_{WZ} + 2q^2 V_{WZ}^2\right)\Phi.$$

In components

$$\int d^4\theta\, \Phi^\dagger e^{2qV_{WZ}}\Phi \;=\; \eta^{\mu\nu}\left(\nabla_\mu\phi\right)^*\left(\nabla_\nu\phi\right) - i\bar\psi\bar\sigma^\mu\nabla_\mu\psi$$
$$+i\sqrt{2}q\left(\phi^*\lambda\psi - \phi\bar\lambda\bar\psi\right) \qquad (6.102)$$
$$+|F|^2 + qD\phi^*\phi,$$

where,

$$\nabla_\mu\phi \;=\; \partial_\mu\phi + iqA_\mu\phi$$
$$\nabla_\mu\psi \;=\; \partial_\mu\psi + iqA_\mu\psi.$$

The supersymmetric generalisation of QED requires at least two chiral superfields Φ_+ and Φ_- so that we may write a gauge invariant mass term. Therefore, we have

$$\text{chiral superfields} \qquad \Phi_\pm \to e^{\mp 2ie\Lambda}\Phi_\pm,$$
$$\text{vector superfield} \qquad e^{2V} \to e^{-2i\Lambda^\dagger}e^{2V}e^{2i\Lambda}.$$

The lagrangian in superfields is

$$\mathcal{L}_{SQED} \;=\; \frac{1}{2}\int d^2\theta\, W^\alpha W_\alpha + \int d^4\theta\left(\Phi_+^\dagger e^{2eV}\Phi_+ + \Phi_-^\dagger e^{-2eV}\Phi_-\right)$$
$$-m\left(\int d^2\theta\, \Phi_+\Phi_- + \int d^2\bar\theta\, \Phi_+^\dagger\Phi_-^\dagger\right), \qquad (6.103)$$

and in components

$$\mathcal{L}_{SQED} \;=\; \frac{1}{2}D^2 - \frac{1}{4}F_{\mu\nu}F^{\mu\nu} - i\bar\lambda\bar\sigma^\mu\partial_\mu\lambda$$
$$+\eta^{\mu\nu}(\nabla_\mu\phi_+)^*(\nabla_\nu\phi_+) + \eta^{\mu\nu}(\nabla_\mu\phi_-)^*(\nabla_\nu\phi_-)$$
$$-i\bar\psi_+\bar\sigma^\mu\nabla_\mu\psi_+ - i\bar\psi_-\bar\sigma^\mu\nabla_\mu\psi_- + m\left(\psi_+\psi_- + \bar\psi_+\bar\psi_-\right)$$
$$+i\sqrt{2}e\left(\phi_+^*\lambda\psi_+ - \phi_+\bar\lambda\bar\psi_+ - \phi_-^*\lambda\psi_- - \phi_-\bar\lambda\bar\psi_-\right)$$
$$-m\left(\phi_+F_- + \phi_-F_+ + \phi_+^*F_-^* + \phi_-^*F_+^*\right)$$
$$+|F_+|^2 + |F_-|^2 + eD\left(\phi_+^*\phi_+ - \phi_-^*\phi_-\right). \qquad (6.104)$$

Exercise: *Re-express the above lagrangian in terms of the Dirac spinor $\Psi_D = \begin{pmatrix} \psi_+ \\ \bar{\psi}_- \end{pmatrix}$. What are the additional fields in the lagrangian compared to QED?*

Notice that the lagrangian of supersymmetric QED (6.104) contains two complex auxiliary fields $F_\pm$ and one real auxiliary field D. Their (algebraic) equations of motions are

$$
\begin{aligned}
D + e\left(\phi_+^* \phi_+ - \phi_-^* \phi_-\right) &= 0, \\
F_\pm^* &= m\phi_\mp.
\end{aligned}
$$

(The RHS of the first equation above is $-\kappa$ in case the Fayet-Iliopoulos term (6.69) is present.) These equations can be solved to determine the classical scalar potential

$$
\begin{aligned}
\mathcal{V}(\phi_\pm) &= \left(|F_+|^2 + |F_-|^2\right) + \frac{1}{2}D^2 \\
&= m^2 \left(|\phi_+|^2 + |\phi_-|^2\right) \qquad (6.105) \\
&\quad + \frac{1}{2}e^2 \left(|\phi_+|^2 - |\phi_-|^2\right)^2.
\end{aligned}
$$

The two terms above are called the F- and D-term respectively.

In the massless case, we only have the D-term. Therefore, there are infinitely many degenerate vacua labelled by, upto gauge equivalence,

$$
\langle \phi_+ \rangle = a = \langle \phi_- \rangle,
$$

for any complex number a.

In any vacuum with $a \neq 0$, gauge symmetry is broken by the (super) Higgs mechanism. The gauge superfield becomes massive by absorbing one chiral superfield degree of freedom from the matter. One of the chiral superfield degrees of freedom, however, still remains massless. A gauge invariant description of this massless degree of freedom can be given in terms of

$$
X = \Phi_+ \Phi_-. \qquad (6.106)
$$

In the vacuum, $\langle X \rangle = a^2$ is an arbitrary complex number. This means that there is no superpotential for the chiral superfield X, at least at the classical level:

$$W_{cl}(X) = 0.$$

The above is known as the *D-flatness condition.*

The vacuum expectation value $\langle X \rangle$ gives a gauge invariant parametrisation of the moduli space of classical vacua. This space has a singularity at the origin, *i.e.* at $\langle X \rangle = 0$, corresponding to the fact that at this point the U(1) gauge symmetry is unbroken, and all the original microscopic degrees of freedom are massless there (see Fig.6.1(b)).

The degeneracy of the classical vacua is accidental. There is no symmetry that relates different vacua parametrised by different choices for a, indeed they are inequivalent theories. This degeneracy may therefore be lifted in the quantum theory by a dynamically generated superpotential $W_{eff}(X)$.

Recall that X is gauge invariant because $\phi_\pm$ carry equal and opposite charges

$$\phi_\pm \to e^{\mp ie\lambda}\phi_\pm \qquad \Rightarrow \qquad X \to X.$$

This freedom in defining X can formally be extended to a complexification of the gauge group $U^{\mathbf{C}}(1) \approx \mathbf{C}^*$, under which

$$\phi_+ \to \rho\phi_+, \qquad \phi_- \to \rho^{-1}\phi_-; \qquad (\rho \in \mathbf{C}^*).$$

(In the above, we have extended the gauge parameter to take non-zero complex values.) The moduli space of vacua was described by setting the (D-term of the) classical superpotential to zero, and quotienting by the gauge symmetry. This turns out to be exactly equivalent to quotienting by the compexified gauge group. Thus, the space of chiral superfields modulo the complexified gauge group may be parametrised by the gauge invariant polynomials of chiral superfields[14]. (In a more general situation, there may be relations between the possible gauge invariant polynomials.)

6.8   Supersymmetric QCD — classical theory

Much of what we learnt in Lecture 6.7 generalises to the case of non-abelian gauge theories. In the present lecture, we shall discuss that generalisation.

Consider a chiral superfield $\Phi = \{\Phi_i\}$ which belong to some representation R of a Lie algebra $\mathbf{g}$:

$$\Phi_i \rightarrow \Phi_i{}' = \left(e^{-2i\Lambda}\right)_i{}^j \Phi_j = \left(e^{-2i\Lambda^a t_i^{aj}}\right) \Phi_j, \tag{6.107}$$

where, $\|t_{ij}^a\|$ with $a = 1, 2, \cdots, \dim \mathbf{g}$, $i, j = 1, 2, \cdots, \dim R$, are matrices in the representation R. The gauge invariant kinetic term for the field Φ is

$$\left(\Phi^\dagger e^{2V} \Phi\right) = \Phi^{\dagger i} \left(e^{2V^a t^a}\right)_i{}^j \Phi_j = \mathrm{tr}_R \left(e^{2V^a t^a} \Phi\Phi^\dagger\right).$$

This is gauge invariant because the tensor product of the representations R, its conjugate $\bar{R}$ and the adjoint contains the singlet.

More generally the lagrangian of a Wess-Zumino type model interacting with a non-abelian gauge field is

$$\mathcal{L} = \frac{1}{8\pi}\mathrm{Im}\left[\tau \int d^2\theta\, \mathrm{tr}\,(W^\alpha W_\alpha)\right] + \int d^4\theta\left(\Phi_I^\dagger e^{2V}\Phi_I\right)$$

$$- \int d^2\theta\left(m_{IJ}\Phi_I\Phi_J + g_{IJK}\Phi_I\Phi_J\Phi_K\right) + \mathrm{h.c.} \tag{6.108}$$

The 'mass' term is allowed only when $R_I = \bar{R}_J$. For a single chiral superfield, this in only possible for SU(2) (of all SU(n)'s). Similarly the g_{IJK} term is allowed if $R_I \otimes R_J \otimes R_K$ contains the singlet. For example, a single chiral superfield in the doublet of SU(2) can have a mass term but not a cubic self coupling.

Let us now look at the D-terms in the lagrangian

$$\mathcal{L}_D = \frac{1}{2g^2}\,\mathrm{tr}\,D^2 + \phi^* D\phi$$

$$= \frac{1}{2g^2}\,D^a D^b\,\mathrm{tr}\,(t^a t^b) + D^a \phi^{*i} t_i^{aj} \phi_j$$

$$= \frac{1}{2g^2}\,C(R)D^a D^a + D^a\,\mathrm{tr}_R\left(t^a(\phi\phi^*)\right), \tag{6.109}$$

where we have used the normalisation (6.59). The equation of motion for D^a is therefore

$$D^a = -\frac{g^2}{C(R)}\, \mathrm{tr}_R\left(t^a(\phi\phi^*)\right), \tag{6.110}$$

which leads to a scalar potential

$$\mathcal{V}_D = \left[\frac{g^2}{C(R)}\, \mathrm{tr}_R\left(t^a(\phi\phi^*)\right)\right]^2. \tag{6.111}$$

This potential must vanish in a vacuum which preserves supersymmetry. Thus, for unbroken supersymmetry, we require

$$\mathrm{tr}_R\left(t^a(\phi\phi^*)\right) = 0, \tag{6.112}$$

the D-flatness condition.

The supersymmetric version of QCD that we shall consider has

- N_f chiral superfields in the fundamental representation $\mathbf{N}_c$ of the gauge group $\mathrm{SU}(N_c)$:

$$Q_i^A, \quad A = 1, 2, \cdots, N_f;\ i = 1, 2, \cdots, N_c.$$

  (The colour index will not always be displayed explicitly.) There is a global $\mathrm{U}(N_f)$ *flavour* symmetry between the Q's which transform in the $\mathbf{N}_f$ representation. The conjugate anti-chiral superfield $Q^\dagger$ belongs to $\overline{\mathbf{N}}_f$ of $\mathrm{U}(N_f)$ and $\overline{\mathbf{N}}_c$ of $\mathrm{SU}(N_c)$. We shall think often think of $||Q_i^A||$ and $||Q_{\tilde{A}}^{\dagger \tilde{i}}||$ as $N_c \times N_f$ matrices.

- N_f chiral superfields $\tilde{Q}_{\tilde{A}}$ in the anti-fundamental $\overline{\mathbf{N}}_c$ of the gauge group $\mathrm{SU}(N_c)$. These transform as $\overline{\mathbf{N}}_f$ of the flavour group. The corresponding conjugate fields are denoted by $\tilde{Q}^\dagger$.

Their dynamics is specified by the lagrangian

$$
\begin{aligned}
\mathcal{L}_{SQCD} \;=\;& \frac{1}{8\pi}\mathrm{Im}\left[\tau\int d^2\theta\,\mathrm{tr}\,(W^\alpha W_\alpha)\right] \\
&+ \int d^4\theta\left[Q^\dagger_{\tilde{A}}e^{2V}Q^A + \tilde{Q}_{\tilde{A}}e^{2V}\tilde{Q}^{\dagger A}\right] \qquad (6.113) \\
&- \int d^2\theta\, m_A\, Q^A Q_{\tilde{A}} - \int d^2\bar\theta\, m^*_A\,\tilde{Q}^{\dagger A}Q^\dagger_{\tilde{A}}.
\end{aligned}
$$

This theory has a global $\mathrm{U}(N_f)$ *flavour* symmetry. In the mass-less case, $m_A = m^*_A = 0$ there are actually two flavour rotations corresponding to the fact that Q and $\tilde{Q}$ can be transformed independently in flavour space, leading to a $\mathrm{U}_\ell(N_f)\times\mathrm{U}_r(N_f)$ global symmetry for the classical theory. In the following, we shall consider the massless case.

The *D*-flatness conditions are

$$
\sum_{A=1}^{N_f}\left[\mathrm{tr}_{\mathbf{N}_c}\left(t^a\phi_{Q^A}\phi^*_{Q^A}\right) - \mathrm{tr}_{\overline{\mathbf{N}}_c}\left(t^a\phi^*_{\tilde{Q}^A}\phi_{\tilde{Q}^A}\right)\right] = 0, \qquad (6.114)
$$

for all $a = 1, 2, \cdots, N_c^2 - 1$. Since the representations $\mathbf{N}_c$ and $\overline{\mathbf{N}}_c$ are isomorphic, the two traces are the same and the D-flatness conditions reduce to

$$
\sum_{A=1}^{N_f}\mathrm{tr}_{\mathbf{N}_c}\left[t^a\left(\phi_{Q^A}\phi^*_{Q^A} - \phi^*_{\tilde{Q}^A}\phi_{\tilde{Q}^A}\right)\right] = 0, \quad \text{for all } a.
$$

By Schur's lemma, this is possible when

$$
\sum_{A=1}^{N_f}\phi_{Q^A}\phi^*_{Q^A} - \phi^*_{\tilde{Q}^A}\phi_{\tilde{Q}^A} = c\,\mathbf{1}_{N_c\times N_c}, \qquad (6.115)
$$

where, c is a constant independent of i (the colour index).

To characterise classical vacua, we need to differentiate between two cases

- $N_f < N_c$:

  By a gauge and flavour rotation, we can 'diagonalise' ϕ_Q. This implies that the constant on the RHS of (6.115) must vanish. Upto gauge and flavour rotations, the solution to the D-flatness conditions are therefore given by

$$\langle \phi_Q \rangle = \langle \phi_{\tilde{Q}}^* \rangle = \begin{pmatrix} a_1 & 0 & \cdots & 0 \\ 0 & a_2 & & \\ \vdots & & \ddots & \\ 0 & & & a_{N_f} \\ 0 & & & 0 \\ \vdots & & & \vdots \\ 0 & & & 0 \end{pmatrix}. \qquad (6.116)$$

  The gauge invariant composite superfields whose lowest components parametrise the moduli space of vacua are

$$M^A{}_{\tilde{B}} = Q^A \tilde{Q}_{\tilde{B}}, \quad A, \tilde{B} = 1, 2, \cdots, N_f, \qquad (6.117)$$

  called the *meson* superfields. The vacuum expectation value of $\langle M^A_{\tilde{B}} \rangle = |a_A|^2 \delta_{A\tilde{B}}$. When this is non-zero, *i.e.* $\langle M^A_{\tilde{B}} \rangle \neq 0$, the gauge group is broken. The most generic behaviour is $SU(N_c) \to SU(N_c - N_f)$. (Notice that given the vev of M, one can determine, upto gauge and flavour rotations, the vevs of Q and $\tilde{Q}$ and vice versa.)

- $N_f \geq N_c$:

  An arbitrary A-independent constant is now allowed, so upto gauge and flavour symmetry, a solution to the D-flatness conditions is

$$\langle \phi_Q \rangle = \begin{pmatrix} a_1 & 0 & \cdots & 0 & 0 & \cdots & 0 \\ 0 & a_2 & & & & & \\ \vdots & & \ddots & & & & \\ 0 & & & a_{N_c} & 0 & \cdots & 0 \end{pmatrix},$$

$$\langle \phi^*_{\tilde{Q}} \rangle \;=\; \begin{pmatrix} \tilde{a}_1 & 0 & \cdots & 0 & 0 & \cdots & 0 \\ 0 & \tilde{a}_2 & & & & & \\ \vdots & & \ddots & & & & \\ 0 & & & \tilde{a}_{N_c} & 0 & \cdots & 0 \end{pmatrix} , \quad (6.118)$$

together with the restriction $|a_A|^2 - |\tilde{a}_A|^2 = c$ (a constant independent of A).

Once again we have the meson chiral superfields

$$M^A{}_{\tilde{B}} \;=\; Q^A \tilde{Q}_{\tilde{B}},$$

$$||\langle M^A_{\tilde{B}} \rangle|| \;=\; \begin{pmatrix} a_1 \tilde{a}^*_1 & & & & & \\ & a_2 \tilde{a}^*_2 & & & & \\ & & \ddots & & & \\ & & & a_{N_c} \tilde{a}^*_{N_c} & & \\ & & & & 0 & \\ & & & & & \ddots \\ & & & & & & 0 \end{pmatrix} .$$

This time, however, it is not possible to determine a and $\tilde{a}$ from the knowledge of $\langle M \rangle$. Moreover, there are new gauge invariant composite operators. For example, in the case $N_f = N_c$, there are two such operators

$$\begin{aligned} B &= \epsilon_{A_1 A_2 \ldots A_{N_c}} Q_1^{A_1} \cdots Q_{N_c}^{A_{N_c}} \\ &= \frac{1}{N_c!} \epsilon_{A_1 A_2 \ldots A_{N_c}} \epsilon^{B_1 B_2 \ldots B_{N_c}} Q_{B_1}^{A_1} \cdots Q_{B_{N_c}}^{A_{N_c}}, \\ \tilde{B} &= \epsilon_{\tilde{A}_1 \tilde{A}_2 \ldots \tilde{A}_{N_c}} \tilde{Q}_1^{\tilde{A}_1} \cdots \tilde{Q}_{N_c}^{\tilde{A}_{N_c}} \qquad (6.119) \\ &= \frac{1}{N_c!} \epsilon_{\tilde{A}_1 \tilde{A}_2 \ldots \tilde{A}_{N_c}} \epsilon^{\tilde{B}_1 \tilde{B}_2 \ldots \tilde{B}_{N_c}} \tilde{Q}_{\tilde{B}_1}^{\tilde{A}_1} \cdots \tilde{Q}_{\tilde{B}_{N_c}}^{\tilde{A}_{N_c}}, \end{aligned}$$

called the *baryons*. The moduli space of vacua is parametrised by $\langle M \rangle$, B and $\tilde{B}$. However, they are not all independent, but related through the relation

$$\det ||\langle M \rangle|| = B \tilde{B}. \qquad (6.120)$$

Likewise, *e.g.* for $N_f = N_c + 1$, we have $2N_f$ baryons

$$
\begin{aligned}
B_A &= \epsilon_{AA_1A_2...A_{N_c}} Q_1^{A_1} \cdots Q_{N_c}^{A_{N_c}} \\
\tilde{B}^{\tilde{A}} &= \epsilon^{\tilde{A}\tilde{A}_1\tilde{A}_2...\tilde{A}_{N_c}} \tilde{Q}^1_{\tilde{A}_1} \cdots \tilde{Q}^{N_c}_{\tilde{A}_{N_c}},
\end{aligned}
\tag{6.121}
$$

which satisfy the following relations

$$
\begin{aligned}
(\det M)\left(M^{-1}\right)^{\tilde{B}}_A - B_A \tilde{B}^{\tilde{B}} &= 0, \\
M^A_{\tilde{B}} B_A = M^A_{\tilde{B}} \tilde{B}^{\tilde{B}} &= 0.
\end{aligned}
\tag{6.122}
$$

This pattern is easily generalisable to higher values of N_f.

Now we would like to know how the quantum theory behaves. We shall approach this question by analysing the low energy effective action involving the degrees of freedom which are light at the energy scale in which we are interested.

We shall assume that supersymmetry is unbroken, (*i.e.* we shall work above the possible supersymmetry breaking scale). This symmetry of effective action will be made manifest by working in terms of superfields. Matter fields are combined into chiral superfields Φ_i (and their conjugates $\Phi_i^\dagger$), while gauge field is described by real/vector superfield V or its descendent chiral spinor superfield W_α.

As we have seen, the moduli space of classical vacua of a theory is parametrised by the vacuum expectation values of (the lowest componenent of) the chiral superfields. The classical superpotential for these fields either vanish, or are determined from the classical lagrangian. We would like to know whether quantum corrections generate an effective (super)potential or change its form from the classical one.

Let us write the effective superpotential as

$$
\int d^2\theta \, W_{\text{eff}}(\{\Phi_i\}, \{g_I\}, \Lambda),
$$

where $\{g_I\}$ are the coupling constants and Λ is the dynamically generated QCD scale. Recall that Λ is determined from

$$\Lambda^{b_0} = \mu^{b_0} \exp\left(-\frac{8\pi^2}{g^2(\mu)}\right),$$

where $b_0 = 3C_2(G) - C_2(R)$ is the coefficient in the lowest order β-function

$$\beta(g) \equiv \mu\frac{dg}{d\mu} = -\frac{b_0}{16\pi^2}\,g^3.$$

However, in supersymmetric gauge theories, it is more convenient to *complexify* Λ by defining

$$\Lambda^{b_0} = \mu^{b_0} \exp\left(-\frac{8\pi^2}{g^2(\mu)} + i\theta_{YM}\right) = \mu^{b_0} \exp\left(2\pi i\tau(\mu)\right). \quad (6.123)$$

(See (6.98) for the definition of τ.)

We have already noticed that the superpotential must be a holomorphic function of the chiral superfields Φ_i, (no explicit dependence on $\Phi_i^\dagger$). Seiberg[16] proposed that the dependence of W_{eff} on $\{g_I\}$ (and Λ when applicable) is also holomorphic. This can be motivated by thinking of the couplings $\{g_I\}$ as the vacuum expectation value of some background chiral superfields $G_I(= g_I + \cdots)$. This conjecture is further motivated from string theory where the coupling constants are actually vacuum expectation values of some chiral superfields. However, within the context of field theory alone its justification is *a posteriori*.

Moreover, if the theory possesses some symmetry in the absence of the superpotential W, *i.e.* the symmetry is broken by the terms in W, then one can formally recover the symmetry by assigning suitable transformation laws on $\{g_I\}$ and Λ, such that $W_{eff}(\{\Phi\}, \{g_I\}, \Lambda)$ is invariant under the combined transformations of $\{\Phi\}$, $\{g_I\}$ and Λ.

With these assumptions, it turns out that it is often possible to determine the effective superpotential *exactly*.

6.9   Quantum corrections & effective action I

In the previous lecture we sketched an argument, (due to Seiberg), to determine the effective superpotential of a quantum theory. Let us recall that strategy.

1. Ascertain all the symmetries of a theory in the absence of the superpotential. Assign suitable (formal) transformation properties to the coupling constants such that the symmetry remains valid even in the presence of the superpotential.

2. Regard the quantum corrected effective superpotential to be a holomorphic function not only of the light chiral superfields, but also of the coupling constants $\{g_I\}$ (and Λ in the case of supersymmetric QCD).

3. The effective superpotential should match the results of perturbation theory in the appropriate limit.

These requirements are often stringent enough to determine the effective superpotential. The superpotential so determined will also include non-perturbative corrections.

Parenthetically, let us add that the kinetic term of the gauge fields

$$\int d^2\theta \, \mathrm{Im} \left(\tau_{eff}(\{\Phi\}, \{g_I\}, \Lambda) \, W^\alpha W_\alpha \right)$$

is also holomorphic. More precisely, τ_{eff} is holomorphic in its arguments. However, in many cases it so happens that the gauge symmetry is broken and the theory is in a Higgs or a confining phase. In case gauge symmetry remains unbroken and one is in the Coulomb phase, the same arguments may be applied to determine τ_{eff}.

As an illustration, let us apply these ideas to determine the quantum effective superpotential of the Wess-Zumino model (6.75). The classical superpotential (6.85) is

$$W(\Phi) = \frac{1}{2} m\Phi^2 + \frac{1}{3} g\Phi^3.$$

There is a U(1) R-symmetry that we encountered earlier in eqns.(6.86) and (6.88). It rotates the fermionic component of the superspace

$$\theta \to \theta' = e^{-i\alpha}\theta,$$

and under this transformation the action is invariant if

$$W(\Phi) \to e^{2i\alpha}\,W(\Phi).$$

Since the superpotential has R-charge $+2$, Φ, m and g transform under $U_R(1)$ as follows:

$$\begin{aligned}
\Phi(x,\theta) &\to \Phi(x, e^{-i\alpha}\theta) \\
m &\to e^{2i\alpha}m \\
g &\to e^{2i\alpha}g.
\end{aligned} \tag{6.124}$$

The kinetic term in (6.75) is also invariant another U(1) transformation — let us call it S-symmetry:

$$\Phi(x,\theta) \to e^{i\beta}\Phi(x,\theta). \tag{6.125}$$

The superpotential (6.85) breaks this invariance. However, we can formally recover it by assigning the following transformation rules to the couplings

$$\begin{aligned}
m &\to e^{-2i\beta}m, \\
g &\to e^{-3i\beta}g.
\end{aligned} \tag{6.126}$$

We now demand that $W_{e\!f\!f}$ is invariant under both $U_R(1)$ as well as $U_S(1)$, and also that it is holomorphic in m and g (and of course Φ). (In case m and g are real, we shall formally regard them as complex and set them to their real values at the end.) The most general form of $W_{e\!f\!f}$ consistent with these requirements is

$$\begin{aligned}
W_{e\!f\!f}(\Phi, m, g) = g^a m^b \Phi^c &\xrightarrow{R} e^{i\alpha(2a+2b)}W_{e\!f\!f}(\Phi, m, g) \\
&\xrightarrow{S} e^{i\alpha(-3a-2b+c)}W_{e\!f\!f}(\Phi, m, g).
\end{aligned}$$

Invariance under the two U(1)'s determines (say) a and b in terms of c. The most general form W_{eff} is then

$$
\begin{aligned}
W_{eff}(\Phi, m, g) &= \sum_{n=0}^{\infty} A_n g^{n-2} m^{3-n} \Phi^n \\
&= m\Phi^2 \sum_{n=0}^{\infty} A_n \left(\frac{g\Phi}{m}\right)^n \qquad (6.127) \\
&= m\Phi^2 f\left(\frac{g\Phi}{m}\right), \qquad (6.128)
\end{aligned}
$$

where, A_n are arbitrary constant coefficients and $f(\cdot)$ is an arbitrary function defined by the Taylor series above.

It remains to determine the function f, or equivalently the coefficients A_n. To that end, notice that for small values of g, perturbation theory is applicable. We compare this with the Taylor series

$$
W_{eff} \overset{g \to 0}{=} m\Phi^2 \left(A_0 + A_1 \frac{g\Phi}{m} + \sum_{n=2}^{\infty} A_n \left(\frac{g\Phi}{m}\right)^n \right)
$$

to find that $A_0 = 1/2$ and $A_1 = 1/3$. It follows from the properties of superspace Feynman rules[10] that the higher terms in the sum come from the tree level process shown in Fig.6.3. (Loops are not possible as it would increase powers of g but not of Φ.)

For $n \geq 2$, the Feynman diagram in Fig.6.3 is not one particle irreducible, and hence it cannot contribute to the effective superpotential. Therefore the effective superpotential is exactly the same as the classical superpotential (6.85)

$$
W_{eff}(\Phi, m, g) = W(\Phi, m, g) = \frac{1}{2} m\Phi^2 + \frac{1}{3} g\Phi^3.
$$

In other words, the superpotential of the Wess-Zumino model is not renormalised by quantum (and non-perturbative) corrections. This important result, first derived in perturbation theory, is called the *non-renormalisation theorem.*

---

[10] A complete discussion of superspace perturbation theory is beyond the scope these lectures. However, all we need to know for this is the fact that the propagator for chiral superfield at zero momentum $\langle \Phi\Phi \rangle (k=0) \sim m^{-1}$.

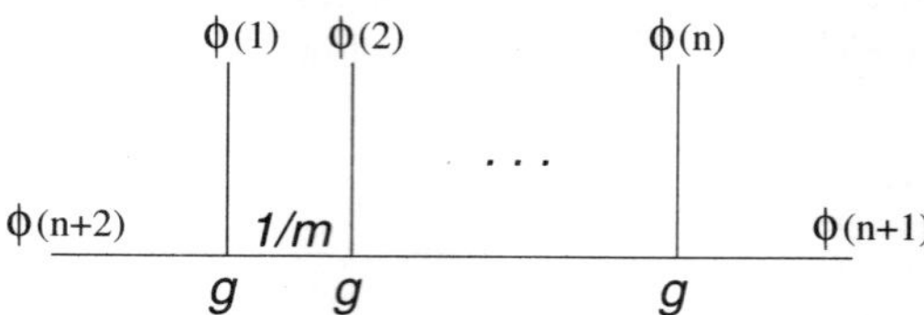

Figure 6.3: Superspace Feynman diagram contributing to the sum in Eq.(6.127).

There is, however, a caveat to the above argument. It was already known, (through the analysis of perturbation theory of superfields), that in case there are massless chiral superfields, the non-renormalisation theorem is invalidated. Indeed it was found[17] that the effective potential in the Wess-Zumino model with $m = 0$ gets a contribution $g^3(g^*)^2\Phi^3$ which comes from the two-loop diagram shown in Fig.6.4. This is clearly non-holomorphic.

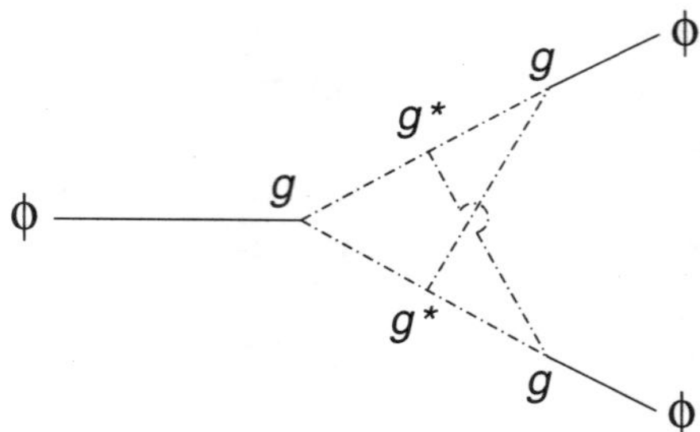

Figure 6.4: A two loop diagram which gives a non-holomorphic contribution to the 1PI effective superpotential. (A solid line denotes a $\Phi\Phi$ propagator while a dot-dashed line stands for a $\Phi^\dagger\Phi$ or a $\Phi\Phi^\dagger$ propagator.)

Similar violation of holomorphy was found in SQCD at two loops. In this case the problem is even more serious due to supersymmetric relation between the stress-tensor $T_{\mu\nu}$ and the axial current j_μ^5. The latter does not receive any correction beyond one loop. A resolution to this apparent paradox was suggested by Shifman and Vainshtein[18], who argued that one needs to distinguish between two commonly used notions of effective action (or effective potential).

- From the microscopic lagrangian $\mathcal{L}(\{\phi\})$, one defines a (Euclidean) generating functional for the correlation functions

$$
\begin{aligned}
Z[\{J\}] &= e^{-F[\{J\}]} \\
&= \int [\mathcal{D}\phi]\, \exp\left(-\int d^4x \left(\mathcal{L}[\{\phi\}] + \sum J\phi\right)\right),
\end{aligned}
$$

  which is a functional of the external sources $\{J\}$. A Legendre transform of this generating functional is a functional of the effective fields (called classical fields) $\{\phi_{cl}\}$

$$
\Gamma[\{\phi_{cl}\}] = -F[\{J\}] - \sum \int d^4x\, J(x)\phi_{cl}(x),
$$

  where

$$
\frac{\delta}{\delta J(x)} F[\{J\}] = -\phi_{cl}(x),
$$

$$
\frac{\delta}{\delta \phi_{cl}} \Gamma[\{\phi_{cl}\}] = -J(x).
$$

  $\Gamma[\{\phi_{cl}\}]$ generates all the one particle irreducible graphs of the theory, and is called the (1PI) effective action. Notice that in defining this effective action one integrates out *all* the degrees of freedom.

- The alternative notion is that of a Wilsonian effective action. To define this, let us first Fourier transform to momentum space $\{\phi(x)\} \rightarrow \{\phi(k)\}$. Now divide the degrees of freedom into low energy and high energy modes[11]

$$
\phi(k) = \phi_<(k) + \phi_>(k),
$$

---

[11]This is really meaningful in Euclidean continuation.

where

$$\phi_<(k) \;=\; \begin{cases} \phi(k) & \text{for } \mu \leq |k| \leq M_{UV}, \\ 0 & \text{for } 0 \leq |k| < \mu, \end{cases}$$

$$\phi_>(k) \;=\; \begin{cases} 0 & \text{for } \mu \leq |k| \leq M_{UV}, \\ \phi(k) & \text{for } 0 \leq |k| < \mu, \end{cases}$$

where M_{UV} is some UV cut-off. The Wilsonian effective action is defined via

$$\int [\mathcal{D}\phi]\, e^{-S[\{\phi\}]} \;=\; \int \prod_k [\mathcal{D}\phi_<(k)] \int \prod_k [\mathcal{D}\phi_>(k)]\, e^{-S[\{\phi\}]}$$

$$= \int \prod_{0 \leq |k| < \mu} [\mathcal{D}\phi_<(k)] \exp\left(-S_W[\{\phi_<\}]\right)$$

Thus in the Wilsonian effective action, one only integrates over the high energy degrees of freedom. In particular the IR region is excluded from the integrals.

Since the violation of holomorphy comes from the IR region, the Wilsonian effetcive action is free of this problem, while the 1PI effective action may, (and as we saw in general it so does), suffer from a *holomorphic anomaly*. Seiberg's arguments are therefore applicable to the Wilsonian effective action. (However, when there is no massless particle, the two notions are almost identical.) There are other limitations of the 1PI effective action. For example, the source term added to the superpotential may lead to supersymmetry breaking. (See Ref.[15] for such subtleties.)

Finally we are ready to discuss quantum corrections to *massless* supersymmetric QCD. To this end we need to determine the global symmetries of the theory. Recall from the last lecture that when $m = 0$, there are two independent flavour symmetry

$$U_\ell(N_f) \times U_r(N_f) \;\approx\; SU_\ell(N_f) \times SU_r(N_f) \times U_\ell(1) \times U_r(1)$$

$$\approx\; SU_\ell(N_f) \times SU_r(N_f) \times U_B(1) \times U_A(1),$$

where $U_B(1)$ and $U_A(1)$ are the diagonal and anti-diagonal parts of the two $U(1)$ subgroups respectively. The conserved charge arising from $U_B(1)$ is called the *baryon number*, and this is a good symmetry of the quantum theory. The axial $U_A(1)$ on the other hand, acts oppositely on the left- and right-moving quarks, and is *anomalous*. We shall normalise so as to assign baryon number $+1$ (respectively -1) to the superfields Q ($\tilde{Q}$). The various $U(1)$ quantum numbers of the quark superfields are as follows:

	Q	$\tilde{Q}$	$Q^\dagger$	$\tilde{Q}^\dagger$
L	$+1$	0	-1	0
R	0	-1	0	$+1$
B	$+1$	-1	-1	$+1$
A	$+1$	$+1$	-1	-1

The anomaly of the axial $U_A(1)$ leads to a shift of the θ-parameter: $\theta_{YM} \to \theta_{YM} - n\alpha$, where n is the coefficient of the anomaly term in the conservation law, or in other words, the number of fermion zero modes contributing to the anomaly. In this case, $n = 2N_f$, which leads to

$$\exp(2\pi i \tau) \overset{A}{\to} e^{2iN_f\alpha} \exp(2\pi i \tau).$$

In addition, we have the R-symmetry of supersymmetry discussed in lectures VI and the present one. There is no superpotential, (we are consideing the massless case[12]), the superfields Q and $\tilde{Q}$ have zero R-charge. If we write

$$Q = \phi_Q + \theta q_L + \cdots, \qquad \tilde{Q} = \phi_{\tilde{Q}} + \theta \tilde{q}_L + \cdots,$$

the $U_R(1)$ transformation properties are as follows (see Eq.(6.87)):

$$\phi_Q \to \phi_Q, \qquad\qquad \phi_{\tilde{Q}} \to \phi_{\tilde{Q}},$$
$$q_L \to e^{-i\alpha} q_L, \qquad\qquad \tilde{q}_L \to e^{-i\alpha} \tilde{q}_L,$$
$$\lambda \to e^{i\alpha} \lambda.$$

---

[12]The $m \neq 0$ case can be thought of as arising from a Higgs mechanism. This happens naturally if the $N = 1$ supersymmetric theory is embedded in the $N = 2$ theory, which has a superpotential term $\tilde{Q}\Phi Q$, where Φ is the Higgs superfield valued in the adjoint.

This symmetry is also anomalous, leading to

$$\exp(2\pi i\tau) \xrightarrow{R} e^{2i(N_c - N_f)\alpha} \exp(2\pi i\tau).$$

The two anomalous U(1) symmetries may be combined to an anomaly free $U_{\mathcal{R}}(1)$, the generator $\mathcal{R}$ of which is

$$\mathcal{R} = R + \frac{N_f - N_c}{N_f}\, A. \tag{6.129}$$

The following table shows the relevant symmetry properties of the various fields.

Field	Flavour	$U_B(1)$	$U_A(1)$	$U_R(1)$	$U_{\mathcal{R}}(1)$
L-quark q_L	$(\mathbf{N}_f, \mathbf{1})$	$+1$	$+1$	-1	$-\frac{N_c}{N_f}$
squark ϕ_Q	$(\mathbf{N}_f, \mathbf{1})$	$+1$	$+1$	0	$\frac{N_f - N_c}{N_f}$
R-quark q_R	$(\mathbf{1}, \overline{\mathbf{N}}_f)$	-1	$+1$	-1	$-\frac{N_c}{N_f}$
squark $\phi_{\tilde{Q}}$	$(\mathbf{1}, \overline{\mathbf{N}}_f)$	-1	$+1$	0	$\frac{N_f - N_c}{N_f}$
gluino λ	$(\mathbf{1}, \mathbf{1})$	0	0	$+1$	$+1$
meson $M^A_{\tilde{B}}$	$(\mathbf{N}_f, \overline{\mathbf{N}}_f)$	0	$+2$	0	$2\frac{N_f - N_c}{N_f}$
scale Λ	$(\mathbf{1}, \mathbf{1})$	0	$\frac{2N_f}{3N_c - N_f}$	$\frac{2N_c - 2N_f}{3N_c - N_f}$	0

$$\tag{6.130}$$

In the above we have used the relation (6.123) between the QCD scale Λ and the combination τ of the coupling constant g and the θ-angle.

At present we are discussing the case $N_f < N_c$ — the only gauge invariant composite fields are the mesons.

The meson superfield matrix $M^A_{\tilde{B}}$ transforms as $(\mathbf{N}_f, \overline{\mathbf{N}}_f)$ under global flavour symmetry. However, the effective superpotential ought to be a flavour singlet and hence can only be a (holomorphic) function of $\det M$, the determinant of the meson matrix[13]. It can also depend (holomorphically) on τ or equivalently, the complexified QCD scale Λ. We therefore start with

$$W_{\mathit{eff}}(M, \Lambda) = (\det M)^a\, \Lambda^b$$

---

[13] The terms $\operatorname{tr} M, \operatorname{tr} M^2$, etc. are not invariants of independent flavour rotations $SU_\ell(N_f) \times SU_r(N_f)$.

and require that W_{eff} has the right U(1) charges. This fixes a and b:

$$W_{\mathit{eff}}(M, \Lambda) = C_{N_f N_c} \left(\frac{\Lambda^{3N_c - N_f}}{\det M} \right)^{1/(N_c - N_f)}, \qquad (6.131)$$

where $C_{N_f N_c}$ is a constant which cannot be determined by the arguments used so far. Assuming that this constant does not vanish, we find that a non-trivial superpotential is generated by quantum corrections. In the next section, we shall determine the value of this constant to be

$$C_{N_f N_c} = N_c - N_f. \qquad (6.132)$$

The effective superpotential (6.131) is called the Affleck-Dine-Seiberg superpotential. Notice that it has the correct (mass) dimension. We should also add that the superpotential is non-perturbative and hence does not violate any perturbative non-renormalisation theorem.

Let us analyse the behaviour of this superpotential. In order to find the supersymmetric vacua, we set the F-term to zero. The F-term

$$\begin{aligned}
F_{AB}^* &= \frac{\partial}{\partial M_{AB}} W_{\mathit{eff}} \qquad\qquad (6.133) \\
&= (N_c - N_f) \left(\frac{\Lambda^{3N_c - N_f}}{\det M} \right)^{1/(N_c - N_f)} \left(M^{-1} \right)_{AB}
\end{aligned}$$

implies that the bosonic potential tends to zero as $\langle M_{AB} \rangle \sim a$ tends to infinity (see Fig.6.5).

Thus, we come to the strange conclusion that there is no stable supersymmetric vacuum (at finite vaccum expectation values of the meson superfields). This is inspite of the fact that the classical theory we started with had an infinitely degenerate vacua.

Although this seems strange, it is the only possibility consistent with supersymmetry. The alternative based on expectation from ordinary QCD is chiral symmetry breaking and confinement. In this case the quarks would condense in pairs

$$\langle q_{LA} \tilde{q}_{LB} \rangle \neq 0.$$

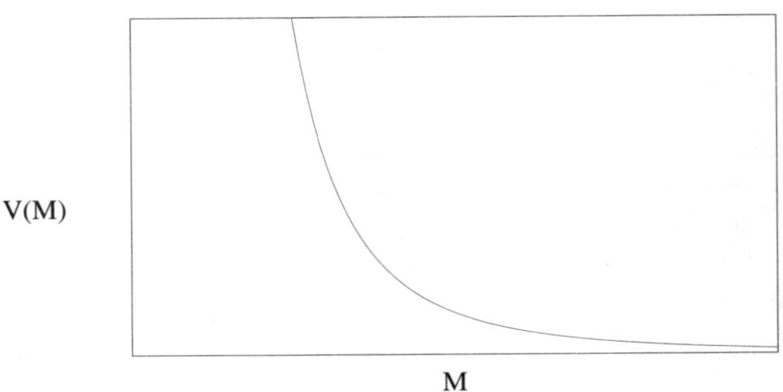

Figure 6.5: Runaway potential of the 'meson' fields.

However, this is the F-term of M_{AB}, a non-zero value for which would lead to supersymmetry breaking. Therefore a QCD like vacuum with quark pair condensate is unstable compared to any vacuum that allows for unbroken supersymmetry[14].

The gaugino bilinear $\lambda\lambda$ can, however, acquire a vacuum expectation value consistent with supersymmetry. For example, in pure super-Yang-Mills theory with $N_f = 0$, the effective superpotential is

$$W_{\textit{eff}}(\Lambda) = c_{N_c}\Lambda^3 = c_{N_c}\mu^3 \exp\left(\frac{2\pi i \tau_{\textit{eff}}}{N_c}\right). \qquad (6.134)$$

Now, $\lambda\lambda$ is the scalar component of the composite chiral superfield $W^\alpha W_\alpha$ which couples to τ. The vacuum expectation value $\langle\lambda\lambda\rangle$ can therefore be found by differentiating the effective lagrangian

---

[14]Recall that we are considering quark superfields which are massless. The situation changes drastically if these are massive. With a mass term $\mathrm{tr}(mM)$ for the mesons in the effective superpotential, one can show that a supersymmetric vacuum is possible at a finite value of M. This moves off to infinity as m is taken to zero.

$\mathcal{L}_{\textit{eff}}$ with respect to F_τ, the F-term of τ.

$$
\begin{aligned}
\langle\lambda\lambda\rangle &= 16\pi i\,\frac{\partial}{\partial F_\tau}\int d^2\theta\,W_{\textit{eff}}(\tau)\\[2mm]
&= 16\pi i\,\frac{\partial}{\partial\tau}\,W_{\textit{eff}}(\tau)\\[2mm]
&= -\frac{32\pi^2}{N_c}c_{N_c}\mu^3\exp\left(\frac{2\pi i\tau}{N_c}\right),
\end{aligned}
\tag{6.135}
$$

or, $\langle\lambda\lambda\rangle\sim e^{-8\pi^2/N_c g^2}$ (for $\theta_{YM}=0$), is the non-perturbative gaugino condensate.

Recall that the R-symmetry shifts $\lambda\to e^{i\alpha}\lambda$, hence $\theta_{YM}\to\theta_{YM}+2N_c\alpha$. When α is a multiple of π/N_c, theta angle shifts by 2π, and a $\mathbf{Z}_{2N_c}$ subgroup of $U_R(1)$ remains unbroken. However the vacuum expectation value $\langle\lambda\lambda\rangle$ is invariant under $\lambda\to-\lambda$ ($\alpha=\pi$), breaking $\mathbf{Z}_{2N_c}$ spontaneously to a $\mathbf{Z}_2$ subgroup. This leads to N_c inequivalent vacuum states in pure super-Yang-Mills theory parametrised by the vev of gaugino bilinear

$$
\langle\lambda\lambda\rangle=\exp\left(\frac{2\pi i m}{N_c}\right),\quad m=0,1,\cdots,N_c-1.
\tag{6.136}
$$

Exercise: *(Holomorphic decoupling) Consider a theory of two chiral superfields given by the superpotential*

$$
W(\Phi_1,\Phi_2)=\frac{1}{2}m\Phi_2^2+g\Phi_1^2\Phi_2.
$$

Analyse the classical moduli space of this theory and show that at a generic point the field Φ_2 is heavy. Find the effective superpotential for the light field Φ_1 by 'integrating out' Φ_2. (In the classical theory, this means that one ignores the kinetic term for the heavy field, and solves its resulting algebraic equation of motion. Now apply Seiberg's analysis to determine the low energy effective superpotential assuming that the only relevant degrees of freedom are those of the light field Φ_1.

Compare the two results.

Identify the diagram that 'renormalises' the superpotential, (it is actually a tree diagram), and see that it does not violate the non-renormalisation theorem.

6.10   Quantum corrections & effective action II

We begin this section by determining the constant in the Affleck-Dine-Seiberg superpotential for supersymmetric gauge theories with $N_f < N_c$. Then we shall move on to the case $N_f \geq N_c$.

The strategy used to find the constants $C_{N_f N_c}$ in (6.131) is in relating the theory with N_f flavours to that with $N_f - 1$ flavours through *holomorphic decoupling*. The idea of decoupling is quite simple and is illustrated in an exercise in the previous section. Let us start with the theory with N_f flavours and add (by hand) a mass term to the last, *i.e.*, N_fth quark superfield $M_{N_f N_f}$:

$$W_{\mathit{eff}}^{(N_f)}(M, \Lambda) = C_{N_c N_f} \left(\frac{\Lambda^{3N_c - N_f}}{\det M} \right)^{1/(N_c - N_f)} + m M_{N_f N_f}.$$

(6.137)

The equations of motion for the mesons with only one index N_f are

$$\frac{\partial W^{(N_f)}}{\partial M_{N_f A}} \bigg|_{A \neq N_f} = -\frac{C_{N_c N_f}}{N_c - N_f} \left(\frac{\Lambda^{3N_c - N_f}}{\det M} \right)^{1/(N_c - N_f)} \left(M^{-1} \right)_{N_f A} = 0$$

(6.138)

and similar equations for the components $M_{A N_f}$. The potential is minimised when all these components (along the last row and column) vanish. Therefore, the meson matrix of the theory reduces to a block-diagonal form with an $(N_f - 1) \times (N_f - 1)$ block and $M_{N_f N_f}$ as the last diagonal element. Thanks to this, the determinant of the of the meson matrix factorises:

$$\det \|M_{(N_f)}\| = \det \|(M_{(N_f - 1)}\| \times M_{N_f N_f}.$$

(6.139)

It remains to solve for $M_{N_f N_f}$ from its equation of motion, obtained by minimising $W^{(N_f)}$:

$$\frac{\partial W^{(N_f)}}{\partial M_{N_f N_f}} = -\frac{C_{N_c N_f}}{N_c - N_f} \left(\frac{\Lambda^{3N_c - N_f}}{\det M} \right)^{1/(N_c - N_f)} \left(M^{-1} \right)_{N_f N_f} + m = 0.$$

$$(6.140)$$

Using (6.139) and (6.140), the solution is

$$M_{N_f N_f} = \left[\frac{C_{N_c N_f}}{m(N_c - N_f)} \left(\frac{\Lambda^{3N_c - N_f}}{\det M_{(N_f - 1)}} \right)^{\frac{1}{N_c - N_f}} \right]^{\frac{N_c - N_f}{N_c - N_f + 1}}$$

$$(6.141)$$

Substituting above in (6.137) in turn, one finds the effective superpotential

$$W_{\textit{eff}} = (N_c - N_f + 1) \left(\frac{C_{N_c N_f}}{N_c - N_f} \right)^{\frac{N_c - N_f}{N_c - N_f + 1}} \left(\frac{m \Lambda_{(N_f)}^{3N_c - N_f}}{\det M_{(N_f - 1)}} \right)^{\frac{1}{N_c - N_f + 1}}$$

$$(6.142)$$

This resembles the effective superpotential of the theory with $N_f - 1$ flavours. Indeed, the match is exact provided we make the following identifications

$$C_{N_c (N_f - 1)} = (N_c - N_f + 1) \left(\frac{C_{N_c N_f}}{N_c - N_f} \right)^{\frac{N_c - N_f}{N_c - N_f + 1}} \qquad (6.143)$$

$$\Lambda_{(N_f - 1)}^{3N_c - N_f + 1} = m \, \Lambda_{(N_f)}^{3N_c - N_f}. \qquad (6.144)$$

The first of these equations can be rewritten as

$$\left(\frac{C_{N_c N_f}}{N_c - N_f} \right)^{N_c - N_f} = \left(\frac{C_{N_c (N_f - 1)}}{N_c - N_f + 1} \right)^{N_c - N_f + 1}, \qquad (6.145)$$

which is a recursion relation for the coefficients in the Affleck-Dine-Seiberg superpotential. It is easy to see that $C_{N_c N_f} = N_c - N_f$ in (6.132) is a solution to (6.145). A more systematic solution would need a boundary condition. This comes from the theory with $N_f = N_c - 1$, in which it is possible to calculate the effective superpotential exactly from instanton calculation. This yields the value $C_{N_C(N_c-1)} = 1$.

The requirement (6.144) is also consistent with the running of the coupling constant from the renormalisation group equation:

$$\frac{8\pi^2}{g^2(\mu)} = - (3N_c - N_f) \ln \left(\frac{\Lambda_{(N_f)}}{\mu} \right). \qquad (6.146)$$

In the theory described by the effective superpotential (6.137), the number of 'light' flavour degrees of freedom is N_f above the scale m. But below this scale it is $N_f - 1$. Hence in this regime it obeys the Eq.(6.146) with $N_f \rightarrow N_f - 1$. By continuity, the coupling constants should match at the transition scale $\mu = m$. Demanding that this happens is precisely the condition (6.144).

In summary, we have determined the effective superpotential governing the dynamics of supersymmetric QCD with small number of flavours $N_f < N_c$. Curiously, this superpotential does not have allow for a supersymmetric vacuum at any finite vev of the mesons, which are the relevant gauge invariant degrees of freedom. The only exception is pure SYM theory, in which there are no mesons, and which have N_c supersymmetric vacua parametrised by the vev of the gluino condensate.

Let us now turn to the cases where the number of flavour is larger than the number of colour: $N_f \geq N_c$. The story is much more complex here and we shall have to content ourselves with a summary of the main results and only a brief mention of some of the arguments.

Consider the case of equal number of flavours and colours $N_f = N_c$ first. The effective superpotential (6.131) for $N_f < N_c$ cannot evidently be extrapolated to this case. Indeed, if we look back at the table (6.130), we find that all the R-charges of the fields Q, $\bar{Q}$, M, B and of the scale Λ vanish in this case. It is, therefore, impossible to write a superpotential, which has R-charge $+2$ in terms of these.

Recall that classically this theory has a moduli space of vacua parametrised by the meson matrix $||M||$ and the baryons B, $\tilde{B}$

subject to the constraint (6.120): $\det\langle M\rangle = B\tilde{B}$. Schematically, the geometry of this space is that of a cone. At large values of $\langle M\rangle$, we are in a Higgs phase. However, at the origin of this space, all fields are massless and gauge symmetry is restored. Thus the origin is a 'singular' point. (See Fig.6.6.)

Seiberg argued that quantum mechanically, the relation (6.120) gets modified to

$$\det M - B\tilde{B} = \Lambda^{2N_c}. \tag{6.147}$$

There is no symmetry that prevents this modification — both sides have the same R-charge (zero) and dimension ($2N_c$). Moreover, it is consistent with holomorphic decoupling when we add a mass term $mM_{N_cN_c}$ to the last meson. To see how this works, we introduce an auxiliary chiral superfield Φ of R-charge and dimension 2, to write a superpotential

$$W_{\mathit{eff}} = \frac{1}{\Lambda^{2N_c-1}}\,\Phi\left(B\tilde{B} - \det M + \Lambda^{2N_c}\right) + m\,M_{N_cN_c}. \tag{6.148}$$

The equation of motion of Φ implements the relation (6.147). At this point we can simply repeat the steps we followed in the case $N_f < N_c$. The F-flatness conditions for the superpotential can be solved easily. We find that

$$\begin{aligned}
B = \tilde{B} &= 0, \\
\Phi &= \frac{m\Lambda^{2N_c-1}}{\det M_{(N_c-1)}}, \\
M_{N_cN_c} &= \frac{\Lambda^{2N_c}}{\det M_{(N_c-1)}}.
\end{aligned}$$

Substituting these values in (6.148), one gets the effective superpotential of the theory with $N_f = N_c - 1$. Finally, there are other, more sophisticated, checks in favour of Seiberg's argument that we are not in a position to get into.

The quantum moduli space defined by Eq.(6.147) is no longer singular (see Fig.6.6). In the asymptotic regions (where the vev of M is large), quantum corrections are small and the moduli space resembles the classical description. The theory is in the

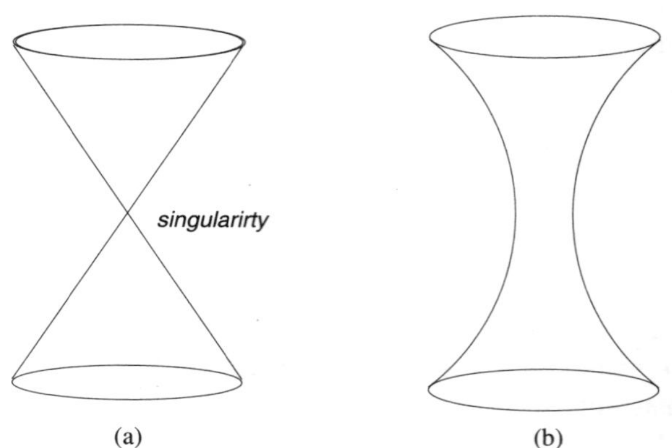

Figure 6.6: Moduli space of supersymmetric vacua of SQCD with $N_f = N_c$ (a) classical (6.120) and (b) quantum (6.147).

Higgs phase. However, the situation is quite different for small values of M. There is no singularity, vevs of M never vanish and consequently do not become massless. The gauge theory is *confining*. We see that the Higgs and the confining phases are continuously connected in the quantum moduli space.

As a matter of fact, there is no gauge invariant way of dinstiguishing these two phases in a theory with fundamental matter (quarks). The phases are characterised by the potential (between a quark-antiquark pair), which grows linearly in the confining phase and is a constant in the Higgs phase:

$$V_C(r) \quad \propto \quad r, \qquad (6.149)$$

$$V_H(r) \quad = \quad \text{constant}. \qquad (6.150)$$

However, the former is unstable to pair production. As the separation between the quark and the antiquark increases the energy of the flux string grows. Beyond a point it is more favourable to break the string by producing a quark-antiquark pair. Thus the potential, although increasing linearly for small separation, saturates beyond a point and becomes a constant.

For the next case $N_f = N_c+1$, Seiberg argued that the effective superpotential

$$W_{eff} = \frac{1}{\Lambda^{2N_c-1}} \left(\det M - B_A M_{AB} \tilde{B}_B \right) \tag{6.151}$$

is generated. It is easy to see that W_{eff} has the correct dimension and R-charge. One can also check that it is consistent with holomorphic decoupling.

To proceeed beyond this, let us first consider the case of very large number of flavour. The β-function vanishes at $N_f = 3N_c$ and is positive beyond this value. Therefore, supersymmetric QCD with number of flavour exceeding $3N_c$ is not asymptotically free. Clearly this is very different from what we have seen so far. This theory is free in the infra-red with the degrees of freedom being the gluons and quarks. It is in a phase where the interaction between *electric* type test charges is a non-abelian analogue of the Coulomb interaction — the potential $V(r)$ falls off as

$$V_E(r) \sim \frac{1}{r \ln(r\Lambda)}. \tag{6.152}$$

This is called the *free electric phase*.

Let us return to the theories in which N_f is bigger than N_c, but not by much. The baryons carry $N_f - N_c \equiv \check{N}_c$ indices. These can therefore be viewed as bound states of $\check{N}_c$ component fields. Seiberg suggested that these be thought of as physical asymptotic states of a *dual* theory with gauge group SU($\check{N}_c$). It also has chiral superfields corresponding to dual quarks $\check{Q}_A$ and antiquarks $\tilde{\check{Q}}_A$ that come in N_f flavours ($A = 1, 2, \cdots, N_f$) and in addition, gauge singlet chiral superfield T_{AB} coupled to the dual (anti-)quarks through the superpotential

$$W = \check{Q}_A T_{AB} \tilde{\check{Q}}_B.$$

The relation between this description and the original one is called
the *nonabelian electric-magnetic duality*. In terms of the dual
theory, the β-function vanishes at $N_f = 3\check{N}_c$, *i.e.*, at $N_f = 3N_c/2$.
For lower values of N_f the dual theory is free in the infra-red with
the dual gluons and quarks being the relevant degrees of freedom.
The potential between electric charges of the dual theory

$$V_M(r) = \frac{\ln(r\Lambda)}{r} \qquad (6.153)$$

is again Coulomb like. By duality this is the behaviour between
the magnetic charges of the original SQCD. In the region $N_c+2 \leq$
$N_f < \frac{3}{2}N_c$, the theory is said to be in the *free magnetic phase*.

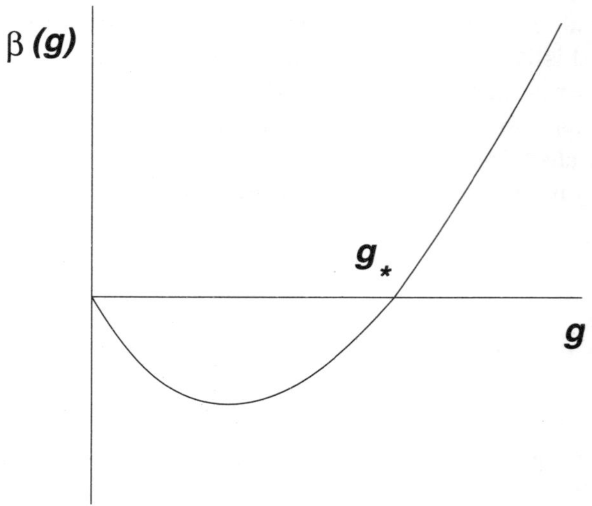

Figure 6.7: β-function of SQCD for $\frac{3}{2}N_c \leq N_f \leq 3N_c$. Notice the
nontrivial fixed point at g_*.

Finally, in the range $\frac{3}{2}N_c \leq N_f \leq N_c$, the β-function has a
more interesting behaviour. The first and the second coefficients

in

$$\beta(g) = -\left(3N_c - N_f\right)\frac{g^3}{16\pi^2} + \left(2N_cN_f - 3N_c^2 - \frac{N_f}{N_c}\right)\frac{g^5}{128\pi^4}$$

are comparable and there is a new fixed point at $g = g_* \neq 0$ (see Fig.6.7). This corresponds to a nontrivial interacting superconformal field theory. At the limiting values $N_f = 3N_c/2$ (respectively $N_f = 3N_c$) the (dual) mesons become free fields. Their (anomalous) dimension is one.

In summary, for $N_c + 2 \leq N_f < 3N_c/2$ and $N_f > 3N_c$ the theory is IR-free and is of magnetic, respectively electric, type (in terms of the original microscopic degrees of freedom). In the intermediate region $\frac{3}{2}N_c \leq N_f \leq N_c$, it is a superconformal field theory corresponding to a nontrivial zero of the β-function.

6.11 Appendix: Fierz identities

There are some identities which involve spinors. Much of it follows from simple angular momentum addition rule.

If we take the product of a spin $s_1 = \frac{1}{2}$ representation[15] with another one $s_2 = \frac{1}{2}$, the result is a direct sum of spin-0 and spin-1 representations:

$$\begin{aligned}
(s_1 = 1/2) \otimes (s_2 = 1/2) &= (s^{(1)} = 0) \oplus (s^{(2)} = 1),\\
\textit{i.e,} \qquad \mathbf{2} \otimes \mathbf{2} &= \mathbf{1} \oplus \mathbf{3}.
\end{aligned}$$

A simpler fact is

$$\mathbf{2} \otimes \mathbf{1} = \mathbf{2}.$$

Since the Lorentz algebra is (almost) a direct sum of two angular momentum algebras (see (6.5)), the above rules for tensoring representations apply to the Lorentz algebra as well. The defining representations here are the $\left(\frac{1}{2}, 0\right)$ or $(\mathbf{2}, \mathbf{1})$ representation called the left-handed spinor ψ_α and the $\left(0, \frac{1}{2}\right)$ or $(\mathbf{1}, \mathbf{2})$ representation called the right-handed spinor $\bar{\chi}^{\dot\alpha}$.

---

[15]The spin-$\frac{1}{2}$ representation is called the defining or fundamental representation of SU(2), as all other representations can be constructed from it.

There are three basic possibilities in combining these representations.

- We can tensor two left-handed spinors

$$(\mathbf{2}, \mathbf{1}) \otimes (\mathbf{2}, \mathbf{1}) = (\mathbf{1}, \mathbf{1}) \oplus (\mathbf{3}, \mathbf{1}).$$

More explicitly in terms of the spinors

$$\psi_\alpha \chi_\beta = \frac{1}{2}\, \epsilon_{\alpha\beta}(\psi\chi) - \frac{1}{2}\, (\psi\sigma_{\mu\nu}\chi)\, \sigma^{\mu\nu}{}_{\alpha\beta}, \qquad (6.154)$$

we get a scalar and a rank-2 self-dual antisymmetric tensor.

Notice that the first term in the RHS above is antisymmetric (in indices α and β) while the second one is symmetric, as is expected from the representations of the angular momentum algebra. The coefficients $\frac{1}{2}\epsilon_{\alpha\beta}$ and $-\frac{1}{2}\sigma^{\mu\nu}{}_{\alpha\beta}$ are the Clebsch-Gordon coefficients relating different representations.

In order to prove (6.154), based on covariance properties one can argue that

$$\begin{aligned}
\psi_\alpha \chi_\beta &= \frac{1}{2}(\psi_\alpha \chi_\beta - \chi_\beta \psi_\alpha) + \frac{1}{2}(\psi_\alpha \chi_\beta + \chi_\beta \psi_\alpha), \\
&= \frac{1}{2}\, A\epsilon_{\alpha\beta} + \frac{1}{2}\, B_{\mu\nu}\sigma^{\mu\nu}{}_{\alpha\beta}.
\end{aligned}$$

The unknown coefficients A and $B_{\mu\nu}$ may be determined by contracting both sides of the above equation by $\epsilon^{\beta\alpha}$ and $\sigma^{\rho\lambda\beta\alpha}$ respectively. For the latter, we need to use the fact that $\sigma_{\mu\nu}$ is self-dual (6.9), and that

$$\mathrm{tr}\,(\sigma^{\rho\lambda}\sigma^{\mu\nu}) = \frac{1}{2}(\eta^{\rho\mu}\eta^{\lambda\nu} - \eta^{\rho\nu}\eta^{\lambda\mu}) + \frac{i}{2}\, \epsilon^{\rho\lambda\mu\nu}.$$

- Similarly, for two right-handed spinors

$$(\mathbf{1}, \mathbf{2}) \otimes (\mathbf{1}, \mathbf{2}) = (\mathbf{1}, \mathbf{1}) \oplus (\mathbf{1}, \mathbf{3}).$$

More explicitly in terms of the spinors

$$\bar{\psi}^{\dot\alpha}\bar{\chi}^{\dot\beta} = \frac{1}{2}\,\epsilon^{\dot\alpha\dot\beta}(\bar\psi\bar\chi) - \frac{1}{2}\,(\bar\psi\bar\sigma_{\mu\nu}\bar\chi)\,\bar\sigma^{\mu\nu\dot\alpha\dot\beta}, \qquad (6.155)$$

where on the RHS we find a scalar and a rank-2 anti-self-dual antisymmetric tensor.

- Finally, combining a left- and a right-handed spinor

$$\begin{aligned}
(\mathbf{2},\mathbf{1})\otimes(\mathbf{1},\mathbf{2}) &= (\mathbf{2},\mathbf{2}), \\
\psi_\alpha\bar\chi_{\dot\beta} &= \frac{1}{2}\,(\psi\sigma_\mu\bar\chi)\,\sigma^\mu_{\alpha\dot\beta}, \qquad (6.156)
\end{aligned}$$

we get a Lorentz vector.

Using these relations, following from basic group theoretic facts, it is easy to derive the following relations.

$$\begin{aligned}
\psi_\alpha\psi_\beta &= \frac{1}{2}\,\epsilon_{\alpha\beta}(\psi\psi) \\
\bar\chi_{\dot\alpha}\bar\chi_{\dot\beta} &= -\frac{1}{2}\,\epsilon_{\dot\alpha\dot\beta}(\chi\chi) \\
(\theta\psi)(\theta\chi) &= -\frac{1}{2}(\theta\theta)(\psi\chi) \\
(\bar\theta\bar\psi)(\bar\theta\bar\chi) &= -\frac{1}{2}(\bar\theta\bar\theta)(\bar\psi\bar\chi) \\
\psi\sigma^\mu\bar\chi &= -\bar\chi\sigma^\mu\psi \qquad (6.157) \\
\psi\sigma^\mu\bar\sigma^\nu\chi &= \chi\sigma^\nu\bar\sigma^\mu\psi \\
(\sigma^\mu\bar\theta)_\alpha(\theta\sigma^\nu\bar\theta) &= \frac{1}{2}\,\eta^{\mu\nu}\theta_\alpha(\bar\theta\bar\theta) - i(\sigma^{\mu\nu}\theta)_\alpha(\bar\theta\bar\theta) \\
(\theta\sigma^\mu\bar\theta)(\theta\sigma^\nu\bar\theta) &= \frac{1}{2}\,\eta^{\mu\nu}(\theta\theta)(\bar\theta\bar\theta) \\
(\theta\psi)(\bar\theta\bar\chi) &= \frac{1}{2}(\theta\sigma^\mu\bar\theta)(\psi\sigma_\mu\bar\chi)
\end{aligned}$$

These are known as *Fierz identities*, and are very useful in simplifying expressions involving many spinor fields.

Bibliography

[1] J. Lykken, *Introduction to supersymmetry*, TASI lectures 1996, (`hep-th/9612114`).

[2] J. Wess and J. Bagger, *Supersymmetry and supergravity*, 2nd Ed (1992), Princeton University Press, Princeton.

[3] M. Peskin, *Duality in supersymmetric Yang-Mills theory*, TASI lectures 1996, (`hep-th/9702094`).

[4] N. Seiberg, *The dynamics of $N = 1$ supersymmetric field theories in four dimensions* in Quantum fields & strings: A course for mathematicians, Eds. P. Deligne *et al*, also available at `http://medan.math.ias.edu/QFT/spring/index.html`.

[5] P. Sohnius, *Phys. Rep.* **128** (1985) 39.

[6] P. West, *Introduction to supersymmetry and supergravity*, World Scientific, Singapore.

[7] Y. Golfand and E. Likhtman, *JETP Lett.* **13** (1971) 323.

[8] J. Wess and B. Zumino, *Phys. Lett.* **B49** (1974) 52.

[9] S. Coleman and Mandula, *Phys. Rev.* **159** (1967) 1251.

[10] R. Haag, J. Łopuszanskí and M. Sohnius, *Nucl. Phys.* **B88** (1975) 257.

[11] A. Salam and J. Strathdee, *Nucl. Phys.* **76** (1974) 477.

[12] L. O'Raifeartaigh, *Nucl. Phys.* **B96** (1975) 331.

[13] P. Fayet and Iliopoulos, *Phys. Lett.* **B51** (1974) 461.

[14] M. Luty and W. Taylor, *Phys. Rev.* **D53** (1996) 3399, (`hep-th/9506098`).

[15] K. Intriligator and N. Seiberg, *Lectures on supersymmetric gauge theories and electric-magnetic duality, Nucl. Phys. Proc. Suppl. B45* (1996) 1, (`hep-th/9509066`).

[16] N. Seiberg, *Phys. Lett.* **B318** (1993) 469.

[17] I. Jack, D. Jones and P. West, *Phys. Lett.* **B258** (1991) 382.

[18] M. Shifman and A. Vainshtein, *Nucl. Phys.* **B277** (1986) 456.

Chapter 7

Introduction to AdS/CFT Duality

Dileep P. Jatkar

7.1 Introduction

In these notes on the AdS/CFT correspondence[1] I have tried to present the material as it was covered in the lectures given at the SERC school. The AdS/CFT correspondence (or duality) has been studied in great detail over last few years and there are several reviews are available in the literature. Some of the useful references are [2, 3, 4][1]. It is not possible to cover the entire subject in these lecture notes, nor is it my intention to do so. I would, in fact, restrict myself to a simpler and modest goal of stating the correspondence and giving some elementary methods of doing checks on the predictions of this correspondence.

I have assumed that the reader has basic knowledge of field theory and general relativity. I have tried to keep these notes self-contained by giving simple examples, whenever needed, by borrowing them from either field theory or relativity.

I thank Ashoke Sen for teaching me the AdS/CFT correspondence, Charan Aulakh for explaining Kaluza-Klein formalism and

---

[1]The list of references is horribly incomplete. The reader is referred to [4] for a larger list of references.

Debashis Ghoshal and Sudhakar Panda for numerous discussions on this topic. I also thank organisers of the SERC School, Kolkata for kind hospitality.

Standard model of particle physics is a gauge theory with $SU(3) \times SU(2) \times U(1)$ gauge group. It gives partial unification of strong, weak and electromagnetic forces. $SU(3)$ gauge theory is a theory of strong interactions, also known as quantum chromodynamics (QCD). Whereas $SU(2) \times U(1)$ is a gauge theory for electroweak interactions. Fermionic matter in this model contains leptons and quarks. Leptons are observed in free state and do not have strong interactions. Quarks, on the other hand, are not found in free state, they are confined. Baryons and mesons which are formed out of three quarks and a quark-antiquark pair respectively, are observed in free state. Most of the experimental results are in good agreement with the predictions of the Standard Model. (For details, see the other lectures in this volume.) It is, however, still incomplete because

1. It does not include the force due to gravity.

2. The phenomenon of quark confinement is quite poorly understood in the standard model.

QCD is an asymptotically free theory, *i.e.*, at high energies perturbation expansion is more and more reliable. Confinement, on the other hand, is a low energy, strong coupling effect.

Historically in 1960's dual resonance model of hadrons was popular. This model qualitatively explained the occurance of several baryonic and mesonic resonances in terms of vibrational excitations of a string. Later it was discovered that baryons and mesons are formed out of quarks and QCD described their high energy behaviour much more convincingly than the ad-hoc string hypothesis of dual models.

The string hypothesis, however, was so appealing that to this date we believe that physics of quark confinement can be uncovered if we can understand "QCD string" better. Another support to the hypothesis of QCD string came from 't Hooft's large N expansion of $SU(N)$ gauge theories. He rearranged perturbation expansion of gauge theories in terms of $1/N$ expansion. This $1/N$ expansion naturally generated two dimensional Riemann surfaces which occur in the perturbation expansion of interacting closed string. The action for a $U(N)$ gauge theory is

$$S = -\frac{1}{4g_{YM}^2} \int d^4x \, \mathrm{Tr}(F_{\mu\nu}^a F^{a\mu\nu}) \tag{7.1}$$

where, $\mu, \nu, \cdots = 1, 2, 3$; $a, b, \cdots = 1, 2, \cdots, N^2$ and

$$F_{\mu\nu}^a = \partial_\mu A_\nu^a - \partial_\nu A_\mu^a + f^{abc} A_\mu^b A_\nu^c. \tag{7.2}$$

The gauge field A_μ^a can be written in terms of $N \times N$ matrices

$$(A_\mu)_j^i = A_\mu^a (T^a)_j^i, \tag{7.3}$$

where T^a are generators of $U(N)$ in fundamental representation, *i.e.*, $i, j = 1, \cdots N$ and

$$\begin{aligned}
(F_{\mu\nu})_j^i = F_{\mu\nu}^a (T^a)_j^i &= (\partial_\mu A_\nu)_j^i - (\partial_\nu A_\mu)_j^i \\
&+ (A_\mu)_k^i (A_\nu)_j^k - (A_\nu)_k^i (A_\mu)_j^k
\end{aligned} \tag{7.4}$$

and

$$\mathrm{Tr}(F_{\mu\nu}^a F^{a\mu\nu}) = \mathrm{Tr}\left((F_{\mu\nu})_j^i (F^{\mu\nu})_k^j\right). \tag{7.5}$$

Recall propagator for the gauge field A_μ^a is denoted by a wavy line.

$$a \,\, \mathord{\sim\!\!\!\sim\!\!\!\sim\!\!\!\sim\!\!\!\sim}\, b \underset{p_\mu}{} \quad = \quad \frac{\delta^{ab}}{p^2 + i\epsilon}\left(\frac{g_{\mu\nu}p^2 - p_\mu p_\nu}{p^2}\right)$$

In the two index notation propagator for $(A_\mu)_j^i$ is denoted by a double line.

$$\xrightarrow{\quad j \qquad p_\mu \qquad k \quad}_{\;i \qquad\qquad\qquad l} \quad = \quad \frac{\delta^i_l \delta^k_j}{p^2 + i\epsilon}\left(\frac{g_{\mu\nu}p^2 - p_\mu p_\nu}{p^2}\right)$$

The arrows on the double line graph do not represent momentum. Double line graphs with arrow provide orientation to the graph generated by them. From group theoretic point of view, the gauge field A_μ belongs to the adjoint representation of the gauge group $U(N)$. Adjoint representation can be obtained by taking tensor product of the fundamental representation and the anti-fundamental representation.

$$N \otimes \bar{N} = N^2. \tag{7.6}$$

In the $N \times N$ matrix notation for $A_\mu \equiv (A_\mu)^i_j$ we will denote the fundamental representation by subscript and anti-fundamental by superscript. The arrow is always directed from fundamental representation index to anti-fundamental representation index.

Let us now redefine the gauge coupling

$$\lambda = N g_{YM}^2. \tag{7.7}$$

This new coupling λ is known as the 't Hooft coupling. In terms of this new coupling, Yang-Mills action can be written as

$$S_{YM} = -\frac{1}{4}\frac{N}{\lambda} \int d^4x\, F^i_{\mu\nu\,j} F^{\mu\nu\,j}_i. \tag{7.8}$$

Notice that each interaction vertex, whether cubic or quartic, has a coupling N/λ and the propagator is proportional to λ/N.

Fig. 7.1(i) is a vacuum diagram with two vertices and three propagators. Notice that it has three (closed) index loops. Every index loop contributes a factor N to the combinatorics of the diagram. Thus combinatorial factor for this diagram is

$$\left(\frac{\lambda}{N}\right)^3 \times \left(\frac{N}{\lambda}\right)^2 \times N^3 = N^2\lambda \tag{7.9}$$

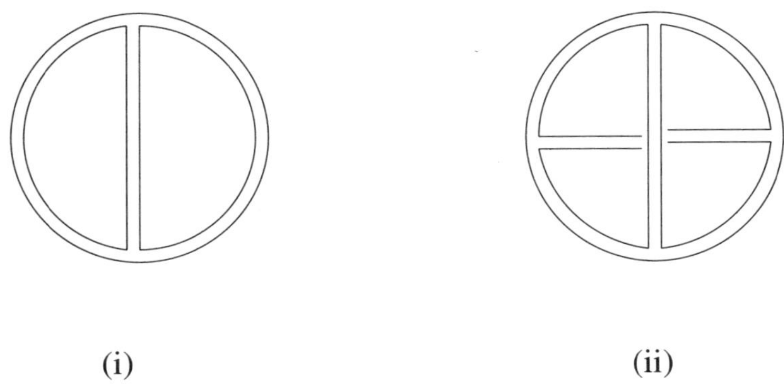

(i)   (ii)

Figure 7.1: Double line representation of Feynman graphs.

Fig. 7.1(ii) is a vacuum diagram with four vertices and six propagators. It also has two index loops. Combinatorial factor for this diagram is

$$\left(\frac{\lambda}{N}\right)^6 \times \left(\frac{N}{\lambda}\right)^4 \times N^2 = N^0 \lambda^2 \qquad (7.10)$$

Using these examples we can write down a general formula for vacuum graphs in this theory. If a diagram has V vertices, E propagators and F loops then this diagram has a coefficient proportional to

$$N^{V-E+F} \lambda^{E-V}. \qquad (7.11)$$

While Fig. 7.1(i) can be drawn on a plane or a sphere, Fig. dblfig(ii) cannot be drawn on a plane without overlapping propagators. Fig. 7.1(ii), however can be drawn on a torus, *i.e.*, on a sphere with one handle. In general, double line graphs can be classified by their topology. More precisely, they are classified by the topology of the simplest two dimensional surface on which it can be drawn.

Compact two dimensional manifolds, generically called Riemann surfaces, are classified by a single topological number called the Euler characteristics.

$$\chi = 2 - 2h \tag{7.12}$$

where h is number of handles. *E.g.*, sphere has no handles, $h = 0$; torus has one handle, $h = 1$; etc. Euler characteristic of a Riemann surface can be computed by drawing a triangular lattice on it. Let a triangular lattice drawn on a Riemann surface has V vertices, E edges and F faces then the Euler number

$$\chi = V - E + F. \tag{7.13}$$

Comparing this with the power of N in the vacuum diagrams, we see that powers of N classify the graphs according to the topology of two dimensional surfaces on with these graphs can be drawn.

$$N^{V-E+F}\lambda^{E-V} = N^{\chi}\lambda^{E-V}. \tag{7.14}$$

In the large N limit only the leading term, *i.e.*, $o(N^2)$ term dominates and terms of $o(1)$ or $o(1/N^2)$ are highly suppressed. Therefore, in the large N limit it appears that we have replaced a perturbative expansion in terms of gauge coupling g_{YM} by another expansion in powers of $1/N$. A natural question one would ask at this point is, what do we gain by doing this?

Clearly, this is a rearrangement of the usual perturbative expansion of the Yang-Mills theory. The advantage, however, is now every term in the $1/N$ expansion contains all order terms with respect to the Yang-Mills coupling g_{YM}. Thus every term in the $1/N$ expansion is expected to capture physics of the Yang-Mills theory to all orders in perturbation theory and hopefully more. For example, if we could write down even the leading order term in the $1/N$ expansion in a closed form, we would get some handle on the nonperturbative physics of Yang-Mills theories.

Let us make a small digression into string theory. Consider propagation of a closed string in space-time. A non-interacting closed string maps out a cylinder. Let us define the interaction as splitting and joining of closed strings. Space-time diagrams of the interacting string are drawn in Fig. 7.2.

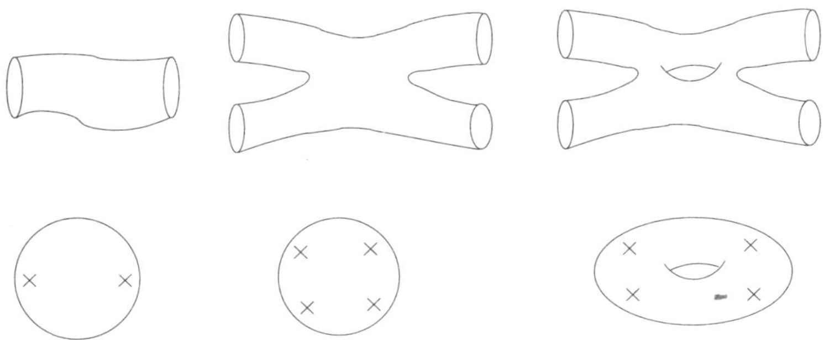

Figure 7.2: Propagation and interactions of closed string and their representation in terms of topologically equivalent Riemann surfaces.

Thus perturbative expansion of interacting closed string is identical to the topological expansion of two dimensional surfaces. Thus as mentioned earlier large N expansion of gauge theories have some hidden information about "QCD string".

What happened to dual resonance models of hadrons?

In the seventies they reappeared with a new interpretation as string theory which naturally incorporates quantum theory of gravity. Recent developments in the supersymmetric string theory has led to concrete developments in relating large N gauge theories to supersymmetric string theory.

Throughout these lectures we will be concentrating on the low energy limit of the superstring theory which is known as the supergravity.

7.2   The large N duality conjecture

In this section we will mention the celebrated Maldacena conjecture. We will, however, only be able to study part of this conjecture. Therefore, we will set a modest goal for us in these lectures, which will be stated after the statement of the conjecture.

The Maldacena Conjecture: Type IIB superstring theory on $AdS_5 \times S^5$ is dual to conformally invariant $\mathcal{N} = 4$ supersymmetric $SU(N)$ gauge theory in four dimensions.

Notice that the two theories which are dual to each other are defined in two different dimensions.

1. Type IIB superstring theory is defined in ten dimensions. It does not contain any Yang-Mills type gauge symmetry in ten dimensions.

2. $\mathcal{N} = 4$ supersymmetric $SU(N)$ gauge theory is defined in four dimensions. It is not a theory of gravity.

In these set of lectures we will restrict ourselves to a smaller task of relating ten dimensional type IIB supergravity to $\mathcal{N} = 4$ supersymmetric gauge theory. Type IIB supergravity appears as a low energy limit of type IIB superstring theory. So the modest goal that we would like to achieve in these set of lectures is,

The Conjecture Simplified: Type IIB supergravity theory on $AdS_5 \times S^5$ is dual to conformally invariant $\mathcal{N} = 4$ supersymmetric $SU(N)$ gauge theory in four dimensions.

Let us clarify some terms which appear in the statement of the conjecture.

1. Type IIB supergravity: This is an $\mathcal{N} = 2$ supersymmetric theory of gravity defined in ten dimensions. In ten dimensions there are two $\mathcal{N} = 2$ supersymmetric theories of gravity depending upon the chirality of two supercharges. If two supercharges have opposite chirality they give rise to what is known as type IIA supergravity. If two supercharges have same chirality then we get type IIB supergravity.

2. $AdS_5 \times S^5$: The ten dimensional space on which we will be studying type IIB supergravity is a product of two five dimensional spaces AdS_5 is five dimensional anti-de Sitter space and S^5 is five dimensional sphere. AdS_5 is a non-compact space and depending upon our choice of signature its boundary is either S^4, a four dimensional sphere or $\tilde{M}^4$,

"compactified" four dimensional Minkowski space. S^5 being a sphere is a compact space.

3. $\mathcal{N} = 4$ supersymmetric $SU(N)$ gauge theory: This is a four dimensional gauge theory with four supersymmetries. Due to very high degree of supersymmetry, not only the gauge fields but all the fields are in the adjoint representation of the gauge group. This is because all the fields in the theory belong to single multiplet of $\mathcal{N} = 4$ supersymmetry algebra. We will be interested in the large N limit of this theory.

We will start our investigation from the supergravity side. Let us briefly discuss some basic concepts of general relativity. This will also help us set up our notations.

In the general theory of relativity, the basic object of interest is the metric $g_{\mu\nu}(x)$ $(0 \leq \mu, \nu \leq 3)$, x^0 is time coordinate and $\vec{x}$ are space coordinates. The metric is symmetric in its indices, $g_{\mu\nu}(x) = g_{\nu\mu}(x)$. In the absence of gravitational field the metric reduces to the Minkowski metric $g_{\mu\nu}(x) = \eta_{\mu\nu}$, where $\eta_{\mu\nu} = \mathrm{diag}(-1, 1, 1, 1)$. If the gravitational field is present then $g_{\mu\nu}(x) \neq \eta_{\mu\nu}$. However, if the gravitational field is weak then we can expand the metric $g_{\mu\nu}(x)$ about the flat Minkowski metric.

$$g_{\mu\nu}(x) = \eta_{\mu\nu} + \kappa h_{\mu\nu}(x), \tag{7.15}$$

where, $\kappa = \sqrt{8\pi G_N}$ and G_N is Newton's constant.

Let us compare this situation with electrodynamics. While, $A_\mu = 0$ corresponds to no electromagnetic field, $A_\mu \neq 0$ means presence of electromagnetic field. A_0 is the electrostatic potential and $\vec{A}$ is magnetic vector potential. Similarly, we can define the gravitational scalar potential as

$$g_{00} - \eta_{00} = 2\phi, \tag{7.16}$$

where ϕ is the gravitational scalar potential in the Newtonian gravity. Acceleration due to gravity is given by $-\vec{\nabla}\phi$. In the non-relativistic limit, where we match the formulae of general relativity with that of the Newtonian gravity, metric components g_{0i} and g_{ij} are small and can be ignored.

Geometric Interpretation

In absence of gravity, the proper time or proper interval between two spacetime events, one at x^μ and other at $x^\mu + \delta x^\mu$ is given by (repeated indices are summed over)

$$d\tau^2 = \eta_{\mu\nu}dx^\mu dx^\nu. \tag{7.17}$$

In the presence of gravity this formula is modified,

$$d\tau^2 = g_{\mu\nu}(x)dx^\mu dx^\nu. \tag{7.18}$$

To study field theory of gravity, first thing we need to do is to define the action functional. To do that let us start with a definition of an affine Christoffel connection

$$\Gamma^\sigma_{\mu\nu} = \frac{1}{2}g^{\sigma\rho}(\partial_\mu g_{\rho\nu} + \partial_\nu g_{\mu\rho} - \partial_\rho g_{\mu\nu}), \tag{7.19}$$

where, $g^{\sigma\rho}$ is inverse metric and $\partial_\mu = \partial/\partial x^\mu$. Indices can be lowered using metric tensor and raised using inverse metric tensor. Using the affine connection we can define the Riemann tensor,

$$R^\mu_{\nu\rho\sigma} = \partial_\sigma\Gamma^\mu_{\nu\rho} - \partial_\rho\Gamma^\mu_{\nu\sigma} + \Gamma^\lambda_{\nu\rho}\Gamma^\mu_{\lambda\sigma} - \Gamma^\lambda_{\nu\sigma}\Gamma^\mu_{\lambda\rho}. \tag{7.20}$$

Ricci tensor and Ricci scalar can be derived from the Riemann tensor as

$$R^\mu_{\nu\mu\sigma} = R_{\nu\sigma}, \qquad R^\mu_\mu = R. \tag{7.21}$$

Under general coordinate transformation, the metric transforms homogeneously.

$$x^\mu \to x'^\mu = f^\mu(x), \tag{7.22}$$

$$g_{\mu\nu}(x) \to \tilde{g}_{\mu\nu}(x') = g_{\rho\sigma}(x)\frac{\partial x^\rho}{\partial x'^\mu}\frac{\partial x^\sigma}{\partial x'^\nu}. \tag{7.23}$$

The affine connection, on the other hand, transforms inhomogeneously,

$$\tilde{\Gamma}^\alpha_{\beta\gamma}(x') = \frac{\partial x'^\alpha}{\partial x^\sigma}\frac{\partial x^\mu}{\partial x'^\beta}\frac{\partial x^\nu}{\partial x'^\gamma}\Gamma^\sigma_{\mu\nu}(x) + \frac{\partial x'^\alpha}{\partial x^\mu}\frac{\partial^2 x^\mu}{\partial x'^\beta \partial x'^\gamma}. \tag{7.24}$$

Quantities that transform homogeneously under coordinate transformations are called tensors. Those which do not transform homogeneously are non-tensors.

If a tensor is non-zero in one coordinate system then it is non-zero in all coordinate systems related to the former by a general coordinate transformation. Similarly if it is zero in some coordinate system then it is zero in all coordinate systems related to the former by a general coordinate transformation.

> **Exercise**: Show that Riemann tensor $R^{\mu}_{\nu\rho\sigma}$, Ricci tensor $R_{\mu\nu}$ and Ricci scalar R transform homogeneously under coordinate transformation.

Non-tensors, on the other hand, can be set to zero in some coordinate system. This can be done by suitably choosing the coordinate transformation such that homogeneous term cancels with the inhomogeneous term.

Now we are in a position to write down the Einstein-Hilbert action for gravity,

$$S_g = -\frac{1}{2\kappa^2} \int d^4x \sqrt{-\det g}\, R, \qquad (7.25)$$

where, $\det g$ is the determinant of the metric. The equation of motion is

$$R_{\mu\nu} - \frac{1}{2} g_{\mu\nu} R = 0. \qquad (7.26)$$

This is the vacuum Einstein equation. It can be simplified by contracting it with $g^{\mu\nu}$ to get

$$R_{\mu\nu} = 0. \qquad (7.27)$$

> **Exercise**: Show that a general coordinate transformation
>
> $$x^{\mu} \to \tilde{x}^{\mu} = \tilde{f}^{\mu}(x), \qquad (7.28)$$
>
> is a symmetry of the action, S_g.

Let us briefly look at the solutions to the vacuum Einstein's equation. First thing we notice from the expression of the Ricci tensor is that it contains at least one derivative of the metric. Therefore, any non-singular metric with constant entries solves the Einstein's equation. In particular, Minkowski space is a solution. Similarly any metric which can be obtained from this by a general coordinate transformation is also a solution. But these are not new solutions as general coordinate transformations is a symmetry of the action as well as equations of motion.

What are the symmetries of the Minkowski space?

Minkowski metric remains form invariant under Lorentz transformations, *i.e.*, if Λ^ρ_μ generates Lorentz transformation then

$$\eta_{\mu\nu} = \Lambda^\rho_\mu \eta_{\rho\sigma} \Lambda^\sigma_\nu. \tag{7.29}$$

Thus Lorentz boost and spatial rotations are symmetries of Minkowski space. Similarly $x^\mu \to x^\mu + a^\mu$ is also a symmetry of the Minkowski space. Thus the total symmetry of the Minkowski space is given by the Poincaré group which consists of translations and Lorentz transformations.

Let us get back to the Einstein-Hilbert action. Without compromising with the principle of general covariance we can add one more term to the action,

$$-2 \int d^4x \sqrt{-\det g}\,\Lambda, \tag{7.30}$$

where, Λ is the cosmological constant. The new equations of motion are

$$R_{\mu\nu} - \frac{1}{2} g_{\mu\nu} R + \Lambda g_{\mu\nu} = 0. \tag{7.31}$$

Depending on the sign of Λ, the solution $g_{\mu\nu}$ represents either a closed compact space or a non-compact space. We will study this equation of motion in five dimensions, *i.e.*, $\mu, \nu = 0, \cdots, 4$.

Consider a six dimensional flat Euclidean space with a metric

$$d\tau^2 = \sum_{i=1}^{6} dy^i dy^i. \tag{7.32}$$

A five dimensional sphere, S^5, of radius R can be embedded in this space by the equation

$$\sum_{i=1}^{6}(dy^i)^2 = R^2. \tag{7.33}$$

This equation is invariant under six dimensional rotation group, $SO(6)$, which mixes the coordinates y^i. In other words, $SO(6)$ is the symmetry group of five sphere, S^5. Induced metric on S^5 due to this embedding can be obtained by substituting for one of the y's in terms of remaining ones.

$$d\tau^2 = \sum_{i=1}^{5}\left(dy^i dy^i + \frac{(y^i dy^i)^2}{R^2 - \sum_{j=1}^{5}(y^j)^2}\right). \tag{7.34}$$

This metric solves the Einstein's equation with cosmological constant, $\Lambda = -6/R^2$.

Let us now consider flat six dimensional space with $(4,2)$ signature. The metric on this space is written as

$$d\tau^2 = -dudv + \sum_{i,j=0}^{3} \eta_{ij}dx^i dx^j, \quad \eta_{ij} = \mathrm{diag}(-1,1,1,1). \tag{7.35}$$

Five dimensional anti-de Sitter (AdS) space of radius r is embedded as a hypersurface in this space,

$$uv - \sum_{i,j=0}^{3} \eta_{ij}x^i x^j = r^2. \tag{7.36}$$

Let us define $u = x_{-1} + x_4$ and $v = x_{-1} - x_4$, then the equation for the AdS space becomes

$$x_{-1}^2 + x_4^2 - \sum_{i,j=0}^{3} \eta_{ij}x^i x^j = r^2. \tag{7.37}$$

This equation is invariant under $SO(4,2)$ transformations which are generalised rotations in six dimensional space with $(4,2)$ signature. The hypersurface equation is solved by

$$x^0 = r\cosh\rho\cos t, \quad x_{-1} = r\cosh\rho\sin t, \tag{7.38}$$

$$x^a = a \sinh \rho \, \Omega^a, \quad a = 1, 2, 3, 4, \quad \sum_{a=1}^{4} (\Omega^a)^2 = 1, \qquad (7.39)$$

where, $\rho \geq 0$ and $0 \leq t \leq 2\pi$. Induced metric on AdS_5 with this choice of coordinates is

$$d\tau^2 = r^2(-\cosh^2 \rho \, dt^2 + d\rho^2 + \sinh^2 \rho \, d\Omega^2). \qquad (7.40)$$

Due to periodicity of t, this space has closed timelike curves. To avoid complications due to closed timelike curves we could work with the universal cover of this space, which does not have closed timelike curves. However, we will choose to work with the Euclidean AdS_5 instead. Euclidean AdS_5 can be obtained by embedding

$$uv - \sum_{a=1}^{4} (x^a)^2 = r^2, \qquad (7.41)$$

in the six dimensional Minkowski space. This equation is invariant under $SO(5,1)$ transformations. For simplicity we will take $r = 1$. Metric on the six dimensional Minkowski space is

$$d\tau^2 = -du dv + \sum_{a=1}^{4} dx^a dx^a. \qquad (7.42)$$

Let us take $u, v > 0$ and introduce new coordinates

$$z^0 = v^{-1}, \qquad z^a = v^{-1} x^a. \qquad (7.43)$$

Using these new coordinates we can write u in terms of z's and substituting it in the metric we get

$$d\tau^2 = \frac{1}{(z^0)^2} \left[(dz^0)^2 + \sum_{a=1}^{4} (dz^a)^2 \right], \qquad (7.44)$$

where, $z^0 > 0$ and $-\infty < z^a < \infty$. Thus we see that this is a metric on the upper half space and the boundary of this upper half space is R^4 and a point at infinity, which is equal to S^4. There is another way to see that the boundary of Euclidean AdS space is S^4. Let us define new variables, $u = z + x^5$ and $v = z - x^5$. In terms of these new variables, equation for AdS hypersurface becomes

$$z^2 - \sum_{i=1}^{5} x^i x^i = 1. \tag{7.45}$$

For $z > 0$ let us define new coordinates, $x^i = (z + 1)\xi^i$. We can now eliminate z using the hypersurface equation,

$$z = \frac{1 + \sum_{i=1}^{5} \xi^i \xi^i}{1 - \sum_{i=1}^{5} \xi^i \xi^i}. \tag{7.46}$$

The induced metric on AdS space is

$$d\tau^2 = 4 \sum_{i=1}^{5} \frac{d\xi^i d\xi^i}{(1 - \sum_{j=1}^{5} \xi^j \xi^j)^2}. \tag{7.47}$$

Notice, $\xi^i = x^i/(1 + z)$, which means

$$\sum_{i=1}^{5} \xi^i \xi^i = \sum_{i=1}^{5} \frac{x^i x^i}{(1 + z)^2} < \sum_{i=1}^{5} \frac{x^i x^i}{z^2} = \sum_{i=1}^{5} \frac{x^i x^i}{1 + \sum_{j=1}^{5} \xi^j \xi^j} < 1. \tag{7.48}$$

Thus we see that ξ^i are bounded and describe the interior of five dimensional ball. Limiting values of ξ^i which determine the boundary of the space is determined by setting

$$\sum_{i=1}^{5} \xi^i \xi^i = 1. \tag{7.49}$$

This is the equation of S^4, and thus we see that boundary of Euclidean AdS_5 is a four sphere S^4.

We can now put back the radius parameter in the metric

$$d\tau^2 = \frac{r^2}{(z^0)^2} \left[(dz^0)^2 + \sum_{a=1}^{4} (dz^a)^2 \right]. \tag{7.50}$$

This metric solves the Einstein's equation with cosmological constant $\Lambda = 6/r^2$. Notice the boundary S^4 is at infinite distance as the metric tends to infinity as $z^0 \to 0$. The metric, in fact, has a double pole at $z^0 = 0$. Due to this metric on AdS_5 does not really give us the metric on S^4. We can resolve this problem by choosing a function $f(z)$ which has a first order zero at $z^0 = 0$ and write a new metric

$$d\tilde{\tau}^2 = f^2(z)d\tau^2, \tag{7.51}$$

which is non-singular and gives a finite metric on the boundary. Clearly, the choice of $f(z)$ is arbitrary and hence, we can determine the metric on the boundary only up to a local conformal rescaling. Let us now see how symmetries of AdS space appear on the boundary.

1. Consider a transformation $x^a \to R^{ab}x^b$, $u \to u$ and $v \to v$, where $R \in SO(4)$. Coordinates that parametrize the boundary are z^0 and z^a. Under $SO(4)$ rotations

   $$z^0 = v^{-1} \to z^0, \qquad z^a = v^{-1}x^a \to v^{-1}R^{ab}x^b = R^{ab}z^b, \tag{7.52}$$

   which are $SO(4)$ rotations on the boundary S^4.

2. We will now consider the transformation, $x^a \to x^a + vl^a$, $v \to v$ and $u \to u + 2l^ax^a + vl^al^a$, where l^a is some constant vector. This is an $SO(5,1)$ transformationas it leaves invariant the AdS hypersurface equation

   $$uv - \sum_{a=1}^{4} x^ax^a = r^2. \tag{7.53}$$

   These tansformations correspond to translations on the boundary.

   $$z^a \to z^a + l^a, \qquad z^0 \to z^0. \tag{7.54}$$

3. Let us consider the transformation, $u \to \lambda u$, $v \to \lambda^{-1}v$ and $x^a \to x^a$. This transformation corresponds to scaling transformation on the boundary, $z^0 \to \lambda z$ and $z^a \to \lambda z^a$.

4. The transformation $u \to v$ and $v \to u$, shows up as a special conformal transformation on the boundary,

$$z^a \to \frac{z^a}{\sum_{a=1}^{4}(z^a)^2}. \qquad (7.55)$$

Notice, all the transformations written in u, v, and x^a coordinate system are isometries of AdS space. All these isometries appear as conformal transformations on the S^4 boundary of the AdS space.

Let us now look at yet another solution to the Einstein equation, gravitational field due to point mass. The simplest way to solve this problem is to solve the equation without the point mass (delta function) source and then identify parameters of the solution with those of the source. As an example let us first consider electrostatic field due to a point charge. The equation of motion is

$$\nabla^2 \Phi = q\delta(x). \qquad (7.56)$$

Let us first ignore the delta function source and solve the source free equation of motion, $\nabla^2 \Phi = 0$. Solution to this equation is

$$\Phi = \frac{c}{r}, \quad \text{where,} r = \sqrt{x^2 + y^2 + z^2}. \qquad (7.57)$$

This solution, due to presence of point charge has spherical symmetry and is valid everywhere except at $r = 0$, *i.e.*, at the location of the point charge. Using this we can determine electric flux at infinity.

$$\vec{E} = -\vec{\nabla}\Phi = \frac{c}{r^2}. \qquad (7.58)$$

Total electric flux integrated over the surface at infinity is proportional to the charge enclosed inside the surface.

$$\int \vec{E} \cdot d\vec{S} = 4\pi c \equiv q. \qquad (7.59)$$

Thus we find that $c = q/4\pi$ and

$$\Phi = \frac{q}{4\pi r}. \qquad (7.60)$$

We will employ similar technique to obtain gravitational field due to a point mass. Like the point charge, here also we have spherical symmetry. Therefore, the solution will depend only on the radial direction. We will also be seeking time independent solution. A metric consistent with these constraints is

$$d\tau^2 = -A(\rho)dt^2 + B(\rho)d\rho^2 + \rho^2(d\theta^2 + \sin^2\theta d\phi^2). \tag{7.61}$$

Substituting this ansatz into vacuum Einstein equation, we get

$$A(\rho) = \left(1 + \frac{1}{c\rho}\right), \quad B(\rho) = \frac{1}{A(\rho)}, \tag{7.62}$$

where, c is a constant which will be determined by the parameters characterising the point mass source. The solution then looks like

$$d\tau^2 = -\left(1 + \frac{1}{c\rho}\right)dt^2 + \frac{1}{\left(1 + \frac{1}{c\rho}\right)}d\rho^2 + \rho^2(d\theta^2 + \sin^2\theta d\phi^2). \tag{7.63}$$

It is now time to find interpretation for c. For large radial distances

$$g_{tt} = -\left(1 + \frac{1}{c\rho}\right) = -1 - 2\phi(x), \tag{7.64}$$

where $\phi(x)$ is Newtonian gravitational potential.

$$\phi(x) = \frac{1}{2c\rho}. \tag{7.65}$$

For a point particle source of mass M, we expect

$$\phi(x) = -\frac{G_N M}{\rho} \Rightarrow c = \frac{-1}{2G_N M}, \tag{7.66}$$

and the metric is given by

$$\begin{aligned}
d\tau^2 &= -\left(1 - \frac{2G_N M}{\rho}\right)dt^2 + \left(1 - \frac{2G_N M}{\rho}\right)^{-1}d\rho^2 \\
&+ \rho^2(d\theta^2 + \sin^2\theta d\phi^2).
\end{aligned} \tag{7.67}$$

This is the well known Schwarzschild black hole metric in four dimensional spacetime. This solution can be easily generalised to D dimensional black hole solution,

$$d\tau^2 = -\left(1 - \frac{2G_N M}{\rho^{D-3}}\right) dt^2 + \left(1 - \frac{2G_N M}{\rho^{D-3}}\right)^{-1} d\rho^2 + \rho^2 d\Omega_{D-2}^2,$$
(7.68)

where $d\Omega_{D-2}^2$ is a metric on a $D - 2$ dimensional sphere. Such solutions are stable if they carry some conserved charge.

Black hole is a point like solution. We are, in general, interested in extended solutions. We are also interested in getting these solutions not from pure gravity but from a theory with matter coupled to gravity. Before we look at these type of theories let us see how we can couple matter fields to gravity. We will start with coupling of a scalar field to gravity. Action for a scalar field in the Minkowski space is

$$S^{(0)}_{scalar} = -\frac{1}{2} \int d^4x \left(\partial_\mu \phi(x) \partial_\nu \phi(x) \eta^{\mu\nu} + m^2 \phi^2(x) + \lambda \phi^4(x)\right).$$
(7.69)

In the presence of gravity we have

$$\begin{aligned}
S_{scalar} &= -\frac{1}{2} \int d^4x \sqrt{-\det g}\Big(\partial_\mu \phi(x) \partial_\nu \phi(x) g^{\mu\nu}(x) \\
&\quad + m^2 \phi^2(x) + \lambda \phi^4(x)\Big).
\end{aligned}$$
(7.70)

Action for a vector field in the Minkowski space is

$$S^{(0)}_{vector} = -\frac{1}{4g_{YM}^2} \mathrm{Tr} \int d^4x F_{\mu\nu} F_{\rho\sigma} \eta^{\mu\rho} \eta^{\nu\sigma},$$
(7.71)

where, $F_{\mu\nu} = F_{\mu\nu}^a T^a = (\partial_\mu A_\nu^a - \partial_\nu A_\mu^a + f^{abc} A_\mu^b A_\nu^c) T^a$. In the presence of gravity this action becomes

$$S_{vector} = -\frac{1}{4g_{YM}^2} \mathrm{Tr} \int d^4x \sqrt{-\det g}\, F_{\mu\nu} F_{\rho\sigma} g^{\mu\rho}(x) g^{\nu\sigma}(x).$$
(7.72)

Similarly action for second rank antisymmetric tensor field is

$$S^{(0)}_T = -\frac{1}{12} \int d^4x H_{\mu\nu\rho} H_{\alpha\beta\gamma} \eta^{\mu\alpha} \eta^{\nu\beta} \eta^{\rho\gamma},$$
(7.73)

where, $H_{\mu\nu\rho} = 3\partial_{[\mu}B_{\nu\rho]}$. This action in the presence of gravity becomes

$$S_T = -\frac{1}{12}\int d^4x\sqrt{-\det g}\,H_{\mu\nu\rho}H_{\alpha\beta\gamma}g^{\mu\alpha}(x)g^{\nu\beta}(x)g^{\rho\gamma}(x). \quad (7.74)$$

To be more specific we are interested in type IIB supergravity in ten dimensions. The bosonic field content of this theory is a metric g_{MN}, a complex second rank anti-symmetric tensor $B_{MN} = B_{MN}^1 + iB_{MN}^2$, a complex scalar $\tau = a + i\exp(-\phi)$ and a fourth rank self-dual antisymmetric tensor D_{MNPQ}^+. The fermionic field content is a complex chiral gravitino $\psi_M^\alpha = \psi_{1M}^\alpha + i\psi_{2M}^\alpha$ and a complex chiral spinor with same chirality as the gravitino $\lambda^\alpha = \lambda_1^\alpha + i\lambda_2^\alpha$. These fields together fill up a single $N = 2$ supermultiplet of ten dimensional type IIB supergravity.

Due to presence of self-dual field it is not possible to write down a covariant action for this theory. We can, however, write down a covariant action without imposing self duality condition and use this condition at the level of equations of motion. The bosonic part of the action is

$$
\begin{aligned}
S_{IIB} \;=\; & -\frac{1}{2\kappa^2}\int d^{10}x\sqrt{-g}\Big[R + \frac{1}{2}g^{MN}\partial_M\bar\tau\partial_N\tau \\
& + \frac{1}{12}g^{M_1N_1}g^{M_2N_2}g^{M_3N_3}H_{M_1M_2M_3}^+ H_{N_1N_2N_3} \qquad (7.75)\\
& + \frac{1}{240}g^{M_1N_1}\cdots g^{M_5N_5}F_{M_1M_2M_3M_4M_5}F_{N_1N_2N_3N_4N_5}\Big],
\end{aligned}
$$

where, $\kappa = \sqrt{8\pi G_N}$, G_N is ten dimensional Newton's constant, $H_{MNP} = \partial_M B_{NP}+$ cyclic permutations, $F_{M_1\cdots M_5} = \partial_{M_1}D_{M_2\cdots M_5}^+ +$ cyclic permutations $-\frac{5}{4}\kappa\,\mathrm{Im}(B_{M_1M_2}H_{M_3M_4M_5}^*)$.

The equations of motion for bosonic sector following from this action are

$$D^M P_M \;=\; \frac{1}{24}G_{MNP}G^{MNP}$$

$$D^P G_{MNP} \;=\; P^L G_{MNL}^* - \frac{2}{3}i\kappa F_{MNPQR}G^{PQR} \qquad (7.76)$$

$$R_{MN} \;=\; P_M P_N^* + P_M^* P_N + \frac{1}{6}\kappa^2 F_{L_1\cdots L_4 M}F_N^{L_1\cdots L_4}$$

$$+\frac{1}{8}\kappa^2(G_M^{PQ}G^*_{NPQ} + G_M^{*PQ}G_{NPQ} - \frac{1}{6}g_{MN}G^{PQR}G^*_{PQR})$$

where,

$$\begin{aligned}
P_M &= f^2\partial_M\tau \\
G_{MNP} &= f(H_{MNP} - \tau H^*_{MNP}) \\
f &= \frac{1}{\sqrt{1-\tau\bar{\tau}}}
\end{aligned}\tag{7.77}$$

Another useful set of equations are supersymmetry transformations of the fermionic fields.

$$\begin{aligned}
\delta\psi_M &= \left(\nabla_M - \frac{3}{16}G_{MNP}\gamma^{NP} + \frac{1}{48}G^{NPQ}\gamma_{MNPQ}\right. \\
&\quad \left. - \frac{1}{192}iF_{MNPQR}\gamma^{NPQR}\right)\epsilon \tag{7.78} \\
&\quad + \text{terms quadratic in } \lambda, \\
\delta\lambda &= \left(\frac{1}{24}iG_{MNP}\gamma^{MNP} + \frac{1}{2}iP_M\gamma^M\right)\epsilon. \tag{7.79}
\end{aligned}$$

In addition to the equations of motion derived from this action, the field strength of fourth rank antisymmetric tensor satisfies the self-duality relation

$$F_{M_1\cdots M_5} = \epsilon_{M_1\cdots M_{10}}F^{M_6\cdots M_{10}}. \tag{7.80}$$

As is evident from the equations of motion that they are coupled non-linear differential equations involving $g_{MN}, \tau, B_{MN}, D^+_{MNPQ}$ and fermions. These coupled differential equations are solved by the following classical field configuration

$$\begin{aligned}
\tau_{cl} &= a_0 + ig^{-1} \\
g_{MN} &= \text{Metric on } (AdS_5)_r \times (S^5)_r \tag{7.81} \\
F_{mnpqr} &= c\varepsilon_{mnpqr}, \qquad F_{\alpha\beta\gamma\delta\rho} = c\varepsilon_{\alpha\beta\gamma\delta\rho},
\end{aligned}$$

where, a_0 and g are constants, $\{y^m\}$ are coordinates on S^5, $5 \leq m \leq 9$ and $\{x^\alpha\}$ are coordinates on AdS_5. $0 \leq \alpha \leq 4$, $\varepsilon_{mnpqr} = \sqrt{\det g_{s^5}}\epsilon_{mnpqr}$ and $\epsilon_{56789} = 1$ and similarly $\varepsilon_{\alpha\beta\gamma\delta\rho} =$

$\sqrt{-\det g_{AdS_5}}\,\epsilon_{\alpha\beta\gamma\delta\rho}$ and $\epsilon_{01234} = 1$. Flux of F_{mnpqr} through S^5 is quantized

$$\int F_{56789}dy^5 dy^6 dy^7 dy^8 dy^9 = Q \in \mathbb{Z} \tag{7.82}$$

and the constant c is determined in terms of Q. $AdS_5 \times S^5$ is a solution provided radius of both AdS_5 and S^5 are equal,

$$r = (4\pi g Q \alpha'^2)^{\frac{1}{4}}, \tag{7.83}$$

where α' is a fundamental length scale which descends down from type IIB string theory. This solution preserves all 32 supersymmetries of the ten dimensional type IIB theory. Notice this solution contains three parameters g, a_0 and Q.

7.3 $\mathcal{N} = 4$ Super-Yang-Mills theory

So far we have seen that $AdS_5 \times S^5$ background with nonzero flux of fifth rank field strength on S^5 is a solution to type IIB equations of motion with rest all bosonic as well as fermionic fields set to zero except τ which is constant. We also noted that this solution is parametrized by three parameters. In this section we will relate these parameters to those on the $\mathcal{N} = 4$ $SU(N)$ super-Yang-Mills side.

Let us now consider $\mathcal{N} = 4$ supersymmetric $SU(N)$ gauge theory in four spacetime dimensions. The field content of this theory has only one superfield. This superfield contains an $SU(N)$ gauge field A^a_μ, six scalars ϕ^a_m, $m = 1\cdots 6$ and four Weyl fermions λ^a_i, $i = 1\cdots 4$. All these fields belong to adjoint representation of $SU(N)$, $a = 1\cdots N^2 - 1$. The action for this theory is completely fixed by the requirement of $\mathcal{N} = 4$ supersymmetry:

$$\begin{aligned}
S_{YM} &= -\frac{1}{g_{YM}^2}\int d^4x \Big(\frac{1}{4}F^a_{\mu\nu}F^{a\mu\nu} + \frac{1}{2}\sum_{m=1}^{6}(D_\mu\phi^a_m D^\mu\phi^a_m) \\
&\quad + \sum_{i=1}^{4}\bar{\lambda}^a_i \slashed{D}\lambda^a_i + \sum_{i,m}\bar{\lambda}^a_i \phi^b_m \lambda^c_i f^{abc} \\
&\quad + \sum_{m,n}([\phi^a_m, \phi^b_n])^2\Big) + \frac{\theta}{16\pi^2}\int d^4x F^a_{\mu\nu}F^a_{\rho\sigma}\epsilon^{\mu\nu\rho\sigma}.
\end{aligned} \tag{7.84}$$

$\mathcal{N} = 4$ supersymmetric $SU(N)$ Yang-Mills theory is completely parametrized by g_{YM}, θ and N. Notice this theory is also parametrized by three parameters.

Now we are in a position to write down a more refined form of the conjecture:

Type IIB theory on $AdS_5 \times S^5$ is dual to $\mathcal{N} = 4$ supersymmetric $SU(N)$ gauge theory if $g = g_{YM}^2$, $a_0 = \theta/2\pi$ and $Q = N$.

This statement does not clarify how are these two theories related. The precise relation is, the $\mathcal{N} = 4$ supersymmetric $SU(N)$ gauge theory lives on S^4 which is the boundary of AdS_5. Recall that the space S^5 has $SO(6)$ symmetry. This symmetry keeps the metric on S^5 invariant. Similarly, AdS_5 has $SO(5,1)$ symmetry. Thus type IIB supergravity on $Ads_5 \times S^5$ has $SO(6) \times SO(5,1)$ symmetry. We have also seen earlier that the symmetries of AdS_5 induce $SO(5,1)$ conformal symmetries on the boundary S^4. If type IIB on $AdS_5 \times S^5$ is related to $\mathcal{N} = 4$ supersymmetric $SU(N)$ gauge theory on the boundary S^4, then the natural question is how do these symmetries appear in the gauge theory? $\mathcal{N} = 4$ supersymmetric $SU(N)$ gauge theory is conformally invariant as it has no explicit mass parameter and it has vanishing β-function to all orders in perturbation theory. Therefore conformal transformations of the boundary S^4, which forms $SO(5,1)$ group, are symmetry transformations of this theory. Apart from this $\mathcal{N} = 4$ supersymmetric $SU(N)$ gauge theory also has $SO(6)$ symmetry in the form of R-symmetry. To see this let us recall the field content of $\mathcal{N} = 4$ gauge theory. It contains a gauge field A_μ^a, six scalars ϕ_m^a ($m = 1 \cdots 6$) and four spin 1/2 fermions ψ_i^a ($i = 1 \cdots 4$). All these fields fit into a single supermultiplet and hence all belong to the same gauge group representation, namely, the adjoint representation. $\mathcal{N} = 4$ supersymmetry means we have four supercharges Q_α^i and $\bar{Q}^{\dot{\alpha}i}$ ($i = 1 \cdots 4$). Since this theory is completely symmetric with respect to these four supersymmetries, any linear combination of them is also a symmetry of the theory. Therefore

$$Q_\alpha^i \rightarrow U_j^i Q_\alpha^j, \tag{7.85}$$

and similarly for $\bar{Q}^{\dot{\alpha}i}$, is a symmetry of the theory. The transformation matrix U^i_j generically belongs to $U(4)$ but the nontrivial symmetry transformations reside in $SU(4)$ which is isomorphic to $SO(6)$. The supersymmetry charges Q^i_α and $\bar{Q}^{\dot\alpha i}$ belong to the fundamental (**4**) and the anti-fundamental ($\bar{\mathbf{4}}$) representations of $SU(4)$, respectively. These representations are spinorial representations of $SO(6)$. The $SU(N)$ gauge field is a singlet of this global symmetry, four fermions transform in the spinorial representation and six scalars are in the vector representation of $SO(6)$ which is a 4×4 antisymmetric matrix representation of $SU(4)$. Transformation rule for the scalars is

$$\phi^a_m \equiv \phi^a_{[ij]}, \qquad \phi^a_{[ij]} \rightarrow U^{[k}_{[i} U^{l]}_{j]} \phi^a_{[kl]}. \tag{7.86}$$

Under these transformation, action of $\mathcal{N} = 4$ supersymmetric $SU(N)$ gauge theory remains invariant.

Thus we see that symmetries of type IIB supergravity on $AdS_5 \times S^5$ are identical[2] to that of $\mathcal{N} = 4$ SUSY $SU(N)$ Yang-Mills for any N.

Since $\mathcal{N} = 4$ SUSY gauge theory doesnot have any explicit mass parameter all the correlation functions in the theory will have power law behaviour. That is,

$$\langle \Phi(x)\Phi(y) \rangle = \frac{1}{(x-y)^{2\Delta}}, \tag{7.87}$$

where Δ is scaling dimension of the field Φ. Suppose Φ is a scalar field in four dimensions then its Lagrangian density is given by

$$\mathcal{L} = -\frac{1}{2}\partial_\mu\Phi\partial^\mu\Phi. \tag{7.88}$$

In four dimensions, the field Φ has engineering mass dimension 1. This scaling dimension of the free field gets modified due to interactions. $\mathcal{N} = 4$ SUSY $SU(N)$ gauge theory is classically conformally invariant and remains conformally invariant to all orders

---

[2]Historically this fact was known as early as 1985 to Günaydin and Marcus Class. Quant. Gravity, **2**, L11 (1985). However, at that time it was not clear how and why these two theories are related.

in perturbation theory if there is no spontaneous symmetry breaking. Therefore, in this theory scaling dimensions of all the fields are identical to their classical mass dimensions.

In any gauge theory all observables are gauge invariant objects. Let us construct gauge invariant operators corresponding to these observables in $\mathcal{N} = 4$ SUSY $SU(N)$ gauge theory. Let us again recall that the gauge field A_μ^a, fermions ψ_i^a and scalars ϕ_m^a with $i = 1, \cdots 4$ and $m = 1, \cdots 6$ are all in the adjoint representation of the gauge group $SU(N)$. While the gauge field A_μ^a is invariant under the global $SO(6)$ symmetry, fermions belong to the representation **4** which is a spinor of $SO(6)$ and scalars belong to **6** which is the fundamental representation of $SO(6)$. In what follows, Tr represents trace over $SU(N)$ indices only. All other indices on fermions and scalars will be suppressed.

The following table is a partial list of gauge invariant operators in $\mathcal{N} = 4$ SUSY SU(N) gauge theory. We would like to know how type IIB theory on $AdS_5 \times S^5$ produces this result.

Spin	Operators	Δ	$SO(6)$ Rep.
0	$\mathrm{Tr}[\phi^n]$ $(n \geq 2)$	$2, 3, 4, \cdots$	$\mathbf{20}, \mathbf{50}, \cdots$
0	$\mathrm{Tr}[F_{\mu\nu}F^{\mu\nu}\phi^n]$ $(n \geq 0)$	$4, 5, 6, \cdots$	$\mathbf{1}, \mathbf{6}, \mathbf{20}, \cdots$
0	$\mathrm{Tr}[F_{\mu\rho}F^{\rho}_{\nu}F^{\nu}_{\sigma}F^{\sigma\mu}\phi^n]$ $(n \geq 0)$	$8, 9, 10, \cdots$	$\mathbf{1}, \mathbf{6}, \mathbf{20}, \cdots$
0	$\mathrm{Tr}[\psi\psi\phi^n]$ $(n \geq 0)$	$3, 4, 5, \cdots$	$\mathbf{10}, \mathbf{45}, \cdots$
0	$\mathrm{Tr}[F_{\mu\nu}F^{\mu\nu}\psi\psi\phi^n]$ $(n \geq 0)$	$7, 8, 9, \cdots$	$\mathbf{10}, \mathbf{45}, \cdots$
2	$\mathrm{Tr}[T_{\mu\nu}\phi^n]$ $(n \geq 0)$	$4, 5, 6, \cdots$	$\mathbf{1}, \mathbf{6}, \mathbf{20}, \cdots$
Ant.	$\mathrm{Tr}[F_{\mu\nu}\phi^n]$ $(n \geq 1)$	$3, 4, 5, \cdots$	$\mathbf{6}, \mathbf{20}, \cdots$
Ant.	$\mathrm{Tr}[F_{\mu\nu}\psi\psi\phi^n]$ $(n \geq 0)$	$7, 8, 9, \cdots$	$\mathbf{10}, \mathbf{45}, \cdots$
Ant.	$\mathrm{Tr}[F_{\mu\rho}F^{\mu\rho}F_{\sigma\nu}\phi^n]$ $(n \geq 0)$	$6, 7, 8, \cdots$	$\mathbf{1}, \mathbf{6}, \mathbf{20}, \cdots$
$\frac{1}{2}$	$\mathrm{Tr}[\psi\phi^n]$ $(n \geq 1)$	$\frac{5}{2}, \frac{7}{2} \cdots$	$\mathbf{20}^*, \mathbf{50}^*, \cdots$
$\frac{1}{2}$	$\mathrm{Tr}[\psi F_{\mu\nu}F^{\mu\nu}\phi^n]$ $(n \geq 0)$	$\frac{11}{2}, \frac{13}{2} \cdots$	$\mathbf{4}, \mathbf{20}, \cdots$
$\frac{1}{2}$	$\mathrm{Tr}[\psi\sigma_{\mu\nu}F^{\mu\nu}\phi^n]$ $(n \geq 0)$	$\frac{7}{2}, \frac{9}{2} \cdots$	$\mathbf{4}^*, \mathbf{20}^*, \cdots$

Type IIB supergravity equations of motion are also solved by the following field configuration

$$
\begin{aligned}
\tau_{cl} &= a_0 + ig^{-1}, \qquad B_{MN} = 0, \\
F_5 &= dt dx^1 dx^2 dx^3 dx^4 df^{-1},
\end{aligned}
\tag{7.89}
$$

$$
d\tau^2 = \frac{1}{\sqrt{f(\rho)}} \sum_{\mu,\nu=0}^{3} \eta_{\mu\nu} dx^\mu dx^\nu + \sqrt{f(\rho)}(d\rho^2 + \rho^2 d\Omega_5^2),
$$

where $f(\rho) = 1 + (r^4/\rho^4)$ and $r^4 = 4\pi g Q \alpha'^2$. This is a three-brane solution as the metric does not depend on time and three space dimensions. This in a way is a generalisation of the Schwarzschild metric. While Schwarzschild solution corresponds to a point like object, this one is an extended object with extension in three spatial directions. This is a stable solution as it carries a conserved charge with respect to fifth rank field stength F_5. This charge can be computed by integrating F_5 over a five dimensional sphere S^5 transverse to the three-brane.

Let us now consider a limit $\rho \ll r$. In this limit $f(\rho) \gg 1$ and hence we can safely ignore the factor 1 in the expression for $f(\rho)$. With this approximation the metric takes the form

$$
\begin{aligned}
d\tau^2 &= \frac{\rho^2}{r^2} \left(\sum_{\mu,\nu=0}^{3} \eta_{\mu\nu} dx^\mu dx^\nu \right) + \frac{r^2}{\rho^2} \left(d\rho^2 + \rho^2 d\Omega_5^2 \right) \\
&= \left(\frac{\rho^2}{r^2} \sum_{\mu,\nu=0}^{3} \eta_{\mu\nu} dx^\mu dx^\nu + \frac{r^2}{\rho^2} d\rho^2 \right) + r^2 d\Omega_5^2.
\end{aligned}
\tag{7.90}
$$

An expression in the paranthesis is a metric on AdS_5 with radius r and with Lorentzian signature. The term outside the paranthesis is a metric on S^5 with radius r. Thus as we go very close to the three brane, metric transforms into $AdS_5 \times S^5$. Notice $\rho \to 0$ is a boundary of AdS_5 (in addition to this there is a point at infinity, $\rho = \infty$.) which is at infinite distance away. The metric on the boundary is the usual Minkowski metric.

7.4 Collective coordinates

It is easy to see from the form of the three brane solution that the brane is located at $x^4 = x^5 = x^6 = x^7 = x^8 = x^9 = 0$ or equivalently at $\rho = 0$. However, any point in $(x^4, \cdots, x^9)$ space is as good. In other words translations of three brane in $(x^4, \cdots, x^9)$ space does not cost any energy.

How does the three brane know about the translation symmetry in the transverse direction?

Low energy excitations on the three brane contain six massless collective coordinates, one each for translation in $(x^4, \cdots, x^9)$. From the field theory point of view these six collective coordinates are six scalar fields living on the three brane.

What happens to supersymmetry?

Claim: Three brane solution breaks half the supersymmetry of type IIB theory. That is, out of 32 supersymmetry charges, 16 are preserved by the solution.

In four dimensions, 16 supercharges implies $\mathcal{N} = 4$ supersymmetry. We have also seen that collective coordinates give rise to six scalar fields in the 3+1 dimensions spanned by the three brane. This along with the fact that we have $\mathcal{N} = 4$ supersymmetry uniquely fixes the theory on the 3+1 dimensional space spanned by the three brane. When the charge carried by the solution $Q = N$ then the corresponding theory in 3+1 dimensions is $\mathcal{N} = 4$ supersymmetric $SU(N)$ Yang-Mills theory.

Let us now take stock of what we have done so far. We are studying the conjecture of Maldacena, *i.e.*, Type IIB string theory on $AdS_5 \times S^5$ is dual to $\mathcal{N} = 4$ supersymmetric $SU(N)$ gauge theory in four dimensions. To study this conjectured relationship between a theory of gravity and the Yang-Mills theory we studied field content of type IIB supergravity and wrote down the equations of motions for these fields. We then saw that metric on $AdS_5 \times S^5$ with nontrivial five form field strength on the five sphere is a solution to the type IIB supergravity equations of motion. This solution has $SO(5,1) \times SO(6)$ symmetry with 32 supersymmetries. This solution is parametrized by g, the string coupling, a_0, value of the RR scalar and Q, integral of the five form flux through S^5.

$\mathcal{N} = 4$ SUSY $SU(N)$ in four dimensions has a single multiplet. Field content of this multiplet and the action for those fields is completely determined by supersymmetry. This theory is parametrized by the Yang-Mills coupling constant g_{YM}, θ, the vacuum angle and N.

We then refined the statement of the conjecture and carried out a few checks. In particular, we saw that both type IIB theory on $Ads_5 \times S^5$ and $\mathcal{N} = 4$ SUSY $SU(N)$ have same symmetries, *i.e.*, $SO(5,1) \times SO(6)$ with 32 supersymmetries.

On the gauge theory side we wrote down the spectrum of gauge invariant operators in the $\mathcal{N} = 4$ $SU(N)$ superconformal field theory and on the type IIB side, we showed that three-brane metric with nontrivial five form F_5 flux is a solution to type IIB equations of motion. This solution breaks 1/2 supersymmetry of type IIB theory. This implies the theory living on the world volume of three-brane is $\mathcal{N} = 4$ $SU(N)$ gauge theory. In the limit $\rho \ll r$ we recover the $AdS_5 \times S^5$ geometry with gauge theory define on the boundary of AdS_5. In the process we also recover remaining 16 supercharges in the form of conformal supersymmetries.

To gather further evidence for this conjecture we will study the geometry in more detail. To compare the spectrum of gauge invariant operators with the spectrum of fields on the AdS side we need to compute Kaluza-Klein spectrum of type IIB theory compactified to AdS_5. This will be done by analysing spectrum of type IIB theory[3] on S^5. Each field in the type IIB theory will give rise to an infinite tower of fields with mass

$$\text{mass} \sim \frac{n}{r}, \qquad n \geq 0,$$

where r is the radius of S^5. Suppose $\{\chi_i\}$ be a set of all the fields propagating on AdS_5. This contains infinite towers obtained from Kaluza-Klein reduction of type IIB on S^5. Let us define partition

---

[3]As mentioned earlier we will concentrate only on the massless sector of type IIB theory, *i.e.*, we will work in the supergravity approximation.

function of type IIB on AdS_5 as

$$Z_{IIB} = \int \prod_i [d\chi_i] \exp\left(-S[\{\chi_i\}]\right). \tag{7.91}$$

Since AdS_5 has a boundary we need to specify boundary condition for every χ_i. Let $\chi_i \to \chi_i^B$ at the boundary then

$$Z_{IIB}[\{\chi_i^B\}] = \int \prod_i [d\chi_i] \exp\left(-S[\{\chi_i\}]\right)|_{\chi_i \to \chi_i^B}. \tag{7.92}$$

A partial list of gauge invariant operators in the superconformal field theory is summarised in table 1. Let us denote them collectively as $\mathcal{O}_i$ The generating functional for correlation functions of these observables is

$$\begin{aligned}
Z_{SCFT}[\{J_i\}] &= \int [d\Phi] \exp\Big(-S[\Phi] \\
&+ \sum_i \int_{S^4} d^4x \, \sqrt{\det g_{s^4}} \, J_i(x)\mathcal{O}^i(x)\Big),
\end{aligned} \tag{7.93}$$

where Φ collectively denotes all the fields in $\mathcal{N} = 4$ superconformal field theory. An n-point correlation function of these gauge invariant operators is obtained by differentiating Z_{SCFT} by appropriate sources.

$$\frac{\delta^n Z_{SCFT}}{\delta J_{i_1}(x_1)\cdots\delta J_{i_n}(x_n)}\Big|_{J_i=0\forall i} = \langle \mathcal{O}_{i_1}(x_1)\cdots\mathcal{O}_{i_n}(x_n)\rangle. \tag{7.94}$$

With this formulation we can state the conjecture in the following form:

There is a one to one correspondence between χ_i and $\mathcal{O}_i$ such that

$$Z_{IIB}[\{\chi_i^B\}] = Z_{SCFT}[\{\chi_i^B\}]. \tag{7.95}$$

That is, boundary values of the fields in type IIB on AdS_5 are such that $\chi_i^B = J_i$.

7.5 Kaluza-Klein reduction

We will start with te classical background, *i.e.*, metric on $AdS_5 \times S^5$, fifth rank field strength $F_{mnpqr} = c\epsilon_{mnpqr}$ and $F_{\alpha\beta\gamma\delta\eta} = c\epsilon_{\alpha\beta\gamma\delta\eta}$ and $\tau = a_0 + ig^{-1}$, and we will study fluctuation of all the fields about this background. In other words, we will linearise IIB equations of motion in this background and seek solutions to the linearised equations of motion.

Since S^5 is a compact space, we will write all the fields in ten dimensional type IIB supergravity on $AdS_5 \times S^5$ in terms of spherical harmonics on S^5. We will then evaluate their spectrum using the Kaluza-Klein mechanism.

To understand the Kaluza-Klein mechanism, let us consider a 2+1 dimensional theory. Suppose we have periodic boundary condition in one direction, say, in y-direction.

$$y = y + 2\pi l \tag{7.96}$$

If l is very large the space will appear to be two dimensional, but if l is very small, smaller than the resolution of the most powerful microscope then the space will appear to be one dimensional. In effect we will end up with a 1+1 dimensional theory starting from a 2+1 dimensional theory.

More concretely, let us consider free scalar field theory in 2+1 dimensions.

$$S = -\frac{1}{2} \int dt d^2x \left(\eta^{\mu\nu} \partial_\mu \phi \partial_\nu \phi + m^2 \phi^2 \right). \tag{7.97}$$

Since y direction is periodic, we can expand ϕ as

$$\phi(x, y, t) = \frac{1}{\sqrt{2\pi l}} \sum_{n=-\infty}^{\infty} \phi_n(x, t) \exp(in\frac{y}{l}). \tag{7.98}$$

Since ϕ is a real scalar field, it imposes reality condition on the modes of ϕ. The zero mode, $\phi_0(x, t)$ is real and

$$\phi_{-n}(x, t) = \phi_n^*(x, t). \tag{7.99}$$

Substituting this in the action gives

$$
\begin{aligned}
S &= \frac{1}{2} \int dt dx \sum_{n=-\infty}^{\infty} \left[\partial_t \phi_{-n} \partial_t \phi_n - \partial_x \phi_{-n} \partial_x \phi_n \right.\\
&\qquad\qquad\qquad \left. - \frac{n^2}{l^2} \phi_{-n} \phi_n - m^2 \phi_{-n} \phi_n \right]\\
&= \frac{1}{2} \int dt dx \left[\partial_t \phi_0 \partial_t \phi_0 - \partial_x \phi_0 \partial_x \phi_0 - m^2 \phi_0 \phi_0 \right]\\
&\quad + \int dt dx \sum_{n=1}^{\infty} \left[\partial_t \phi_n^* \partial_t \phi_n - \partial_x \phi_n^* \partial_x \phi_n - \left(m^2 + \frac{n^2}{l^2} \right) \phi_n^* \phi_n \right]
\end{aligned}
$$

Thus we see that we have an infinite number of scalar fields in $1-1$ dimensions. ϕ_0 is a real scalar field with mass m and ϕ_n, for each n is a complex scalar field with mass $\sqrt{m^2 + (n/l)^2}$. As $l \to 0$ all ϕ_n's become infinitely massive and decouple from low energy theory leaving us with only ϕ_0. In this limit we get a $1+1$ dimensional free scalar field theory with mass m.

We will use the same mechanism to reduce type IIB supergravity on S^5 and get the resulting theory on AdS_5. We will first linearise the bosonic equations of motion in this background. Let us recall the background about which we will be linearising the fields is

$$
\begin{aligned}
g_{MN} &= \left(\bar{g}_{\alpha\beta}, \bar{g}_{mn} \right), & \tau &= \text{constant},\\
F_{mnpqr} &= c\epsilon_{mnpqr}, & F_{\alpha\beta\gamma\delta\epsilon} &= c\epsilon_{\alpha\beta\gamma\delta\epsilon},
\end{aligned}
\tag{7.100}
$$

where $\alpha, \beta, \cdots$ are AdS_5 indices and $m, n, \cdots$ are S^5 indices and c is a constant. Linearisation of field equations is done as follows. We expand all the fields about their background values

$$
\begin{aligned}
g_{\alpha\beta} &= \bar{g}_{\alpha\beta} + h_{\alpha\beta}(x, y), & g_{mn} &= \bar{g}_{mn} + h_{mn}(x, y),\\
g_{n\alpha} &= 0 + h_{m\alpha}(x, y), & \tau &= \tau_0 + \hat{\tau}(x, y),\\
B_{\alpha\beta} &= 0 + \hat{B}_{\alpha\beta}(x, y), & B_{mn} &= \hat{B}_{mn}(x, y),\\
B_{m\alpha} &= 0 + \hat{B}_{m\alpha}(x, y), & D_{\alpha\beta\gamma\delta} &= \bar{D}_{\alpha\beta\gamma\delta} + \hat{D}_{\alpha\beta\gamma\delta}(x, y),\\
D_{\alpha\beta\gamma m} &= 0 + \hat{D}_{\alpha\beta\gamma m}(x, y), & D_{\alpha\beta mn} &= 0 + \hat{D}_{\alpha\beta mn}(x, y).
\end{aligned}
$$

We will also linearise the equations on motion (7.76),

$$
\begin{aligned}
D^M \partial_M \hat{\tau} &= 0 \\
R_{MN} &= \frac{1}{6} \kappa^2 c \epsilon_{MPQRS} \hat{F}_N^{PQRS} \\
\hat{F}_{M_1 M_2 \cdots M_5} &= \epsilon_{M_1 M_2 \cdots M_{10}} \hat{F}^{M_6 \cdots M_{10}} \\
D^M \partial_{[M} \hat{B}_{NP]} &= -\frac{2i}{3} c \epsilon_{NPQRS} D^Q \hat{B}^{RS}.
\end{aligned}
\tag{7.101}
$$

Using the ansatz for massless fields and substituting it in eq.(7.101) we get linearised equations of motion decomposed in terms of components living on AdS space and on sphere. We can now look for solutions to these equations of motion. To solve equations of motion for all the fluctuations is a long and tedius exercise. We will carry it out only for a scalar field. Before we get into that let us look at the tensorial structure of the fluctuations.

1. $h_{\alpha\beta}(x, y)$ are components of the metric fluctuations with indices on AdS_5. It is a symmetric second rank tensor on AdS_5. It is, however, a function of both coordinates on AdS_5 and S^5. Since it has no S^5 index on it, it is a scalar on S^5.

2. $h_{\alpha m}(x, y)$ are components of metric fluctuations with one AdS_5 index and one S^5 index. It is a vector on AdS_5 as well as on S^5.

3. $h_{mn}(x, y)$ are components of metric along the S^5 direction. These components are split into two parts, $h_{(mn)}(x, y)$ which is a traceless symmetric part and $h_m^m(x, y)$ which is the trace part.

   (a) $h_{(mn)}(x, y)$ is a symmetric second rank tensor on S^5 and a scalar on AdS_5 as it has no AdS_5 index.

   (b) $h_m^m(x, y)$ is a scalar on S^5 as well as on AdS_5.

In the full analysis similar tensorial decomposition is done for $\hat{B}_{\alpha\beta}$, $\hat{B}_{\alpha m}$, $\hat{B}_{mn}$ as well as for the components of D_{MNPQ}^+.

Let us now consider linearised equation of motion for $\hat{\tau}$.

$$D^M \partial_M \hat{\tau} = 0. \tag{7.102}$$

Since the metric decomposes into mdirect sum of metric on AdS_5 and metric on S^5, the laplacian decomposes into the Laplacian on AdS_5 and the Laplacian on S^5. Thus the equation of motion can be written as

$$(\Box_x + \Box_y)\hat{\tau}(x, y) = 0, \tag{7.103}$$

where $\Box_x$ is the Laplacian on AdS_5 and $\Box_y$ is the Laplacian on S^5. If we have an operator $\mathcal{O}$ which is a sum of two commuting operators $\mathcal{O}_1$ and $\mathcal{O}_2$ then eigenfunctions of $\mathcal{O}$ are product of eigenfunctions of $\mathcal{O}_1$ and $\mathcal{O}_2$. Thus,

$$\hat{\tau}(x, y) = \sum_{k=0}^{\infty} \tau^k(x) Y^k(y). \tag{7.104}$$

Substituting this ansatz in the equations of motion we get

$$\sum_{k=0}^{\infty} (\Box_x + \Box_y)\tau^k(x) Y^k(y) = 0, \tag{7.105}$$

which implies for each k

$$Y^k(y)\Box_x \tau^k(x) + \tau^k(x)\Box_y Y^k(y) = 0. \tag{7.106}$$

Now recall, $\Box_y$ is the Laplacian on S^5, hence solutions $Y^k(y)$ are spherical harmonics. Since $Y^k(y)$ doesnot have any S^5 index, these are scalar spherical harmonics. The infinite tower of scalar spherical harmonics on S^5 are parametrized by an index k. The spectrum of these scalar harmonics on S^5 is,

$$\Box_y Y^k(y) = -k(k+4)Y^k(y), \quad k = 0, 1, 2, \cdots \tag{7.107}$$

This implies equation of motion for $\tau^k(x)$ on AdS_5 is

$$\Box_x \tau^k(x) - k(k+4)\tau^k(x) = 0. \tag{7.108}$$

Before we solve this equation, let us look at the spherical harmonics on S^5. When we looked at the tensor decomposition of the fields in the $Ads_5 \times S^5$ background we saw that we not only need spectrum of scalars on S^5 but also that of vectors, symmetric second rank tensors as well as antisymmetric second rank tensors. To determine these we need to study vector as well as second rank tensor harmonics on S^5.

In the flat space, equation of motion for a vector field is

$$\partial^\mu \partial_\mu A_\nu + \partial^\mu \partial_\nu A_\mu = 0. \tag{7.109}$$

The second term in the equation of motion can be removed using the gauge condition. In the curved space this equation get modified, with ordinary derivatives being replaced by covariant derivatives,

$$D^\mu D_\mu A_\nu + D^\mu D_\nu A_\mu = 0 \tag{7.110}$$

where, D_μ is the covariant derivative with affine connection. Since covariant derivatives do not commute, we cannot follow the procedure used in the flat space to eliminate the second term using the gauge condition. We can, however, write

$$D^\mu D_\nu A_\mu = D_\nu D^\mu A_\mu + [D^\mu, D_\nu] A_\mu, \tag{7.111}$$

and set the first term to zero using the gauge condition. The second term is proportional to R_ν^μ, and hence the equation of motion becomes

$$\Box A_\nu + R_\nu^\mu A_\mu = 0. \tag{7.112}$$

Putting $R_n^m = -4\delta_n^m / r^2$ for S^5 we get the equation

$$(\Box - 4) A_n = 0, \tag{7.113}$$

where we have set $r = 1$ for simplicity, but it can be reincorporated back in the equation at any stage. Similarly for the second rank antisymmetric tensor harmonics the equation is

$$(\Box - 6) B_{mn} = 0, \tag{7.114}$$

and for second rank symmetric tensor spherical harmonics we get

$$(\Box - 10)h_{(mn)} = 0. \tag{7.115}$$

We have denoted scalar spherical harmonics by Y^k. We will denote vector, antisymmetric and symmetric tensor harmonics by Y^k_m, $Y^k_{[mn]}$ and $Y^k_{(mn)}$ respectively. The spectrum of these spherical harmonics are

$$
\begin{aligned}
(\Box - 4)Y^k_m &= -(k+1)(k+3)Y^k_m, \\
(\Box - 6)Y^k_{[mn]} &= -(k+2)^2 Y^k_{[mn]}, \\
(\Box - 10)Y^k_{(mn)} &= -(k^2 + 4k + 8)Y^k_{(mn)},
\end{aligned}
\tag{7.116}
$$

where, $k = 1, 2, 3, \cdots$ in the first two equations and $k = 2, 3, 4, \cdots$ in the third. Let us now get back to τ equation of motion on AdS_5. Instead of writing the scalar harmonics spectrum being $-k(k+4)$ we will denote it by $-m_k^2$, *i.e.*,

$$(\Box_x - m_k^2)\tau^k(x) = 0, \tag{7.117}$$

where $\Box_x$ is the Laplacian on AdS space. The metric on AdS_5 that we will use is

$$d\tau^2 = \frac{1}{Z^2}\left(dZ^2 + \sum_{a=1}^{4} dZ^a dZ^a\right), \tag{7.118}$$

we have again set the radius of AdS_5 to be equal to 1. Recall, range of coordinates in this metric are $-\infty < Z^a < \infty$ for $a = 1, 2, 3, 4$ and $0 < Z < \infty$, with $Z = 0$ being the boundary. The metric is singular at $Z = 0$. We will regularise it by taking $Z = \epsilon$ as the boundary and then take $\epsilon \to 0$ in the end. Using these coordinates equation of motion for τ becomes

$$Z^2\left[\frac{\partial^2}{\partial Z^2} - \frac{3}{Z}\frac{\partial}{\partial Z} + \sum_{a=1}^{4}\frac{\partial^2}{\partial(Z^a)^2}\right]\tau^k - m_k^2\tau^k = 0. \tag{7.119}$$

Since the metric on AdS_5 and hence the Laplacian is independent of Z^a coordinates, we can get rid of $\partial^2/\partial(Z^a)^2$ by Fourier transformation

$$\tau^k(Z, Z^a) = \frac{1}{(2\pi)^2}\int d^4p \exp\left(ip_a Z^a\right)\tilde{\tau}^k(Z, p_a), \tag{7.120}$$

where $p^a p_a = p^2$. Substituting this in the equations of motion gives

$$\frac{d^2 \tilde{\tau}^k(Z, p)}{dZ^2} - \frac{3}{Z} \frac{d\tilde{\tau}^k(Z, p)}{dZ} - \left(p^2 + \frac{m_k^2}{Z^2} \right) \tilde{\tau}^k(Z, p) = 0. \quad (7.121)$$

To solve this equation let us first substitute $\tilde{\tau}^k(Z) = Z^2 \varphi^k(Z)$ and then redefine $\rho = pZ$ to get

$$\frac{d^2 \varphi^k}{d\rho^2} + \frac{1}{\rho} \frac{d\varphi^k}{d\rho} - \left(1 + \frac{\nu_k^2}{\rho^2} \right) \varphi^k = 0, \quad \nu_k = \sqrt{4 + m_k^2}. \quad (7.122)$$

We can solve this equation in different limits. Let us first consider large ρ limit. In this limit the equation of motion becomes

$$\frac{d^2 \varphi^k}{d\rho^2} - \varphi^k = 0, \quad (7.123)$$

and the solution is

$$\varphi^k(\rho) = A \exp(\rho) + B \exp(-\rho). \quad (7.124)$$

Regularity of the solution at $rho \to \infty$ implies $A = 0$. In the $\rho \to 0$ limit, the equation of motion becomes

$$\frac{d^2 \varphi^k}{d\rho^2} + \frac{1}{\rho} \frac{d\varphi^k}{d\rho} - \frac{\nu_k^2}{\rho^2} \varphi^k = 0. \quad (7.125)$$

If we take $\varphi^k = \rho^\lambda$ then

$$\lambda(\lambda - 1) + \lambda - \nu_k^2 = 0. \quad (7.126)$$

This is solved by $\lambda = \pm \nu_k$ and hence the solution is

$$\varphi^k = D\rho^{-\nu_k}(1 + \rho + \cdots) + E\rho^{\nu_k}(1 + \rho + \cdots) \quad (7.127)$$

Due to the singularity of the metric at $\rho = 0$, the regular solution is not consistent with the regularity condition at $\rho \to \infty$. Therefore the solution as $\rho \to 0$ is $\varphi \sim D\rho^{-\nu_k}$. Putting this back in the formula for $\tilde{\tau}^k$ we get

$$\tilde{\tau}^k(Z) \sim DZ^2(pZ)^{-\nu_k} \sim Dp^{-\nu_k} Z^{2-\nu_k}. \quad (7.128)$$

Therefore at $Z = \epsilon$,

$$\tilde{\tau}^k(\epsilon) \sim \epsilon^{2-\nu_k}. \tag{7.129}$$

Thus boundary behaviour of the fields $\{\chi_i\}$ on the AdS_5 belonging to τ^k is

$$\tau_B^k(Z^a, \epsilon) \sim \epsilon^{2-\nu_k} \tau_0^k(Z^a). \tag{7.130}$$

As mentioned earlier these boundary values of the fields on the AdS_5 space become sources in the super Yang-Mills theory partition function. Note that these sources would couple only to a subset of scalar gauge invariant operators of the super Yang-Mills theory.

Incorporating these sources in the super Yang-Mills partition function we can write

$$Z_{SYM}\left[\{\epsilon^{2-\nu_k}\tau_0^k(Z^a)\}\right] = \int [d\Phi]\, \exp\left(-S[\Phi]\right.$$
$$\left. + \sum_k \int d^4Z \sqrt{\det g_4}\, \epsilon^{2-\nu_k} \mathcal{O}_k(Z)\tau_0^K(Z)\right).$$

Recall that the four dimensional metric $g_{ab}^{(4)} = \epsilon^{-2}\delta_{ab}$ at $Z = \epsilon$. Therefore the determinant is $\det g_{ab}^{(4)} = \epsilon^{-8}$. Incorporating this in the super Yang-Mills partition function yields

$$Z_{SYM}\left[\{\epsilon^{2-\nu_k}\tau_0^k(Z^a)\}\right] = \int [d\Phi]\, \exp\left(-S[\Phi]\right.$$
$$\left. + \sum_k \int d^4Z \epsilon^{-2-\nu_k} \mathcal{O}_k(Z)\tau_0^K(Z)\right).$$

We can now use this partition function to evaluate correlation functions of some of the gauge invariant operators in the gauge theory. These correlation functions obtained by differentiating with respect to the sources are gauge invariant by construction. Let us look at the two point correlation function.

$$\frac{\delta^2 Z_{SYM}}{\delta\tau_0^k(Z^a)\delta\tau_0^{k'}(Z'^a)}\bigg|_{\substack{\forall k \\ \tau_0^k=0}} = \frac{1}{\epsilon^{4+\nu^k+\nu^{k'}}} \left\langle \mathcal{O}_k(Z^a)\mathcal{O}_{k'}(Z'^a)\right\rangle_{g_{ab}=\frac{1}{\epsilon^2}\delta_{ab}}$$

$$\tag{7.131}$$

A nonsingular metric can be obtained by making the coordinate change $Z^a = \epsilon x^a$. Recall, the operator $\mathcal{O}_k(Z^a)$ has scaling dimension Δ_k when the four dimensional metric $g_{ab} = \delta_{ab}$. Therefore, we can express the two point correlation in terms of x^a coordinates

$$\left\langle \mathcal{O}_k(Z^a)\mathcal{O}_{k'}(Z'^a) \right\rangle_{g_{ab}=\frac{1}{\epsilon^2}\delta_{ab}} = \left\langle \mathcal{O}_k(\epsilon^{-1}Z^a)\mathcal{O}_{k'}(\epsilon^{-1}Z'^a) \right\rangle_{g_{ab}=\delta_{ab}}. \tag{7.132}$$

Since the super Yang-Mills theory defined at the boundary is conformally invariant, it implies

$$\left\langle \mathcal{O}_k(\epsilon^{-1}Z^a)\mathcal{O}_{k'}(\epsilon^{-1}Z'^a) \right\rangle_{g_{ab}=\delta_{ab}}$$
$$= a\epsilon^{\Delta_k+\Delta_{k'}} \left\langle \mathcal{O}_k(Z^a)\mathcal{O}_{k'}(Z'^a) \right\rangle_{g_{ab}=\delta_{ab}}.$$

By rescaling the metric it can be written as

$$\left\langle \mathcal{O}_k(Z^a)\mathcal{O}_{k'}(Z'^a) \right\rangle_{g_{ab}=\delta_{ab}}$$
$$= a\epsilon^{-\Delta_k-\Delta_{k'}} \left\langle \mathcal{O}_k(Z^a)\mathcal{O}_{k'}(Z'^a) \right\rangle_{g_{ab}=\epsilon^{-2}\delta_{ab}}.$$

We can now compare this result with that derived from Z_{SYM} and read out the scaling dimensions

$$\Delta_k = 2 + \nu_k = 2 + \sqrt{4 + m_k^2}. \tag{7.133}$$

Thus for scalar, gauge invariant operators Δ_k can be determined by inserting the mass spectrum of the scalar spherical harmonics. Recall, the scalar spherical harmonics on S^5 have the mass spectrum

$$m_k^2 = k(k + 4), \tag{7.134}$$

thus the scaling dimensions are

$$\Delta_k = 2 + \sqrt{4 + k(k + 4)} = k + 4, \quad k \geq 0. \tag{7.135}$$

On the type IIB side we have been studying dilaton-RR scalar modulus and its S^5 spherical harmonics. The lowest component of this complex scalar is $SO(6)$ singlet and higher harmonics go as **6**, **20**, $\cdots$. The lowest level operator, corresponding to $k = 0$, has scaling dimension 4. It is easy to determine the gauge invariant operators in gauge theory with this data for Δ_k and $SO(6)$ representations,

$$\text{Tr}(F^{\mu\nu}F_{\mu\nu}\phi^k)(z^a), \quad k \geq 0. \tag{7.136}$$

The formulas for Δ_k for vector, second rank antisymmetric tensor and second rank symmetric tensor gauge invariant operators in terms m_k^2 are:

$$\text{vectors:} \quad \Delta_k = 2 + \sqrt{1 + m_k^2},$$

$$\text{rank 2 antisymm. tensor:} \quad \Delta_k = 2 + m_k, \tag{7.137}$$

$$\text{rank 2 symm. tensor:} \quad \Delta_k = 2 + \sqrt{4 + m_k^2}.$$

There are a few caveats in relating the mass formulae for the spherical harmonics to the scaling dimensions. This is because linearised equations of motion are coupled and it is necessary to diagonalise the mass matrix for every coupled set. In case of fermions, one needs to deal with the Killing spinor equations. These equations are written in the local Lorentz frame. We have not dealt with these coordinates in these lectures, so our discussion of fermionic sector is at best sketchy.

Bibliography

[1] J.M. Maldacena, *The large N limit of superconformal field theories and supergravity*, Adv. Theor. Math. Phys. **2**, 231 (1998) [Int. J. Theor. Phys. **38**, 1113 (1999)] [hep-th/9711200].

[2] E. Witten, *Anti-de Sitter space and holography*, Adv. Theor. Math. Phys. **2**, 253 (1998) [hep-th/9802150].

[3] S.S. Gubser, I.R. Klebanov and A.M. Polyakov, *Gauge theory correlators from non-critical string theory*, Phys. Lett. **B428**, 105 (1998) [hep-th/9802109].

[4] O. Aharony, S.S. Gubser, J.M. Maldacena, H. Ooguri and Y. Oz, *Large N field theories, string theory and gravity*, Phys. Rept. **323**, 183 (2000) [hep-th/9905111].

Chapter 8

New Scenarios for Physics Beyond the Standard Model

Sandip P. Trivedi

8.1 Introduction

The Standard Model is a remarkably successful theory, consistent with all known experimental facts[1]. Yet there is good reason to believe that it is incomplete, and that new physics must come into play close to a TeV. Hopefully this new physics will be accessible at the Large Hadron Collider, which will begin gathering data a few years from now.

We believe that the Standard Model is incomplete because it suffers from a serious problem, called the hierarchy problem. There are two important energy scales in the Standard Model. The Planck scale M_{Pl}, which governs the strength of gravity, and, the electroweak scale M_{ew}, which governs the strength of the weak interactions. Their ratio is very small:

$$\frac{M_{ew}}{M_{Pl}} = 10^{-17}.$$

(8.1)

The large hierarchy of scales is very unnatural and it's existence is called the hierarchy problem.

---

[1] A modest extension can incorporate neutrino masses.

Let us explain this a little more. The electroweak scale is related to the mass of the Higgs field. Quite generally, one can argue that in any effective field theory containing a scalar field, valid below the cutoff scale Λ_c, quantum effects drive the mass of the scalar to be of order, Λ_c. If the Standard Model is valid all the way up to the Planck scale this would mean that the Higss mass and therefore the scale of electroweak symmetry breaking would naturally be of order the Planck scale. Miraculous cancellations would have to occur in order to get a ratio as small as in Eq.(8.1).

This is clearly unsatisfactory and leads us to believe that new physics must come in at a scale not much above the electro-weak scale. In previous attempts to solve the hierarchy problem, like supersymmetry or technicolor, new particle physics degrees of freedom, with mass at the TeV scale, were used to explain why the electroweak scale is so much smaller than the Planck scale.

More recently, new scenarios, motivated by string theory, have been constructed to solve the hierarchy problem. The novel feature in these scenarios is that the behavior of gravity is altered and the fundamental scale associated with it, say the string scale, is taken to be of order a TeV. The modifications in gravity, especially the presence of extra dimensions in spacetime, is then used to explain why the four dimensional Planck scale, M_{Pl}, is so much larger than the electroweak scale.

There are two scenarios of this type which go by the name of the ADD scenario and the Randall Sundrum scenario. These lead to experimental signatures which can be used to distinguish them from supersymmetry and technicolor and also from each other. The collider signatures of these models will allow us to test them at the LHC. In addition, in the ADD case, some of the signatures can be tested in experiments which probe for the modifications of gravity at distance scales of order a millimeter.

At the very least, these new scenarios are an interesting foil to the more conventional ones like supersymmetry. If true, they could lead to dramatic consequences in the next round of collider experiments. Indeed, the discovery of extra dimensions in space-time would be truly revolutionary. It would probably also answer our fervent prayers to test string theory experimentally.

These lectures were delivered at the *XVI SERC School on Theoretical High Energy Physics* held at the Harish-Chandra Research Institute, Allahabad, during February-March, 2001. They are targeted at beginning to middle-level graduate students, with a passing familiarity with General Relativity. The lectures are planned as follows. In Sec. 2, we discuss the ADD model and some of its consequences. The attempt is on getting across the basic ideas and we rely almost entirely on Newtonian gravity to do so. Sec. 3, deals with the RS model and requires some knowledge of Einstein's equations. In Sec. 3.3 we discuss the Golberger-Wise mechanism for stabilising the Radion moduli, and conclude in Sec. 3.4. The emphasis of these lectures is theoretical. The phenomenological aspects were the central focus of a companion set of lectures given by Sreerup Raichaudhury.

It is my pleasure to thank Debashis Ghoshal for help in preparing these notes; Sreerup Raychaudhuri for coordinating our respectively lectures and related discussion; Ashoke Sen for his comments; and most of all to the participants of the school for their enthusiasm and interest.

A word about references. In the last few years this subject has seen an intense burst of research activity and a huge literature has been spawned as a result. Fortunately, the seminal papers are themselves very readable and a good starting point. For the ADD scenario these are [1, 2] and [3] respectively. For the Randall-Sundrum (RS) scenario these are [4, 5, 6].

Finally, a few comments about notation. We will work in conventions where $\hbar = 1$, $c = 1$, and use the metric, $\eta^{\mu\nu} = (-1, 1, 1, 1)$.

8.2 The ADD scenario

The model we will discuss here was first proposed by Arkani-Hamed, Dimopoulos and Dvali [1, 2, 3]. The central features are as follows. In this model space-time is taken to have more than four dimensions. The extra dimensions are curled up into a small compact manifold, as in Kaluza Klein compactifications. The gravity degrees of freedom live in all the spacetime dimensions, while the Standard Model degrees of freedom live on a $3+1$ dimensional domain wall, which does not extend in the compact directions. In string theory *D-branes* are a natural candidate for such domain walls, and accordingly the domain wall is sometimes called a *brane* in the literature. However, it is worth noting here that the Standard Model domain wall in the ADD scenario need not necessarily be a D-brane.

We now turn to understanding these features in more detail.

8.2.1   Gravity in extra dimensions

We begin by first understanding how gravity works in extra dimensions. Newtonian gravity is a good place to start. Newton's law says that the gravitational force between two point masses, M and m at a distance $\vec{r}$ apart, in $3+1$, dimensions takes the form:

$$\vec{F} = -\frac{G_N M m}{r^2}\,\hat{r}. \tag{8.2}$$

The inverse square law is analogous to Coulomb's Law in Electrostatics and in fact one can associate a Newtonian potential, ϕ, with the point mass M which solves Poisson equation:

$$\nabla^2 \phi = 4\pi G_N \rho(\vec{r}) = 4\pi G_N M \delta^3(\vec{r}), \tag{8.3}$$

and takes the form

$$\phi = -\frac{G_N M}{r}. \tag{8.4}$$

The force Eq.(8.2) is given in terms of ϕ by

$$\vec{F} = -m\,\vec{\nabla}\phi. \tag{8.5}$$

In $d+1$ dimensions one can similarly construct a Newtonian potential, as a solution of Poisson equation. It takes the form,

$$\nabla^2 \phi = 4\pi G_N^{d+1} M \delta^d(\vec{r}) \tag{8.6}$$

and is given by

$$\phi = -\frac{4\pi}{(d-2)V_{S^{d-1}}}G_N^{d+1}\frac{M}{r^{d-2}}. \tag{8.7}$$

Here G_N^{d+1} is the gravitational constant in $d+1$ dimensions[2]. $V_{S^{d-1}}$ stands for the volume of a units $d-1$ dimensional sphere, and is given by

$$V_{S^{d-1}} = \frac{2\pi^{d/2}}{\Gamma(d/2)}. \tag{8.8}$$

The resulting force on a point mass, m, is then given by Eq.(8.5). The Newtonian potential must be dimensionless, this means that G_N^{d+1} must have dimensions of

$$[G_N^{d+1}] = [\text{Length}]^{d-1}. \tag{8.9}$$

Notice that the gravitational force law falls off more rapidly as the dimensionality of spacetime increases.

Now let us ask what happens when the extra dimensions present are compactified. We start with the case when there is one additional direction compactified on a circle of length $L = 2\pi R$. The resulting potential can be obtained by solving Poisson's equation using the method of images.

For now, we note the following intuitively obvious feature about the the resulting potential. For $r \ll L$ we expect that it takes the form Eq.(8.7), with $d = 4$, while for $r \gg L$, it should take the form in $3+1$ dimensional spacetime, Eq.(8.4). What is the relation between the $4+1$ dimensional gravitational constant G_N^5, that governs the short distance behaviour, and the $3+1$ dimensional one, G_N, that governs the long distance behaviour ? On dimensional grounds we would expect that this takes the form:

$$G_N^5 \sim G_N^4 L.$$

---

[2]In our notation the $3+1$ dimensional Newton constant is denoted as G_N, and the Planck mass by M_{Pl}, where $M_{Pl}^2 = 1/G_N$. More generally, in $d+1$ dimensions, the gravitational constant is denoted with a superscript, as G_N^{d+1}, and the Planck mass by M_{Pl}^{d+1}.

It is worth fixing the constant of proportionality in the relation above. The simplest way to do this is to use Gauss' law in analogy with Electrostatics. From the potential we can construct the field strength for the mass M, $\vec{F} = -\vec{\nabla}\phi$. We see from Eq.(8.6) that the total flux of this field strength crossing a surfaces at infinity, is $\Phi = 4\pi G_N^5 M$. This flux can also be calculated by taking the long distance behavior of the potential Eq.(8.4). This gives $\Phi = 4\pi G_N M L$. Equating the two gives the relation :

$$G_N^5 = G_N L. \tag{8.10}$$

If there are p extra directions compactified on circles of equal radius, R, a similar argument shows that

$$G_N^{4+p} = G_N (2\pi R)^p. \tag{8.11}$$

More generally, for a compactification of volume V, the factor $(2\pi R)^p$, on the RHS of Eq.(8.11) is replaced by V.

8.2.2   The size of extra dimensions

We are now ready to ask the following important question: How big can the extra dimensions be? Recall that our central motivation here is to try and solve the hierarchy problem. To do so we will take the the fundamental scale associated with gravity to be about a TeV. So we set Newton's constant $G_N^{4+p} = (1\text{TeV})^{-(p+2)}$. G_N has an observed value $(10^{19}\text{ GeV})^{-2}$. Then from Eq.(8.11) we learn that the size of the extra dimensions is given by

$$L = 2\pi R = \left[\frac{G_N^{4+p}}{G_N}\right]^{1/p} \tag{8.12}$$

For $p = 1$ one finds that $L \simeq 10^{13}$m this is clearly ruled out by experiments since the inverse square law has been tested well on these scales. For $p = 2$ one finds $L \simeq 1$, this is an interesting scale which is being currently probed by experiments which look for deviations from the inverse square law. Finally, string theory is known to require $p = 6$, which gives, $L \simeq 10$ fermi. This is a very short distance scale and experiments which directly look for deviations from the inverse square law at such short distances are not practical.

It is worth emphasising the central lesson we have learnt above in words. We see from Eq.(8.11) that G_N, the four dimensional Newton's constant, can be much smaller than one in TeV units, if L, the size of the compactification, can be made much bigger than the (inverse) TeV scale. The essential idea in the ADD scenario then is to take the fundamental scale of gravity to be of order the electroweak scale, and to explain the small value of G_N (or equivalently the large value of M_{Pl}) as arising due to the large volume of the compactified extra dimensions.

8.2.3 The standard model lives on a domain wall

There is one immediate question that might have come to the mind of the reader. The other forces of nature like electromagnetism are very well tested on distance scales of order a millimeter or fermi. Departures from the inverse square law for these forces are certainly ruled out. To prevent any such deviations, in the ADD scenario the Standard Model degrees of freedom are taken to live on a $3+1$ dimensional domain wall which does not extend in the p extra compact directions. As was mentioned in the introduction, D-branes in string theory are a natural candidate for such domain walls.

8.2.4 The extra particles in the ADD scenario

Let us turn next to understanding the additional excitations, beyond the Standard Model particles, present in the ADD scenario. Clearly any attempt to test the scenario depends crucially on understanding the spectrum of these new particles and how they couple to the Standard Model degrees of freedom.

As we will see below there are two mass scales which characterise the spectrum of the additional particles that are present. These are the TeV scale, and the scale $1/L$, where L is the compactification size, Eq.(8.11). We see from the discussion above that these are very different in magnitude.

As was mentioned above, we are taking the fundamental scale associated with gravity to be of order 1 TeV. It is then reasonable to believe that extra degrees charged under the Standard Model will enter in this mass range. In fact this is what happens in string theory, where the fundamental scale is of order the string tension and higher Regge excitations of the string with Standard Model quantum numbers have a mass of this order.

Particles which have a mass of order $1/L$ are only gravitationally coupled. This has to do with the fact that only gravity in the ADD scenario lives in the extra dimensions. We will see below that these extra excitations arise due to higher vibrational modes, *i.e.*, higher harmonics, of the graviton in the extra dimensions. We will also see that typically in collider processes many of these higher harmonic modes are excited, so that even though each of them is only gravitationally coupled, cumulatively they can have a significant impact on such processes.

8.2.5   Kaluza-Klein harmonics

In this section we will understand in more detail what the higher harmonics of the graviton are, their masses and couplings. These higher vibrational modes are also called Kaluza-Klein modes, after Kaluza and Klein who were amongst the first to consider compactifications of higher dimensional gravity.

We start with a much simpler problem, of a massless free scalar field theory in $4 + 1$ dimenions where the extra direction is compactified on a circle of length $L = 2\pi R$. We denote the extra direction by y below, x^μ, $\mu = 0, \cdots 3$, denote the non-compact directions. The action is given by

$$S = -\frac{1}{2} \int d^4x dy \ (\partial_\mu \phi \partial^\mu \phi + \partial_y \phi \partial_y \phi \tag{8.13}$$

Now we can Fourier decompose the field ϕ as follows:

$$\phi(x, y) = \sum_n \phi_n(x^\mu) e^{ik_n y}, \tag{8.14}$$

where $k_n = 2\pi n/L$. Inserting this into the action above gives:

$$S = -\frac{1}{2} \int d^4 L \sum_n \left(\partial_\mu \phi_{-n} \partial^\mu \phi_n - k_n^2 \phi_{-n} \phi_n \right). \tag{8.15}$$

We see that a single field in $4 + 1$ dimensions can be thought of as an infinite collection of fields in $3 + 1$ dimensions. The zero mode $n = 0$, is a massless field in $3+1$ dimensions, but the higher harmonics are massive, with a mass $m_n^2 = k_n^2 = (2\pi n/L)^2$. This mass grows linearly with n, and for fixed n grows inversely with L.

We have analysed a simple toy scalar field above. But it should be clear that a similar analysis is possible in any translationally invariant field theory compactified on a circle[3]. In particular our comments above are applicable for the gravitational field as well. The action in this case is the Einstein-Hilbert action,

$$S = -\frac{1}{16\pi G_N^{d+1}} \int d^4 x d^p y \sqrt{-g} R. \tag{8.16}$$

It is not difficult to show that compactifying the extra dimensions on circles is a consistent solution of higher dimensional gravity. And that the resulting spectrum for excitations is of the Kaluza Klein type we found above. The only new feature is that due to the extra polarisations of the graviton along the compactified directions, the $3+1$ dimensional theory has additional spin zero and spin one fields, and their higher harmonics, besides the graviton.

To avoid the complications of dealing with the Einstein-Hilbert action, we will illustrate the higher harmonics of gravity by working in the Newtonian approximation below. This will also allow us to understand an important feature, namely the strength with which the KK harmonics couple to the Standard Model matter and their cumulative effect.

---

[3]The idea of Kaluza-Klein compactifications is more general than a circle or a product of circles and applies to manifolds which break translational invariance in the compactified directions too. The spectrum of the higher harmonics in the more general case is more complicated, but on dimensional grounds, one can see that the masses of the Kaluza-Klein harmonics must go like $m_{KK} \sim 1/L$.

8.2.6 The Kaluza-Klein modes of the graviton and the Newtonian potential

We argued above that if we start with p extra dimensions which are compactified with size L, the Newtonian potential at large distances $r \gg L$ behaves as in the $3+1$ dimensional case, Eq.(8.4), while at short distances, $r \ll L$ it behaves as in $p+4$ dimensions, Eq.(8.7) . One expects that the the cross over happens due to the Kaluza-Klein harmonics coming into play. Let us try to understand this better.

For simplicity we consider the case where there is one extra dimension, which, as above, we denote by y. The Standard Model matter will be taken to live on a brane at $y = 0$. Our starting point then is Poisson's equation for the potential Eq.(8.6). Since we will be mainly interested in understanding the gravitational effects of Standard Model matter, we take the density ρ to be a delta function in the y direction with non-vanishing support at the Standard Model brane. This gives us:

$$\nabla^2 \phi = 4\pi G_N^5 \rho(x)\delta(y). \tag{8.17}$$

Now let us decompose the Newtonian potential in terms of its Fourier components, as in Eq.(8.14).

From Eq.(8.17) each mode then satisfies the equation,

$$\nabla_x^2 \phi_n(x) - \left(\frac{2\pi n}{L}\right)^2 \phi_n(x) = \frac{4\pi G_N^5}{L}\rho(x). \tag{8.18}$$

Using Eq.(8.11) to express the five dimensional gravitational constant in terms of the four dimensional one, gives,

$$\nabla_x^2 \phi_n(x) - \left(\frac{2\pi}{L}n\right)^2 \phi_n(x) = 4\pi G_N \rho(x). \tag{8.19}$$

We see that each of the modes couples with four dimensional gravitational strength to the Standard Model matter. The zero mode $n = 0$, is massless and gives rise to the Newtonian potential in $3+1$ dimensions, while the higher harmonics give rise to the additional Kaluza Klein tower of massive fields.

For a point mass, the density is,

$$\rho(x) = M\delta^3(\vec{r}). \tag{8.20}$$

Since the higher modes are massive it is easy to see that at distances $r \gg L$ they give rise to an exponential decaying potential of the form

$$\phi_n = -\frac{G_N M}{r}e^{-m_n r}, \tag{8.21}$$

where $m_n = \frac{2\pi n}{L}$. For $r \gg 1/m_n$ this contribution to the potential rapidly decays, while for $r \ll 1/m_n$ it is approximately equal to that from the zero mode Eq.(8.4).

We can now estimate the total potential at a distance $r \ll L$ by summing over the KK modes. The number of modes which contribute is approximately $N = 2(L/2\pi r)$. The factor of two arises because both modes with positive and negative mode numbers contribute. Each mode gives an approximation contribution equal to the zero mode. As a result the full potential can be approximated to be

$$\phi \simeq -\frac{G_N M}{r}2\left(\frac{L}{2\pi r}\right) \simeq -\frac{G_N L}{\pi}\frac{M}{r^2}. \tag{8.22}$$

Using Eq. (8.11), we see that this can be written as,

$$\phi \simeq -\frac{1}{\pi}\frac{G_N^5}{r^2}. \tag{8.23}$$

In fact, our estimate above even gets the coefficient right, luckily. On comparing with the five dimensional potential Eq.(8.7) with $V_{S^3} = 2\pi^2$, we find agreement with Eq.(8.23).

The exercise above illustrates an important general feature about the coupling of the higher harmonics to the Standard Model degrees of freedom. Each KK mode individually only couples with (four) dimensional gravitational strength to the Standard Model matter. However, at distance scales much smaller than the compactification scale L, many of these modes are typically excited and the cumulative effect of their interactions becomes of order the higher dimensional gravitational coupling. Note that this is much stronger, as we had mentioned above, of order the TeV scale. This feature, which we have seen above in the Newtonian potential, is true much more generally and plays a crucial role in the collider signatures of this model.

A few more comments are worth making at this point.

Notice that the other fundamental forces, *e.g.*, electromagnetism, which only live on the brane, do not change their behaviour at distances smaller than L and their potential continues to go like $1/r$. This means that once we go distances a little below the compactification scale gravity begins to grow quickly in strength in comparison to the other forces. By the time one gets to about a TeV in energy, gravity has caught up with the other forces and became of comparable size. In fact in a string theory scenario all the forces get unified at this scale.

Notice also that at the TeV scale gravity gets strong. The Newtonian potential Eq.(8.23) becomes of order one, since G_N^5 and r are both of order one in TeV units. At this scale one expects interesting non-perturbative effects in gravity to come into play. This includes the formation of black holes. If the ADD scenario is really correct, physics at the TeV scale is going to be truly exciting!

8.2.7   Concluding comments

In this lecture we have explored the ADD scenario as an interesting new attempt at solving the hierarchy problem. The experimental consequences of this scenario will be examined in more detail in the lectures of S. Raychaudhuri in this volume.

Let us end here with a few more comments.

The central idea of the ADD model is to take a compactification with size L, much bigger than 1 in TeV units. This makes the four dimensional Newton constant G_N much smaller than 1 in TeV units. A smaller value of G_N means that gravity is more weakly coupled. In what follows, it will be useful to keep the following picture in mind. Four dimensional gravity arises from the zero mode in the Kaluza-Klein tower. The profile, or "wave function", of the zero mode, is uniformly spread out in the compact directions. In contrast, the Standard Model matter lives on a domain wall which is localised in the compact directions. As a result, for large volume, the overlap between the Standard Model matter and the four dimensional graviton is small, leading to a weak gravitational force.

The hierarchy problem was the main motivation for the ADD model. Let us end by revisiting this issue. As we have just mentioned, solving the hierarchy problem in the ADD model requires us to take the compactification scale to be much bigger than the fundamental scale of gravity, of the order of TeV. But this means there is a large dimensionless number in the model, L TeV. This number takes the value, 10^{16} and 10^5, for two and six extra dimensions respectively. Now, to solve the hierarchy problem we need to explain how this large number arises. In the context of gravity, this is a dynamical question, because the volume should be thought of as a scalar field in the four dimensional effective theory, and the compactification scale L, is related to its expectation value. To date, we have no good ideas on how to generate such a large expectation value for L, without giving as input in some way, a large number in the first place. One has to therefore conclude that, as things stand, the ADD model does not solve the hierarchy problem satisfactorily.

We should caution the reader from taking too pessimistic a view of the model in light of the above observation. We have emphasised the hierarchy problem above. But in the gravitational sector of the Standard Model there is an even bigger naturalness problem, having to do with the smallness of the cosmological constant on the Planck or electorweak scale. This indicates that we are clearly missing something important in our understanding of gravity, and suggests that it is best to keep an open mind at this stage. Perhaps what fixes the cosmological constant to be small might also fix the volume modulus to be large. The experimental consequences of the ADD model, which will be discussed in greater detail in the lectures of S. Raychaudhuri in this volume, are enough motivation to study it carefully.

8.3   The Randall-Sundrum scenario

We now turn to another scenario first proposed by Randall and Sundrum [4, 5]. The model has many features in common with the ADD scenario. The fundamental scale associated with gravity is again taken to be of order a TeV. Extra compactified dimensions play a crucial role in making the four dimensional gravitational constant, G_N, small on this scale. Finally, the Standard Model lives on a $3 + 1$ dimensional brane which does not extend in the extra compact dimensions, while gravity lives in the bulk in all spacetime dimensions.

The main difference is that the extra dimensions are curled up in a manner quite different from what was considered in the ADD case. Rather than a Kaluza-Klein compactification, one has a warped compactification, as we will explain below. The essential idea in the ADD scenario, is to obtain a small value of G_N, by taking the compactification size, L, to be big on the TeV scale. In the Kaluza Klein compactification, the zero mode of the graviton, which gives rise to four dimensional gravity, is, roughly speaking, uniformly spread out in the compact directions. As a result, we saw above that making G_N small enough requires huge values of L. Thus in terms of solving the hierarchy problem the situation in the ADD model is not satisfactory.

In contrast, as we will see below, due to the warped nature of the geometry in the RS case, the graviton zero mode has a profile which is exponentially damped at the location of the Standard Model brane. As a result, for a modest value of L, G_N, can be made adequately small. Further, the required value of L can be be generated by giving modestly small parameters in the model as input. Thus, the Randall-Sundrum scenario looks more promising vis-a-vis the hierarchy problem. We will return to a more detailed critique of the model at the end of this section.

The RS model has experimental signatures quite distinct from the ADD case. There is no closely spaced KK tower of graviton states, with mass splittings of order $1/L$. Instead the higher excitations of the graviton give rise to particles with TeV scale masses.

One final comment. In the ADD case the Newtonian limit was adequate to convey most of the essential ideas, and we avoided a discussion in the context of the General Theory of Relativity. In the RS case though, since the compactification is more complicated, we will have to work with Einstein's equations in analysing the behavior of gravity.

8.3.1 Spacetime in Randall-Sundrum scenario

We start with $4 + 1$ dimensional spacetime. Our notation is as follows: $\hat{\mu} = 0, \cdots, 4$ stand for all direction, compact and non-compact; $\mu = 0, \cdots, 3$ stand for the non-compact dimensions and y stands for the compact direction: $0 \leq y \leq 2\pi R$. The size of the compactification is $L \equiv 2\pi R$.

The Einstein-Hilbert action takes the form:

$$S_{grav} = 2M^3 \int d^4x \, dy \, \sqrt{-g} \, R. \qquad (8.24)$$

M is related to the five dimensional Newton's constant by the relation:

$$2M^3 = \frac{(M_{Pl}^5)^3}{16\pi}. \qquad (8.25)$$

We will sometime loosely call M the five-dimensional Planck scale.

Since the universe around us is approximately Poincaré invariant, we will consider compactifications which preserve Poincare invariance[4]. The most general form of the metric consistent with this symmetry is:

$$ds^2 = f(y)^2 \left[-dt^2 + (dx^1)^2 + (dx^2)^2 + (dx^3)^2 \right] + dy^2. \qquad (8.26)$$

Note the metric components along the x^μ directions, $g_{\mu\nu} = f(y)^2 \eta_{\mu\nu}$, depend on the compact direction y. This is characteristic of warped compactifications. In contrast, in the KK case, these components are independent of the y direction.

---

[4]More accurately, on a time scale of order the Hubble constant, the universe is of slowly expanding Robertson-Walker type. Obtaining such cosmologies in the present context is an interesting question, outside the scope of these lectures.

To obtain a warped compactification one needs spacetime to be curved, this requires some sources for stress-energy. We will consider two sources of this kind. One is a bulk cosmological constant, with action:

$$S_{bulk} = - \int d^4x \, dy \sqrt{-g} \, \Lambda. \qquad (8.27)$$

The second source is localised on two domain walls, located at $y = 0$ and at $y = \pi R$. We will call these the Standard Model brane and the Planck brane respectively. As we will see below, Standard Model degrees of freedom will turn out to live on the Standard model brane, this partly accounts for our nomenclature. The action then has the form,

$$S_{loc} = - \int d^4x dy \sqrt{-g^4} \left(\delta(y) T_{SM} - \delta(y - \pi R) T_{PL} \right). \qquad (8.28)$$

Here T_{Pl} and T_{SM}, refer to the tensions of the Planck and Standard Model branes respectively, and g^4 stands for the determinant of the 4×4 matrix, $g_{\mu\nu}$, with components along the brane world volume.

There are three input parameters that we have introduced so far: Λ, T_{Pl} and T_{SM}. Of these Λ has dimensions $[M]^5$, while the two tensions, T_{Pl} and T_{SM}, have dimensions $[M]^4$. We will see shortly that to obtain a metric of the form Eq.(8.26) as a solution to Einstein's equations, T_{Pl} and T_{SM} must be related very precisely[5] to Λ.

Let us make one comment at this stage. The set up we have considered above, especially the choice for the matter, seems *ad hoc* at the moment. The main motivation for it, as will soon see, is that it gives an example of a warped spacetime with an exponentially damped profile for the graviton. This is the central feature of the RS scenario.

---

[5]In the absence of this precise relationship one will have a more general time-dependent cosmology which will break Poincare invariance in $3 + 1$ dimensions.

The Einstein equations that follow from the action Eqs.(8.24), (8.27) and (8.28), take the form:

$$R_{\hat{\mu}\hat{\nu}} = [2M^3]^{-1} \left(T_{\hat{\mu}\hat{\nu}} - \frac{1}{3} g_{\hat{\mu}\hat{\nu}} T_{\hat{\rho}}^{\hat{\rho}} \right) \tag{8.29}$$

From the metric Eq.(8.26), we have that

$$R_{yy} = -4\frac{f''}{f}, \tag{8.30}$$

where prime denotes derivative with respect to y. Also,

$$R_{\mu\nu} = -\left(\frac{f''}{f} + 3(\frac{f'}{f})^2 \right) g_{\mu\nu}. \tag{8.31}$$

The stress energy is given by

$$T_{yy} = -\frac{\Lambda}{2}, \tag{8.32}$$

and

$$T_{\mu\nu} = -\frac{\Lambda}{2} g_{\mu\nu} - \frac{T_{SM}}{2}\delta(y)g_{\mu\nu} - \frac{T_{Pl}}{2}\delta(y - \pi R)g_{\mu\nu}. \tag{8.33}$$

We will first consider the equations in the bulk, away from $y = 0, \pi R$. The yy-component of the equation gives, from Eqs.(8.30) and (8.32),

$$-4\frac{f''}{f} = \frac{\Lambda}{6M^3}. \tag{8.34}$$

The μ, ν components, from Eqs.(8.31) and (8.33), gives:

$$-\frac{f''}{f} - 3\left(\frac{f'}{f}\right)^2 = \frac{\Lambda}{6M^3}. \tag{8.35}$$

From these we learn that

$$12\left(\frac{f'}{f}\right)^2 = -\frac{\Lambda}{2M^3}. \tag{8.36}$$

This shows that Λ must be negative, so we need a negative bulk cosmological constant. Now solving for f from Eq.(8.36), we can, without loss of generality, take the solution in the region $0 < y < \pi R$ to be,

$$f(y) = e^{ky}, \quad 0 < y < \pi R. \tag{8.37}$$

Here[6], k is a mass scale characterizing the curvature of the spacetime. It is determined by Λ, M to be,

$$k = \sqrt{\frac{|\Lambda|}{24M^3}}. \tag{8.38}$$

In the discussion below, we will often find it useful to express various relations in terms of k, these can always be re-expressed in terms of λ, M using the above equation.

We are now ready to impose the equations at the location of the two branes $y = 0, \pi R$. Since the Einstein equations involve the second derivative of f and the source terms have delta function discontinuities, we see that f must be continuous across the branes and its first derivative with respect to y must have a finite jump. This is analogous to electrostatic problems for the potential, with localised charges.

Continuity of the metric across the two branes determines the metric in the region $\pi R < y < 2\pi R$,

$$f(y) = e^{k(-y+2\pi R)}, \quad \pi R < y < 2\pi R. \tag{8.39}$$

Using the identification, $y \simeq y + 2\pi$, we can write, $f(y)$, in the region, $-\pi R < y < 0$, as,

$$f(y) = e^{-ky}, \quad -\pi R < y < 0. \tag{8.40}$$

The condition on the first derivative of the metric at the SM brane gives,

$$-\frac{4}{f(0)} \left(f'|_{y=\epsilon} - f'|_{y=-\epsilon} \right) = \frac{T_{SM}}{3M^3}, \tag{8.41}$$

---

[6]Choosing the second branch $f = e^{ky}$ is equivalent to switching the Planck and Standard Model branes. An additive constant in the exponent of Eq.(8.37) can be reabsorbed into rescaling x^μ.

which yields

$$T_{SM} = -24M^3 k = -\sqrt{|\Lambda| 24 M^3}. \tag{8.42}$$

Similarly, from the condition on the first derivative of the metric at the Planck brane one finds,

$$T_{Pl} = +24M^3 k = \sqrt{|\Lambda| 24 M^3}. \tag{8.43}$$

Note, as was mentioned above, that for the solution to exist, the tension of the two branes must be precisely tuned in terms of the bulk cosmological constant.

Let us summarise our conclusions from studying the Einstein equations here. We have considered five dimensional gravity in the presence of the a negative bulk cosmological constant term and two branes, the Planck brane and the SM brane with precisely tuned positive and negative tensions taking values Eqs.(8.42) and (8.43) respectively. The resulting spacetime is of warped type Eq.(8.26), with the *warp factor f* being given by Eqs.(8.37) and (8.40) in the regions $0 \leq y \leq \pi R$ and $-\pi R \leq y \leq 0$ respectively.

The bulk solution we have with constant negative cosmological constant is a well known spacetime. It is called anti-de Sitter space (in this case AdS_5). From Einstein's equations we see that it has constant curvature of order k, and it is known to be a highly symmetric space.

There is one interesting feature about the solution above. Although the size of the compactification, R, is a dynamical variable in the problem, since it corresponds to a component of the metric, it is not uniquely determined by the Einstein equations. In fact, the solution we have obtained above is valid for any value of this variable. R is an example of what is called a *moduli field*, and often referred to as the *radion* moduli in the literature. We will have more to say on fixing this modulus below, in the subsection on the Goldberger-Wise variation.

Finally, note that the solution we have is symmetric with respect to a reflection in the the y direction, $y \rightarrow -y$. This is clear from Eqs.(8.37) and (8.40). We can therefore consider a compact-ification, called an *orbifold* in which points related by this symme-try are actually identified. There are two fixed points under this reflection at $y = 0, \pi R$, and the branes are located at these fixed points. However, as is clear from our discussion above, such an identification is not essential for the model.

8.3.2 The energy scales in the RS model and the hierarchy problem

Next, we would like to relate the scales, Λ, M and R, which appear in the solution above to those in the Standard Model. This will allow us to tie the discussion above to the hierarchy problem.

As was mentioned in the previous section, we will take the Standard Model degrees of freedom to live on the Standard Model brane. One feature about the warped geometry we are dealing with, which we will come back to at the end of this subsection, is that while discussing energies or mass scales one needs to be careful about which coordinate system or observer they refer to. In the discussion below, unless we explicitly state otherwise, these mass scales will always refer to measurements made by an observer on the Standard model brane.

We take the five dimensional Planck scale to be the funda-mental scale associated with gravity, and of order the TeV scale. We will also take, the five dimensional cosmological constant, $\Lambda \sim M \sim$ TeV. Strictly speaking, in order to consistently ne-glect corrections to the analysis of Einstein's equations presented here, Λ has to be somewhat smaller[7] than M. Accordingly we will take M to be of order the TeV scale and Λ to be somewhat smaller. It is worth mentioning that the requirement of keeping the corrections small does not require that Λ is vastly smaller than M, so this does not introduce a large hierarchy of scales in the model.

---

[7]Otherwise the curvature k is of order the scale M and higher derivative corrections to the Einstein-Hilbert action, Eq.(8.24), along with other correc-tions due to gravity becoming strong, get important.

We still need to determine R. This is related to the four dimensional Planck scale, M_{Pl}, as follows. We start with a metric of the form:

$$ds^2 = f^2(y)\tilde{g}_{\mu\nu}(x)dx^\mu dx^\nu + dy^2, \qquad (8.44)$$

which is more general than Eq.(8.26), and allows for fluctuations in $\tilde{g}_{\mu\nu}$, corresponding to four dimensional gravity. Substituting this metric in the five dimensional Einstein-Hilbert action Eq.(8.24), one can determine the four dimensional one by carrying out the integral over the y coordinate. We have,

$$\sqrt{-g}R = f^2(y)\sqrt{-\tilde{g}}\tilde{R} + \cdots, \qquad (8.45)$$

where $\tilde{R}$ denotes the curvature scalar made out of the $\tilde{g}_{\mu\nu}$ metric, and the ellipses denote terms involving derivatives of f which are independent of $\tilde{g}_{\mu\nu}$. This gives,

$$
\begin{aligned}
S_{grav} &= -2M^3 \int d^4x\, dy\, \sqrt{-g}R \qquad\qquad (8.46)\\
&= -2M^3 \int d^4x \sqrt{-\tilde{g}}\tilde{R} \int_{-\pi R}^{\pi R} dy\, f^2(y) + \cdots.
\end{aligned}
$$

Substituting for $f(y)$ from Eqs.(8.37) and (8.40), gives[8]

$$S_{grav} = -\frac{2M^3}{k}(e^{2\pi kR} - 1)\int d^4x\sqrt{-\tilde{g}}\tilde{R}. \qquad (8.47)$$

This tells us that the four dimensional Planck scale M_{Pl}, is related to the five dimensional one, by,

$$M_{Pl}^2 \simeq \frac{(M_{Pl}^5)^3}{k}e^{2\pi kR}, \qquad (8.48)$$

where we have used Eq.(8.25) and taken $e^{2\pi kR} \gg 1$. Taking $M_{Pl} = 10^{19}$ GeV and M_{Pl}^5, $k \sim 1$ TeV, gives:

$$kR \sim 12. \qquad (8.49)$$

---

[8]If we had made the identification under $y \leftrightarrow -y$, the integral on the RHS would have half this value.

So we see that for a compactification radius R which is about one order of magnitude bigger than the inverse TeV scale, we obtain the required large value for M_{Pl}. The crucial thing here is that the warp factor $f(y)$ has an exponential profile as a function of y.

The analysis above, assumed that four dimensional gravity corresponds to fluctuations of the form Eq.(8.44). It is worth understanding this better. In fact a simple symmetry argument allows us to identify the four dimensional graviton zero mode. Starting with the metric Eq.(8.26) we can always carry out a gauge transformation

$$x^\mu \to x^\mu + \epsilon^\mu_\nu x^\nu, \tag{8.50}$$

where $\epsilon_{\mu\nu}$ is symmetric and constant. The corresponding change in the metric is then

$$ds^2 = f^2(\eta_{\mu\nu} + h_{\mu\nu})dx^\mu dx^\nu + dy^2, \tag{8.51}$$

where

$$h_{\mu\nu}(x) = \epsilon_{\mu\nu} f(y)^2. \tag{8.52}$$

If the metric we started with is a solution of Einstein equations, doing the coordinate transformation will certainly not change this fact.

It is a small step from here to show that a fluctuation of the metric of the form Eq.(8.51), must solve the equations if

$$h_{\mu\nu}(x) = f(y)^2 \epsilon_{\mu\nu} e^{ik\cdot x}, \tag{8.53}$$

with

$$k^2 = 0, \ \ \epsilon_{\mu\nu}k^\nu = 0, \tag{8.54}$$

and $\epsilon_{\mu\nu}$ being constant. Roughly speaking, the idea is the following. Since we have a solution when the condition Eq.(8.52) is met, the only terms which can survive in the small fluctuation operator are of form

$$(\cdots)\partial_y\partial_\rho h_{\mu\nu} + (\cdots)\partial_\sigma\partial_\rho h_{\mu\nu} = 0, \tag{8.55}$$

where the brackets in front indicate (background) metric dependent coefficients. Now the first term must vanish, since the background metric Eq.(8.26) does not contain off diagonal terms connecting the μ, ν or ρ and the y components. The second term vanishes if the conditions Eq.(8.54) are met.

Thus we see that the required graviton zero mode has the form Eq.(8.53). In particular it has a profile in the compact direction y governed by the function $f(y)$ which is exponentially damped at the Standard Model brane and has most of its support away from it. The fact that the Standard Model matter couples weakly with four-dimensional gravitational coupling strength to the graviton can be now directly traced to this fact by showing that a correctly normalised graviton has an exponentially small value at the Standard Model brane.

A few comments are now in order.

We mentioned above that in the warped background one needs to be careful about which observer or coordinate system we have in mind while referring to particle masses or energy scales. This is because the metric parallel to the non-compact directions suffers from a non-trivial dilation as one moves in the y-direction. It is worth illustrating this with a concrete example.

Consider a particle with mass parameter m located on a brane. By this we mean a mode whose action is given by

$$S = -\frac{1}{2} \int d^4x \sqrt{g^4} \left(g^{\mu\nu} \partial_\mu \phi \partial_\nu \phi + m^2 \phi^2 \right). \qquad (8.56)$$

At the Standard Model brane, where $f(y) = 1$, the particle will indeed have a mass m. However at the Planck brane $y = -\pi R$, one finds using the fact that $f(-\pi R) = e^{\pi k R}$ that the action is given by

$$S_{Pl\,brane} = -\frac{1}{2} \int d^4x \, e^{2\pi k R} \left(\eta^{\mu\nu} \partial_\mu \phi \partial_\nu \phi + e^{2\pi k R} m^2 \phi^2 \right). \qquad (8.57)$$

By considering a canonically normalised field, $\tilde{\phi} = e^{\pi k R} \phi$, we learn that the field has much bigger mass equal to $e^{\pi k R} m$. Taking the particle to be a string excitation for example, so that, $m = M = 1$ TeV, we find from Eq.(8.49) that the mass at the Planck brane is of order the Planck scale.

In short, the warp factor results in energies being red-shifted in a position dependent manner. The Standard Model brane is deep inside the potential well of the ambient gravitational field and as a result all energies on it, when compared to the Planck brane, are hugely red-shifted. In fact this red-shift factor, in going from the Planck to the Standard model brane, is of order M_{ew}/M_{Pl}, the hierarchy of scales with which we began these lectures.

It is also now clear why the Standard Model matter needs to be located on the Standard Model brane. Had we located it on the Planck brane, there would have been an exponential enhancement which would have made four dimensional gravity too strong.

Now that we have understood the different length scales in this scenario, it is time, once again, to revisit the hierarchy question. We see that the hierarchy between the Planck and electroweak scales is exponentially sensitive to the value of R. We had commented earlier, that the value of R is not fixed in the compactification and can take any value, by taking R to be moderately big (in units of M) we obtained the required hierarchy of scales. However, the construction required the brane tensions, T_{SM}, T_{Pl} to be very precisely tuned. One tuning of this sort is to be expected since we are requiring that the four dimensional cosmological constant vanish. However the second fine tunning is extra and undesirable if we are to claim that the hierarchy problem has truly been solved.

In the next section, we will see how to construct a variant of this scenario where this extra fine tuning is avoided. In the resulting model a potential is generated for the radion moduli, R. By giving as input, a moderately small parameter in the model, we will see that the expectation value of R can take the required value for solving the hierarchy problem. This will complete our discussion of the hierarchy problem satisfactorily for the RS model.

One final comment. What about the spectrum of excitations about the solution we have described above? Since the fundamental scale associated with gravity is being taken to be the TeV scale there will certainly be extra states, typically charged under the Standard Model gauge groups, with mass of order a TeV. For example, open string modes living on the SM brane in a string theory construction. From the gravity sector we have already seen how the zero mode giving rise to four dimensional gravity arises. In addition there are higher harmonics. These will be discussed further in the accompanying lectures of Sreerup Raychaudhuri. Here we only note that these higher harmonics are quite different from the ADD case. They are widely space excitations with mass of order a TeV.

There is one additional excitation that comes from the gravity sector, the radion modulus R. Since it can take any value in the compactification above, its potential in the four dimensional theory must vanish and it is a massless field. It is easy to show that the radion couples to SM matter like a Brans-Dicke scalar. A massless Brans-Dicke scalar is well unknown to be phenomenologically unacceptable, this provides another motivation for generating a potential and stabilising this modulus.

8.3.3 The Goldberger-Wise variation

In this section we explore a version of the RS scenario, in which an additional sector is introduced and the radion gets fixed. Moderately small parameters in the additional sector, result in the vacuum expectation value of R, required for solving the hierarchy problem. This section is based on the paper [6], although the treatment below is slightly different in detail.

The additional sector consists of a free scalar field in the bulk which has interactions localised on the Planck and Standard Model brane. Its action takes the form:

$$S = -\frac{1}{2}d^4x dy \sqrt{-g} \int d^4x dy (g^{\hat{\mu}\hat{\nu}}\partial_{\hat{\mu}}\phi\partial_{\hat{\nu}}\phi - m^2\phi^2) \qquad (8.58)$$

in the bulk, with additional interactions on the branes:

$$\begin{aligned}
S_{loc} &= -\int d^4x\,dy\left[\sqrt{-g^4}\lambda(\phi^2 - v_{Pl}^2)^2\delta(y - \pi R)\right.\\
&\left. + \sqrt{-g^4}\,\lambda(\phi^2 - v_{Pl}^2)^2\delta(y)\right], \qquad (8.59)
\end{aligned}$$

where m is the mass of the particle, λ, v_{Pl} and v_{SM} are additional couplings that specify the interactions on the Planck and SM branes[9].

---

[9]Strictly speaking the coupling constant λ could be different on the two branes, we choose it to be the same here for simplicity.

Note that ϕ and λ have mass dimensions $[M]^{3/2}$ and $[M]^{-2}$ respectively, since we are in five dimensions. The stabilisation mechanism we discuss does not require λ, v_{Pl} and v_{SM} to be small on the scale of the five dimensional Planck mass, M, so one is not introducing additional small parameters. For sake of simplifying the analysis though, it will be useful below to take v_{Pl} and v_{SM} to be somewhat small in units of M. This will became more quantitatively precise as we proceed.

We will require one moderately small parameter in order to get the hierarchy of scales. This is the mass of the scalar field, m. It is useful in the following discussion to define,

$$\epsilon = \frac{m^2}{4k^2}, \tag{8.60}$$

where k is the curvature of AdS space, Eq.(8.38). For the stabilisation mechanism to work, we will[10] need $\epsilon < 1$.

The original model we considered above had two additional parameters T_{Pl} and T_{SM}, both of these had to be tuned precisely as we have emphasised above, for the scenario to work. As was mentioned above, one, but not both, of the fine-tunings will be necessary in the model discussed below. For example, in the analysis below, if we allow T_{SM} to vary, T_{Pl} will need to be fine-tuned to a precise value. This is to be expected because we require the four dimensional cosmological constant to vanish.

We now turn to exploring the model introduced above further. It is straightforward to solve the Einstein equations, with the extra stress energy contributions due to the scalar field. We seek Poincare invariant solutions which still satisfy the warped ansatz Eq.(8.26).

If the parameters v_{Pl} and v_{SM} take values sufficiently smaller than one in units of M, as was mentioned above, then the contribution to the bulk stress energy from the scalar field is smaller than the bulk cosmological constant. Thus the solution in the bulk is still of form Eqs.(8.37) and (8.40).

---

[10]Looking ahead ϵ will turn out to be $O(1/30)$ so once again we are not introducing a large hierarchy of scales here.

Next we come to the junction conditions at the two branes. Repeating the analysis leading to Eqs.(8.42) and (8.43), now with the scalar field included, leads to the equations:

$$\lambda(\phi_{SM}^2 - v_{SM}^2)^2 + T_{SM} = T_{SM}^0, \tag{8.61}$$

and

$$\lambda(\phi_{Pl}^2 - v_{Pl}^2)^2 + T_{Pl} = T_{Pl}^0, \tag{8.62}$$

where,

$$T_{Pl}^0 = -T_{SM}^0 = 24M^3k. \tag{8.63}$$

(These would have been the values of the tensions respectively without the scalar field present.) We use the notation $\phi_{SM} \equiv \phi(0)$, $\phi_{Pl} \equiv \phi(\pi R)$, for the values of the scalar field at the two branes respectively.

These can be solved to give:

$$\phi_{SM}^2 = v_{SM}^2 - \frac{\sqrt{T_{SM}^0 - T_{SM}}}{\sqrt{\lambda}}, \tag{8.64}$$

$$\phi_{Pl}^2 = v_{Pl}^2 + \frac{\sqrt{T_{Pl}^0 - T_{Pl}}}{\sqrt{\lambda}}, \tag{8.65}$$

where we have chosen the sign of the discriminant in the two cases so as to meet additional equations that follow.

Next we turn to the scalar field bulk equation,

$$\nabla^2\phi + m^2\phi = 0, \tag{8.66}$$

which has as solution,

$$\phi(y) = Ae^{-(2+\nu)ky} + Be^{(\nu-2)ky}, \quad 0 < y < \pi R, \tag{8.67}$$

where,

$$\nu = \sqrt{4 + \frac{m^2}{k^2}} \simeq 2 + \epsilon. \tag{8.68}$$

In the last step we have used Eq.(8.60) and the approximation, $\epsilon \ll 1$.

From continuity of the scalar field across the two branes, we see that the solution above can be extended to the full space $-\pi R \leq y \leq \pi R$, and takes the form:

$$\phi(y) = Ae^{-(2+\nu)k|y|} + Be^{(\nu-2)k|y|}, \quad -\pi R \leq y \leq \pi R. \qquad (8.69)$$

The constants A, B can be determined in terms of ϕ_{SM}, ϕ_{Pl},

$$A = \left[\phi_{SM} - \phi_{Pl}e^{(2-\nu)\xi}\right]\left[1 + e^{-2\nu\xi} + \cdots\right], \qquad (8.70)$$

$$B = \phi_{Pl}e^{(2-nu)\xi} + \left[\phi_{Pl}e^{(2-\nu)\xi} - \phi_{SM}\right]e^{-2\nu\xi} + \cdots, (8.71)$$

where,

$$\xi = \pi k R, \qquad (8.72)$$

and the ellipses denote terms which are exponentially suppressed for large ξ.

To complete the solution we need to now impose the second junction condition across the branes for the scalar field. This relates the jump in its derivative to the localised interactions Eq.(8.59), and gives rise to two conditions. For any value of T_{SM} (within a range), we can think of these two conditions as determining, R, and the required fine-tuned value of, T_{Pl}.

At the Standard Model brane we get:

$$-k[(2+\nu)A + (\nu-2)B] = 2\lambda(\phi_{SM}^2 - v_{SM}^2)\phi_{SM}. \qquad (8.73)$$

From Eqs.(8.70) and (8.71), this takes the form,

$$k[(2+\nu)(\phi_{SM} - \phi_{Pl}e^{(2-\nu)\xi})] = 2\sqrt{\lambda\left(T_{SM}^0 - T_{SM}\right)}\,\phi_{SM}, \quad (8.74)$$

where on the LHS we have dropped terms which are subleading in ϵ. Similarly at the Planck brane we get,

$$k\left[(2+\nu)\left(\phi_{SM} - \phi_{Pl}e^{(2-\nu)\xi} + \cdots\right)e^{-(2+\nu)\xi} + (\nu-2)\phi_{Pl}\right]$$
$$= 2\sqrt{\lambda(T_{Pl}^0 - T_{Pl})}\phi_{Pl}. \qquad (8.75)$$

Eq.(8.74) can be written as,

$$e^{(2-\nu)\xi} = \left[(2+\nu) - \frac{2}{k}\sqrt{\lambda(T_{SM}^0 - T_{SM})}\right]\frac{\phi_{SM}}{\phi_{Pl}}. \qquad (8.76)$$

We can now explain the essential idea behind the Goldberger-Wise stabilisation mechanism. The crucial factor in Eq.(8.76) is the exponent on the LHS, $(2 - \nu) \simeq \epsilon$. When the parameters in the model, $\{\Lambda, T_{SM}^0, T_{SM}, \lambda, v_{SM}, v_{Pl}\}$ have values of order one in units of M, the RHS wil be of order unity. We see that a moderately small value of ϵ, will then determine kR to be moderately large[11], *e.g.*, of order 12, as is required for a solution to the hierarchy problem[12] Eq.(8.49).

Explicitly solving, Eqs.(8.76) and (8.75), in full generality is not easy. The two equations are coupled together in a non-linear way, since ϕ_{Pl} which appears in Eq.(8.76) is determined by T_{Pl}, which in turn appears in Eq.(8.75). To proceed we make the self-consistent assumption that,

$$\frac{\sqrt{T_{Pl}^0 - T_{Pl}}}{v_{Pl}^2 \sqrt{\lambda}} \ll 1. \tag{8.77}$$

Then from Eq.(8.65) we have that, $\phi_{Pl} \simeq v_{Pl}$. Now substituting into Eq.(8.76) gives,

$$kR = -\frac{1}{\pi} \log \frac{\phi_{SM}}{\phi_{Pl}} \left[(2 + \nu) - \frac{2\sqrt{\lambda(T_{SM}^0 - T_{SM})}}{k} \right]. \tag{8.78}$$

Finally, from Eq.(8.75) we have that

$$2\sqrt{\lambda(T_{Pl}^0 - T_{Pl})} \simeq k\epsilon, \tag{8.79}$$

where we have dropped terms which are exponentially suppressed in $\xi = \pi k R$. This determines the value of T_{Pl}. Taking, for example, $\lambda = 1/M^2$, $\sqrt{T_{SM}^0 - T_{SM}} = kM^{3/2}$, $v_{SM}^2 = kM^2$, gives, from

---

[11]We remind the reader that $\pi k R = \xi$, Eq.(8.72)

[12]As was specified above, we take v_{Pl}, v_{SM} and also Λ, to be somewhat smaller than M, to make the analysis tractable. This does not change the argument above in any essential way. R is only logarithmically sensitive to such variations, the important dependence is on ϵ, in the argument of the exponential factor of Eq.(8.76).

Eq.(8.64), that $\phi_{SM} = \sqrt{kM^2/2}$. For $v_{Pl} = 8kM^2$, $\epsilon = 1/28.4$ we then get, $kR = 12$.

For the specific example above we can also check that all our approximations are indeed met. From Eq.(8.79), with $\lambda = 1/M^2$, $v_{Pl}^2 = 8kM^2$, we have that, $\sqrt{T_{Pl}^0 - T_{Pl}}/(v_{Pl}^2 \simeq \epsilon/16 \ll 1$, so condition (8.77) is indeed met. Also we neglected the contribution of the scalar field to the bulk stress tensor. This contribution goes like $\sim (\partial\phi)^2/2 \sim k^2 v_{Pl,SM}^2 \sim k^3 M^2$. If we take k to be somewhat smaller than M, as we need to, in order to control the corrections to the Einstein equations, then $k^3 M^2 < \Lambda \sim k^2 M^3$. So we see that the scalar contribution is smaller parametrically in k/M.

It is worth summarising the important features of the Goldberger-Wise stabilisation mechanism again, in case they have been obscured by the algebra above. The hierarchy M_{ew}/M_{Pl} is exponentially sensitive to the radius of compactification, R. For a scalar field with a mass somewhat smaller than the string scale, generic interactions with the branes parametrised by couplings which take natural values on the scale M, lead to stabilising the radius R. By taking the mass of the scalar field somewhat smaller than M (or the curvature scale k), one can obtain the required moderately large value of R, to solve the hierarchy problem.

The Goldberger-Wise stabilisation mechanism we have discussed above completes our discussion of the hierarchy problem in the RS scenario. We see that the scenario looks quite promising from this point of view. Unlike the ADD case the large hierarchy can be generating by giving moderately small numbers as input at the starting point.

Let us end with one comment. In our discussion above we worked directly with the five dimensional equations and saw that the radion is fixed. It is also possible, and for many purposes more convenient, to understand this in terms of a potential for the radion in a four dimensional description. While we do not pursue this further here, let us mention that such an analysis also shows that the radion has a mass which, in TeV units, is parametrically small in ϵ. Since the radion has couplings of a Brans-Dicke scalar this can lead to distinct phenomenological consequences.

8.3.4   Concluding remarks and outlook

The hierarchy problem and holography

The hierarchy problem is one of the outstanding issues in the Standard Model and has been deeply studied even before the recent scenarios discussed in these lectures were developed. All the scenarios developed in the past rely on essentially one mechanism, although it is implemented somewhat differently in supersymmetry or technicolor. This mechanism is as follows. One starts at the Planck scale with a slightly relevant coupling, typically a gauge coupling. Under RG evolution this coupling grows only slowly, *e.g.*, logarithmically for a gauge coupling. The electroweak scale (or a related scale like the susy breaking scale) is tied to this coupling becoming strong. This means that due to the slow RG running, M_{ew} and M_{Pl} are separated by several decades in energy and their ratio is small.

We have argued above that the RS scenario with the Goldberger-Wise mechanism, provides a satisfactory solution to the hierarchy problem. It is worth asking whether this is truly a new solution, or whether it is related to the one described above in some manner?

This question takes us to some of the recent developments at the forefronts of string theory. In fact there is good reason to believe that these two mechanisms are equivalent to each other. We have seen that the bulk spacetime in the RS scenario is five dimensional anti-de Sitter space (AdS_5). A few years ago, Maldacena conjectured that string theory in a particular AdS_5 background is equivalent to a gauge theory in $3 + 1$ dimensions. This remarkable correspondence between a five dimensional string theory and a four dimensional gauge theory is an example of a more general phenomenon called *holography* and is expected to be much more generally true, especially in AdS spacetimes. For an introduction, see, for example, [7] and the lectures of Dileep Jatkar in this volume.

A full description of these developments will of course take us well outside the scope of these lectures, but it is worth touching on how they bear on the question we asked above. Let us assume that for the RS model too there is a $3+1$ dimensional hologram. One can imagine the hologram living on the Planck brane. Enough is known about holography to then make the following two statements. First, the extra direction in the five dimensional AdS_5 description is expected to correspond to the energy scale in the hologram. Second, a scalar with small mass (in units of the curvature) in the AdS_5 description, corresponds to an almost marginal operator in the hologram.

This is enough information for us to see that the two mechanisms for solving the hierarchy problem are in fact equivalent under holography. Recall that the essential feature of the Goldberger-Wise mechanism is that it involves a scalar field with a small mass. The distance of the standard model brane is then fixed by the dynamics in terms of this mass, to be moderately large. This results in the huge hierarchy due to the exponential redshift in AdS space. In the hologram, the scalar corresponds to an almost marginal operator, and the small anomalous dimension results in a very slow increase in the strength of the coupling with energy, resulting in a huge hierarchy of scale.

This connection throws additional light on the Goldberger-Wise mechanism, for example it leads one to believe that it is very universal, independent of the detailed nature of the interactions on the branes for the scalar field. However, it also potentially alerts us to a problem in the RS scenario. Traditional attempts to solve the hierarchy problem, without supersymmetry, do not work. These theories go under the name of technicolor, and are well known to have problems with flavor changing neutral currents and precision electro-weak parameters. Our discussion above suggests that the RS model should have a a technicolor hologram (since the RS scenario, as described, is of course not supersymmetric), and this suggests that when one carefully tries to incorporate the standard model degrees of freedom one would run into similar problems.

This suggestion should not prevent us from exploring the phenomenological and model building aspects of the model more thoroughly. Not doing so, would amount to an unduly pessimistic attitude at this point in the development of the field, in our view. It could well be that there are technicolor theories consistent with the various constraints and we have not discovered them so far because of our limited knowledge of the dynamics of non-perturbative gauge theories. The duality with AdS_5 backgrounds might be a useful way to understand this dynamics, and RS like constructions might provide a tractable weakly coupled description of these models[13].

Some concluding comments

Let us end with a few more concluding comments about the RS model. The essential idea of the model is to have a warped compactification with the SM degrees of freedom living on a brane. In the RS model the warp factor has an exponentially suppressed value at the SM brane, this allows us to get the required hierarchy between the electroweak and Planck scales, by giving as input only a moderately small parameter into the model. In contrast in the ADD case, no warping is involved, as a result large extra dimensions are needed to suppress the gravitational interactions, this is not satisfactory from the point of view of resolving the hierarchy problem. On the other hand, as was mentioned above, there are reasons to be worried that more detailed implementations of the RS model will face the same sorts of problems that have plagued technicolor.

We have explored a specific model in the RS scenario, that of branes placed in AdS space. But the idea is much more general. All that is needed is to have a region of the compactification space where the warp factor varies exponentially, even though it might eventually depart from this profile. As long as the SM brane is located at a large enough redshift, the required hierarchy can be obtained. The negative tension of the Standard Model brane, which might have struck the reader as being an unattractive feature, is not essential. These points are borne out by recent string

---

[13]Since these lectures were given, there have been other interesting developments, going by the name of *deconstruction*, also relating technicolor like theories to those with extra dimensions. Unfortunately, they are outside the scope of this review.

constructions. Some of these string constructions have a holographic description in terms of a confining gauge theory, and the hierarchy of scales is tied to a slowly varying gauge coupling, as we mentioned above.

More generally, it is clear that one can interpolate from the ADD scenario to the RS case by considering compactifications where the warp factor goes from being constant to varying exponentially. Presumably, a satisfactory solution to the hierarchy problem would require the warp factor varying in a near exponential profile, for sufficiently long, but it is worth keeping the more general possibility in mind. One can also consider compactifications where the warp factor varies even more rapidly than an exponential. It would be interesting to ask if an analysis of some of the phenomenological aspects can be carried out in a manner independent of the detailed nature of the warp profile.

These lectures are only an introduction. A great deal of research has been carried out in various related directions, even since they were given. Some of the phenomenological aspects are discussed in the companion lectures by S. Raychaudhuri. There has also been progress in understanding some of the cosmological consequences, which we did not touch on at all. Finally, in terms of model building aspects, there is progress in the context of string theory, which we mentioned above, and also from a more field theoretic viewpoint.

In short, the field is wide open. There is much to do, even before the LHC turns on. Jump in and join in the fun!

Bibliography

[1] N. Arkani Hamed, S. Dimopoulos and G.R. Dvali, *The hierarchy problem and new dimension at a millimeter*, Phys. Lett. **B429** (1998) 263–272, [hep-ph/9803315].

[2] I. Antoniadis, N. Arkani-Hamed, S. Dimopoulos and G.R. Dvali, *New dimension at a millimeter to a fermi and superstrings at a TeV*, Phys. Lett. **B436** (1998) 257–263, [hep-ph/9804398].

[3] N. Arkani-Hamed, S. Dimopoulos and G.R. Dvali, *Phenomenology, astrophysics and cosmology of theories with sub-millimeter dimensions and TeV scale quantum gravity*, Phys. Rev. **D59** (1999) 086004, [hep-ph/9807344].

[4] L. Randall and R. Sundrum, *An alternative to compactification*, Phys. Rev. Lett. **83** (1999) 4690–4693, [hep-th/9906064].

[5] L. Randall and R. Sundrum, *A large mass hierarchy from a small extra dimension*, Phys. Rev. Lett. **83** (1999) 3370–3373, [hep-ph/9905221].

[6] W. Goldberger and M.B. Wise, *Modulus stablization with bulk fields*, Phys. Rev. Lett. **83** (1999) 4922–4925, [hep-ph/9907447].

[7] S.P. Trivedi, *Holography, Black Holes and String Theory*, Current Science, **81**, num. 12, pp 1582; see also the other articles in this volume.

Chapter 9

Brane-World Phenomenology

Sreerup Raychaudhuri

These notes are intended to be an introduction to the theory and phenomenology of gravitation in many dimensions, coupling to matter confined to a four-dimensional hypersurface (identified with the visible Universe). The basic framework for this is worked out in the large extra dimensions model of Arkani-Hamed, Dimopoulos and Dvali (ADD). Our focus is on the phenomenological consequences of this scenario. Other aspects of the model are treated in the lectures by Sandip Trivedi in this volume—these should be read in conjunction with those and meant to complement those.

I like to thank the organisers of the XV SERC School on Theoretical High Energy Physics, held at the Harish-Chandra Reseach Institute, Allahabad, in 2001, for the opportunity to deliver the lectures on which these notes are based. Thanks are also due to Rohini M. Godbole and Debashis Ghoshal for their patience and persistence in demanding that these notes be finished. Last, but not the least, thanks are due to Santosh Kumar Rai for typing a large part of the manuscript.

9.1   Gravity coupling to matter

The basic theory of gravity used in the theories under consideration is Einstein's theory of general relativity, enunciated in 1916[1]. In this section we shall see how to couple gravity with matter and how to go to the linear (post-Newtonian) approximation, in which gravity behaves as a gauge theory. This will set the stage for analysing the higher dimensional models.

As with any field theory we start with an action invariant under the basic symmetries, which in this case happens to be invariance under general coordinate transformations. The minimal construction is called the *Einstein-Hilbert action*

$$S = \frac{1}{\kappa^2} \int d^4x \sqrt{g} \; \mathcal{R}, \qquad (9.1)$$

where $\mathcal{R}$ is the Ricci scalar and $\kappa = \sqrt{16\pi G_N} = \sqrt{16\pi}/M_P$. Given this action, we can, by varying $g_{\mu\nu} \rightarrow g_{\mu\nu} + \delta g_{\mu\nu}$ (with fixed endpoints), get the Euler-Lagrange equations:

$$\mathcal{G}_{\mu\nu} \equiv \mathcal{R}_{\mu\nu} - \frac{1}{2} g_{\mu\nu} \mathcal{R} = 0, \qquad (9.2)$$

which are just the (free-field) *Einstein equations*. (To prove this last result, we need $\delta g = g g^{\mu\nu} \delta g_{\mu\nu}$ and $\delta(g_{\mu\nu} g^{\nu\lambda}) = 0$.)

In the presence of matter, the modified Lagrangian density becomes

$$\mathcal{L} = \frac{1}{\kappa^2} \sqrt{g} \left(\mathcal{R} + \mathcal{L}_m \right), \qquad (9.3)$$

where $\mathcal{L}_m$ is the Lagrangian density of the matter fields, in generally covariant form, and obtain the field equations

$$\mathcal{R}_{\mu\nu} - \frac{1}{2} g_{\mu\nu} \mathcal{R} = \frac{\kappa^2}{4\sqrt{g}} \left[\frac{\partial(\sqrt{g}\mathcal{L}_m)}{\partial g_{\alpha\beta}} - \partial_\lambda \frac{\partial(\sqrt{g}\mathcal{L}_m)}{\partial(\partial_\lambda g_{\alpha\beta})} \right] g^{\alpha\mu} g^{\beta\nu}$$

$$= -\frac{\kappa^2}{2} T_{\mu\nu},$$

where

$$T^{\mu\nu} = \frac{1}{2\sqrt{g}} \left[\partial_\lambda \frac{\partial(\sqrt{g}\mathcal{L}_m)}{\partial(\partial_\lambda g_{\mu\nu})} - \frac{\partial(\sqrt{g}\,\mathcal{L}_m)}{\partial g_{\mu\nu}} \right]$$

$$= -\frac{1}{\sqrt{g}}\frac{\partial}{\partial g_{\mu\nu}}\left[\sqrt{g}\mathcal{L}_m\right].$$

This gives a simple prescription to couple gravity to other fields.

The simplest example is to suppose that there is a constant energy density in the background, *i.e.*,

$$\sqrt{g}\,\mathcal{L}_m = \sqrt{g}\,2\Lambda, \tag{9.4}$$

where Λ is called the *cosmological constant*. The factor of 2 is just a convention. We then have

$$T_{\mu\nu} = 2\Lambda g_{\mu\nu} \tag{9.5}$$

and the Einstein equations reduce to

$$R_{\mu\nu} - \frac{1}{2}\,g_{\mu\nu}\mathcal{R} + \Lambda g_{\mu\nu} = 0. \tag{9.6}$$

Given the cosmological principle that spacetime is homogeneous and space is isotropic, this leads to a static, non-expanding Universe for $\Lambda > 0$. Since the universe is known to be expanding almost uniformly, we know that $\Lambda \approx 0$. In fact, recent measurements suggest that Λ may not be exactly zero, but may have a small finite value. However, that will not concern us in these lectures.

The matter-Lagrangian and energy-momentum tensor corresponding to scalar, vector and fermionic fields will be discussed in the subsequent section.

Finally we note that subtracting out $\mathcal{G}^{\mu}_{\mu}$ and $T = T^{\mu}{}_{\mu}$, the Einstein equations become

$$\mathcal{R}_{\mu\nu} = -\frac{1}{2}\kappa^2(T_{\mu\nu} - \frac{1}{2}\,g_{\mu\nu}T) \equiv -\frac{1}{2}\kappa^2\mathcal{S}_{\mu\nu}. \tag{9.7}$$

This form will be useful in the subsequent discussion.

9.2 Linearized gravity

In the usual four-dimensional world, gravity is generally a very weak force, parametrized by the smallness of

$$\kappa = \sqrt{16\pi G_N} = \frac{4\sqrt{\pi}}{M_P} \le 10^{-19}\ (\text{GeV}/c^2)^{-1}. \tag{9.8}$$

In the ADD scenario, however, when we see the full bulk (*i.e.*, at energies approaching the string scale) we see

$$\hat{\kappa} = \sqrt{16\pi \hat{G}_N} = \frac{4\sqrt{\pi}}{M_P} \sim 10^{-3} \ (\text{GeV}/c^2)^{-1}. \tag{9.9}$$

This is many orders of magnitude larger, but still a small quantity, though in a normal process with energy E, the requirement is $\hat{\kappa}E \ll 1$, which is only true for $E \ll 1$ TeV (low energy regime). Thus, in the low energy regime, at least, gravity *is* weak, because $\hat{\kappa}E$ is small.

A weak gravitational field corresponds to a near-flat metric which is close to the Minkowski metric, i.e.

$$g_{\mu\nu}(x) = \eta_{\mu\nu} + \kappa h_{\mu\nu}(x) + \mathcal{O}(\kappa^2), \tag{9.10}$$

the closeness to $\eta_{\mu\nu}$ being determined by the smallness of κ. For reasonably small κ we can safely neglect the higher orders.

Now, if we consider an infinitesimal coordinate transformation:

$$x^\mu \to x'^\mu = x^\mu + \xi^\mu(x), \qquad (\xi^\mu \ll x^\mu) \tag{9.11}$$

so that, for example, $(\partial x'^\mu / \partial x^\alpha) = \delta^\mu_\alpha + \partial_\alpha \xi^\mu$ etc., then the transformation law,

$$g_{\mu\nu}(x) \to g'_{\mu\nu}(x') = \frac{\partial x^\alpha}{\partial x'^\mu} \frac{\partial x^\beta}{\partial x'^\nu} \, g_{\alpha\beta}(x)$$

reduces (in the weak field approximation) to

$$h_{\mu\nu}(x) \to h'_{\mu\nu}(x') = h_{\mu\nu}(x) - \partial_\mu \xi_\nu(x) - \partial_\nu \xi_\mu(x), \tag{9.12}$$

which looks like a gauge transformation, with a vector gauge potential ξ_μ. This is analogous to the electromagnetic gauge transformation law $A_\mu \to A'_\mu = A_\mu + \partial_\mu \xi$ with a scalar gauge potential ξ. Invariance of the action under general co-ordinate transformations now corresponds to gauge invariance of the action under the above transformation.

Using $g_{\mu\nu} = \eta_{\mu\nu} + \kappa\, h_{\mu\nu}$, the Christoffel symbol comes out to be

$$\Gamma^{\mu}_{\nu\lambda} = \frac{1}{2}\,\kappa\,\eta^{\mu\alpha}\left(\partial_{\nu}h_{\lambda\alpha} + \partial_{\lambda}h_{\nu\alpha} - \partial_{\alpha}h_{\nu\lambda}\right), \qquad (9.13)$$

and we can now go on to calculate the Riemann-Christoffel tensor and contract two of its indices to get the Ricci tensor

$$\mathcal{R}_{\mu\nu} = \frac{1}{2}\kappa\left[\partial_{\mu}\partial_{\nu}h + \Box h_{\mu\nu} - \partial_{\mu}\partial_{\lambda}h^{\lambda}{}_{\nu} - \partial_{\lambda}\partial_{\nu}h^{\lambda}{}_{\mu}\right] + \mathcal{O}(\kappa^2),$$
$$(9.14)$$

and the Ricci scalar

$$\mathcal{R} = \kappa\left(\Box h - \partial_{\mu}\partial_{\nu}h^{\mu\nu}\right) + \mathcal{O}(\kappa^2), \qquad (9.15)$$

where $h = h^{\mu}{}_{\mu} = \eta^{\mu\nu}h_{\mu\nu}$.

An important thing to note is that these forms for $\mathcal{R}$ and $\mathcal{R}_{\mu\nu}$ have been written without writing $\eta_{\mu\nu}$ explicitly, which means that we could have taken any constant background metric if we wished, *i.e.*, $g_{\mu\nu} = g^{(0)}_{\mu\nu} + \kappa h_{\mu\nu}$. This is not very relevant to the ADD model, but becomes relevant for the Randall-Sundrum model.

Putting things together, in linearized form, the Einstein equations are

$$\Box h_{\mu\nu} + \partial_{\mu}\partial_{\nu}h - \partial_{\mu}\partial_{\lambda}h^{\lambda}{}_{\nu} - \partial_{\lambda}\partial_{\nu}h^{\lambda}{}_{\mu} = -\kappa\mathcal{S}_{\mu\nu}. \qquad (9.16)$$

However, it is obviously difficult to find a Green's function for the operator on the left. However, one can further reduce the equations to a more tractable form. To this end, we rewrite the LHS of the above equation as

$$\Box h_{\mu\nu} + \partial_{\mu}\left(\frac{1}{2}\partial_{\nu}h - \partial_{\lambda}h^{\lambda}{}_{\nu}\right) + \partial_{\nu}\left(\frac{1}{2}\partial_{\mu}h - \partial_{\lambda}h^{\lambda}{}_{\mu}\right), \qquad (9.17)$$

and choose the *harmonic gauge* (de Donder gauge)

$$\frac{1}{2}\partial_{\nu}h - \partial_{\lambda}h^{\lambda}{}_{\nu} = 0, \qquad (9.18)$$

which requires the gauge potential to satisfy

$$\Box\xi_{\mu} = 0. \qquad (9.19)$$

Einstein equations now reduce to the simple form

$$\Box h_{\mu\nu} = -\kappa \, \mathcal{S}_{\mu\nu}. \tag{9.20}$$

This looks like a wave equation with a source: the gravitational fields $h_{\mu\nu}$ can be solved in the presence of matter[1] sources $\mathcal{S}_{\mu\nu}$ using the usual Green's function technique.

In the absence of matter $\mathcal{S}_{\mu\nu} = 0$, and we recover the wave equation

$$\Box h_{\mu\nu} = 0. \tag{9.21}$$

The free solutions of this can be expanded as plane waves

$$h_{\mu\nu}(x) = \mathcal{N}_k \left[\epsilon_{\mu\nu}(k) e^{ik.x} + \epsilon^*_{\mu\nu}(k) e^{-ik.x} \right], \tag{9.22}$$

where $\epsilon_{\mu\nu}(k)$ is the polarization tensor (or Fourier transform of the gravitational field) and $\mathcal{N}_k$ is a normalisation constant. The polarisation tensor clearly satisfies

$$k^2 \;=\; 0 \qquad \text{(Einstein equation)}$$
$$k_\mu \, \epsilon^{\mu\nu} \;=\; \frac{1}{2} \, k_\nu \, \epsilon^\mu{}_\mu \;=\; 0 \qquad \text{(harmonic gauge)}.$$

---

[1]Though weak-field gravity closely resembles electromagnetic theory, we must remember that gravity is a non-linear (self-coupling) theory because the gravity waves also carry energy and momentum. Thus the energy-momentum tensor for gravity (in the harmonic gauge) is calculated by putting

$$\mathcal{L}_m = \frac{1}{\kappa^2} \sqrt{g} \mathcal{R},$$

and using the prescription given above to find $T_{\mu\nu}$. After simplification the result is

$$\begin{aligned}
T_{\mu\nu} \;=\; & -\frac{1}{4} h_{\alpha\beta} \partial_\mu \partial_\nu h^{\alpha\beta} + \frac{1}{8} h \partial_\mu \partial_\nu h + \frac{1}{8} \eta_{\mu\nu} \left(h^{\alpha\beta} \Box h_{\alpha\beta} - \frac{1}{2} h \Box h \right) \\
& -\frac{1}{4} \left(h_{\mu\alpha} \Box h^\alpha{}_\nu + h_{\nu\alpha} \Box h^\alpha{}_\mu - h_{\mu\nu} \Box h \right) + \frac{1}{8} \partial_\mu \partial_\nu \left(h_{\alpha\beta} h^{\alpha\beta} - \frac{1}{2} h^2 \right) \\
& -\frac{1}{16} \eta_{\mu\nu} \Box \left(h_{\alpha\beta} h^{\alpha\beta} - \frac{1}{2} h^2 \right) - \frac{1}{4} \partial_\alpha \{ \partial_\nu (h_{\mu\beta} h^{\beta\alpha}) + \partial_\mu (h_{\mu\beta} h^{\beta\alpha}) \} \\
& +\frac{1}{2} \partial_\alpha \{ h^{\alpha\beta} (\partial_\nu h_{\mu\beta} + \partial_\mu h_{\nu\beta}) \}.
\end{aligned}$$

Of course this term is absent in the linear (*i.e.*, $\mathcal{O}(\kappa)$) approximation.

Using these, one can show that out of 10 independent components of the 4×4 matrix $h_{\mu\nu}$, only 2 are independent (we use the residual gauge freedom within the harmonic gauge). Details are explained, for example, in the classic text by Weinberg. A plane polarized gravity wave traveling in the 3-direction has $k = (k, 0, 0, k)$ and hence a polarisation tensor

$$h_{\mu\nu} = \begin{pmatrix} 0 & 0 & 0 & 0 \\ 0 & \epsilon_{11} & \epsilon_{12} & 0 \\ 0 & \epsilon_{12} & -\epsilon_{11} & 0 \\ 0 & 0 & 0 & 0 \end{pmatrix},$$

if plane-polarised. If circularly polarised, we get helicity ± 2 states $\epsilon_{11} \pm i\epsilon_{12}$.

In an arbitrary gauge, the linearized Einstein-Hilbert action looks like

$$\frac{1}{\kappa^2} \sqrt{g}\mathcal{R} = \frac{1}{4}\left(\partial^\mu h_{\alpha\beta} \partial_\mu h^{\alpha\beta} + \partial^\mu h \partial_\mu h \right.$$
$$\left. - 2\partial^\mu h_{\mu\alpha} \partial_\nu h^{\nu\alpha} - 2\partial^\mu h_{\mu\alpha} \partial^\alpha h \right),$$

where $h = h_\mu{}^\mu$. This is called the *Fierz-Pauli Lagrangian*, and is invariant under a gauge transformation as in Eq.(9.12). The Euler-Lagrange equations for this, after imposing the de Donder gauge condition, becomes

$$\Box \left(h_{\mu\nu} - \frac{1}{2} h \eta_{\mu\nu} \right) = 0, \tag{9.23}$$

which reduces to the earlier equation $\Box h_{\mu\nu} = 0$ if we demand $h = 0$. This can be done without loss of generality by redefining $h_{\mu\nu} \to h_{\mu\nu} - \frac{1}{4} h \eta_{\mu\nu}$.

Quantization of this gauge field follows on the lines of a usual gauge theory, with addition of gauge-fixing and ghost terms. This involves many subtleties which cannot be discussed in these lectures. We shall simply assume that the gravity waves can be quantized, and that the quanta—called *gravitons*—are spin-2 particles. Feynman rules for these will be read off from the interaction

Lagrangian just as is done for QED. Since the phenomenological applications discussed here will not go beyond the tree-level, it will not be necessary to consider the complexities of the quantum theory in detail (though, of course, these are there).

9.3   Gravitational fields in the ADD model

The large extra dimensions model [2] of Arkani-Hamed, Dimopoulos and Dvali (ADD) has been introduced in the companion lectures by Sandip Trivedi. Here we do not go into the motivations, but straightaway consider the ADD model with the following assumptions:

1. Gravity is weak both in the bulk (4+p-dimensional) and of course, on the domain wall or 3-brane (4-dimensional). This can be justified by the values of κ and $\widehat{\kappa}$ exhibited in Eqs.(9.8) and (9.9) respectively. We can, therefore, make a linearized gravity approximation quite safely.

2. All matter is confined to the brane. By matter we mean all fields in the Standard Model (and possibly its extensions). In the string-theoretic framework, this may require a structure with two or more coincident $D3$-branes.

3. The brane is a thin one, *i.e.*, its thickness, which must be less than 10^{-16}cm ($\sim M_{EW}^{-1}$), is approximated as zero. At the same time, it is heavy and rigid, so that its dynamics may be ignored. This is analogous to the famous Born-Oppenheimer approximation in molecular physics.

4. New physics in the bulk appears at energies $E \sim \widehat{M}_P \sim$ (few) TeV. If this is string theory, then the string tension α' is set by this scale, *i.e.*, $\alpha' \sim \widehat{M}_P^{-2}$. Accordingly we may expect Regge excitations of string theory to appear at these energies.

5. The low-energy effective theory—which has just gravity plus Standard Model—is, therefore, valid only upto $\widehat{M}_P$, which

then acts as a cutoff for this theory. In fact, this is the basic motivation for the model, as it removes the hierarchy problem for the Higgs boson.

In this section we shall closely follow the notations and treatment of Han, Lykken and Zhang[2][3]. We describe bulk coordinates by $\hat{x}^{\hat{\mu}}$ in general, where $\hat{\mu} = 0, 1, \cdots, (4+p)$. We describe brane coordinates by x^{μ}, where $\mu = 0, 1, 2, 3$ (these are just the usual Minkowski coordinates). We assume the compact space to be a p-torus with radii R_c every way, and describe the compact dimensions by y_i, where $i = 1, 2, \cdots, p$.

For linearized gravity in the bulk, we expand the metric as

$$\hat{g}_{\hat{\mu}\hat{\nu}}(\hat{x}) = \eta_{\hat{\mu}\hat{\nu}} + \hat{\kappa}\hat{h}_{\hat{\mu}\hat{\nu}}(\hat{x}), \tag{9.24}$$

where $\eta_{\hat{\mu}\hat{\nu}} = \text{diag.}(1, -1, -1, \cdots, -1)$. The extra p dimensions are all space-like. We can now partition the matrix $\hat{h}_{\hat{\mu}\hat{\nu}}(\hat{x})$ as

$$\hat{h}_{\hat{\mu}\hat{\nu}} = \frac{1}{\sqrt{V_p}} \left(\begin{array}{c|c} h_{\mu\nu} + \phi\eta_{\mu\nu} & A_{\mu i} \\ \hline A_{\nu j} & 2\phi_{ij} \end{array} \right), \tag{9.25}$$

where $\phi = \phi_{ii}$ (traceless if $h_{\mu\nu}$ is traceless) and all fields are functions of $\hat{x}$. The box normalisation factor involves $V_p = (2\pi R_c)^p$, the volume of the p-torus.

We assume this metric obeys the Einstein theory in the $(4+p)$-dimensional bulk and hence can be linearized to give a bulk Fierz-Pauli Lagrangian.

$$\frac{1}{\hat{\kappa}^2}\sqrt{\hat{g}}\hat{\mathcal{R}} = \frac{1}{4}\left(\partial^{\hat{\mu}}\hat{h}_{\hat{\alpha}\hat{\beta}}\partial_{\hat{\mu}}\hat{h}^{\hat{\alpha}\hat{\beta}} + \partial_{\hat{\mu}}\hat{h}\partial^{\hat{\mu}}\hat{h} \right.$$
$$\left. -2\partial^{\hat{\mu}}\hat{h}_{\hat{\mu}\hat{\alpha}}\partial^{\hat{\nu}}\hat{h}^{\hat{\nu}\hat{\alpha}} - 2\partial^{\hat{\mu}}\hat{h}_{\hat{\mu}\hat{\alpha}}\partial^{\hat{\alpha}}\hat{h} \right),$$

where $\hat{h} = \hat{h}^{\hat{\mu}}_{\hat{\mu}}$. This is invariant under a bulk gauge transformation,

$$\hat{x}^{\hat{\mu}} \rightarrow \hat{x}'^{\hat{\mu}} = \hat{x}^{\hat{\mu}} + \hat{\xi}^{\hat{\mu}}(\hat{x}), \qquad (\hat{\xi}^{\hat{\mu}} \ll \hat{x}^{\hat{\mu}})$$
$$\hat{h}_{\hat{\mu}\hat{\nu}}(x) \rightarrow \hat{h}'_{\hat{\mu}\hat{\nu}}(\hat{x}') = \hat{h}_{\hat{\mu}\hat{\nu}}(\hat{x}) - \partial_{\hat{\mu}}\hat{\xi}_{\hat{\nu}}(\hat{x}) - \partial_{\hat{\nu}}\hat{\xi}_{\hat{\mu}}(\hat{x}).$$

---

[2]A very similar treatment was given by Giudice, Rattazzi and Wells[4] at about the same time.

The de Donder gauge condition is now[3]

$$\partial_{\hat{\mu}}\left(\hat{h}_{\hat{\mu}\hat{\nu}} - \frac{1}{2}\hat{h}\eta_{\hat{\mu}\hat{\nu}}\right) = 0. \tag{9.26}$$

To get a feel of what the ϕ and $A_{\mu i}$ fields correspond to, let us take $p = 1$, in which case the pth dimension is just a circle of circumference L and there is just one y coordinate representing the arc length around the circle. The bulk metric becomes

$$\hat{h}_{\hat{\mu}\hat{\nu}} = \frac{1}{\sqrt{L}}\left(\begin{array}{cc} h_{\mu\nu} + \phi\eta_{\mu\nu} & A_\mu \\ A_\nu & 2\phi \end{array}\right), \tag{9.27}$$

where $A_\mu = \hat{h}_{\mu\hat{5}}$, $A_\nu = \hat{h}_{\hat{5}\nu}$ and $\phi = \frac{1}{2}\hat{h}_{\hat{5}\hat{5}}$. Now consider a translation round the circle $y \to y' = y + \xi(x)$ which differs at different points x on the brane (Eq.(9.11)). In the limit of linearised gravity, this corresponds to a five-dimensional gauge potential

$$\hat{\xi}_{\hat{\mu}}(\hat{x}) = \left\{\begin{array}{ll} 0 & \text{for } \hat{\mu} = 0, 1, 2, 3, \\ \xi(x) & \text{for } \hat{\mu} = \hat{5} \end{array}\right. \tag{9.28}$$

Now, the off-diagonal elements of the metric transform as

$$\hat{h}_{\mu 5} \quad \to \quad \hat{h}'_{\mu 5} = \hat{h}_{\mu 5} - \partial_\mu \xi_5 - \partial_5 \xi_\mu$$
$$i.e., \quad A_\mu \quad \to \quad A'_\mu = A_\mu - \partial_\mu \xi, \tag{9.29}$$

which is a U(1) gauge transformation. One can similarly show that ϕ is a scalar[4]. Now observe that in four-dimensions (*i.e.*, in the compactification limit)

$$\begin{aligned} ds^2 &= g_{\mu\nu}dx^\mu dx^\nu = (\eta_{\mu\nu} + \hat{\kappa}\hat{h}_{\mu\nu})dx^\mu dx^\nu \\ &= \left(\eta_{\mu\nu} + \frac{1}{\sqrt{L}}\hat{\kappa}h_{\mu\nu} + \frac{1}{\sqrt{L}}\hat{\kappa}\phi\eta_{\mu\nu}\right)dx^\mu dx^\nu \\ &= \left[(1 + \kappa\phi)\eta_{\mu\nu} + \kappa h_{\mu\nu}\right]dx^\mu dx^\nu \\ &\simeq \left(e^{\kappa\phi}\eta_{\mu\nu} + \kappa h_{\mu\nu}\right)dx^\mu dx^\nu, \end{aligned} \tag{9.30}$$

---

[3]Note that even for $\hat{\mu}, \hat{\nu} = \mu, \nu$, this is not the same as the 4-dimensional de Donder gauge condition because $\hat{h} = \eta^{\hat{\mu}\hat{\nu}}\hat{h}_{\hat{\mu}\hat{\nu}} = \eta^{\mu\nu}(h_{\mu\nu} + \phi\eta_{\mu\nu}) + (-\delta^{ij})\phi_{ij} = h + 4\phi - \phi = h + 3\phi$. One can recover $\hat{h} = h$ only by putting $\phi_{ij} = 0$, which means confining gravity to the brane. (This is a trivial case, since it is as good as having no bulk.)

[4]In this case, since the gauge transformations are general coordinate transformations, there is no distinction between a scalar and a gauge scalar.

where $\kappa = \hat{\kappa}/\sqrt{L}$. It is clear from the above that $h_{\mu\nu}$ represents the ordinary gravitational fields in the linearised approximation. If $h_{\mu\nu} = 0$, then $ds^2 = e^{\kappa\phi}\eta_{\mu\nu}dx^\mu dx^\nu$ *i.e.*, ϕ represents a rescaling of the Minkowski metric or a dilatation. Hence its quanta are called *dilatons* and ϕ is called a dilaton field. A similar relation holds for $p \neq 1$, where $\kappa = \hat{\kappa}/\sqrt{V_p}$. In fact, this allows us to have a large $\hat{\kappa}$ while keeping κ small. This is the original motivation of the ADD model.

In the more general case $p > 1$, we have p 'gauge' fields $A_{\mu i}(x^\mu, y_i)$ i.e. i is like a non-Abelian gauge index. There is still one dilaton field $\phi = \phi_{ii}$. The Euler-Lagrange equations satisfied by these, in the absence of matter, are simply

$$\hat{\Box}\left(\hat{h}_{\hat{\mu}\hat{\nu}} - \frac{1}{2}\hat{h}\eta_{\hat{\mu}\hat{\nu}}\right) = 0,$$
$$\hat{\Box}A_{\mu i} = 0, \qquad (9.31)$$
$$\hat{\Box}\phi_{ij} = 0,$$

where $\hat{\Box} = \Box - \partial^2_{y_1} - \partial^2_{y_2} - \cdots - \partial^2_{y_p}$ is the bulk D'Alembertian operator. Of course, these hold only for free fields, but that is precisely the theory we would like to quantize.

Noting that the fields must be periodic in each of the y_i, i.e. under $y_i \rightarrow y_i + 2\pi R_c$, we now make a *Kaluza-Klein decomposition* of each of these bulk fields into the Fourier modes:

$$h_{\mu\nu}(x, y) = \sum_{\vec{n}} h^{\vec{n}}_{\mu\nu}(x)\, e^{2\pi i \vec{n}.\vec{y}/L},$$
$$A_{\mu i}(x, y) = \sum_{\vec{n}} A^{\vec{n}}_{\mu i}(x)\, e^{2\pi i \vec{n}.\vec{y}/L}, \qquad (9.32)$$
$$\phi_{ij}(x, y) = \sum_{\vec{n}} \phi^{\vec{n}}_{ij}(x)\, e^{2\pi i \vec{n}.\vec{y}/L},$$

where $\vec{y} \equiv (y_1, y_2, \cdots, y_p)$; $\vec{n} \equiv (n_1, n_2, \cdots, n_p)$ and n_i are integers. To keep the fields real (since the metric tensor is real), we must have $h^{-\vec{n}}_{\mu\nu} = (h^{\vec{n}}_{\mu\nu})^*$ etc. We now have, on the brane, towers of Kaluza-Klein modes: $h^{\vec{n}}_{\mu\nu}(x)$ (of spin-2), $A^{\vec{n}}_{\mu i}(x)$ (of spin-1) and $\phi^{\vec{n}}_{ij}(x)$ (of spin-0).

Substitution of these expansions (9.32) into the bulk equations of motion (9.31) lead to four-dimensional equations

$$\left(\Box + M_{\vec{n}}^2\right)\left(h_{\mu\nu}^{\vec{n}} - \frac{1}{2}h^{\vec{n}}\eta_{\mu\nu}\right) = 0,$$

$$\left(\Box + M_{\vec{n}}^2\right) A_{\mu i}^{\vec{n}} = 0, \qquad (9.33)$$

$$\left(\Box + M_{\vec{n}}^2\right) \phi_{ij}^{\vec{n}} = 0, \qquad (9.34)$$

i.e., we get towers of degenerate $h_{\mu\nu}^{\vec{n}}$, $A_{\mu i}^{\vec{n}}$ and $\phi_{ij}^{\vec{n}}$ with masses

$$M_{\vec{n}}^2 = (2\pi)^2 \frac{\vec{n}.\vec{n}}{L^2} = \frac{n_1^2 + n_2^2 + \ldots + n_p^2}{R_c^2}. \qquad (9.35)$$

A similar substitution in the bulk de Donder gauge condition yields two equations:

$$\partial^\mu h_{\mu\nu}^{\vec{n}} - \frac{1}{2}\partial_\nu h^{\vec{n}} + \frac{2\pi i}{L}n_i A_{\nu i}^{\vec{n}} = 0,$$

$$\partial^\mu A_{\mu i}^{\vec{n}} + \frac{i\pi}{L}n_i h^{\vec{n}} + \frac{4\pi i}{L}n_j \phi_{ij}^{\vec{n}} + \frac{2\pi i}{L}n_i \phi^{\vec{n}} = 0. \qquad (9.36)$$

The fields $h_{\mu\nu}^{\vec{n}}$, $A_{\mu i}^{\vec{n}}$ and $\phi_{ij}^{\vec{n}}$ are not invariant under bulk gauge transformations. Though in the special case for $p = 1$ discussed above it looked as if A_μ transformed in a decoupled way from the $h_{\mu\nu}$ and ϕ, this was only for translations around the circle. In general, a coordinate transformation (gauge transformation) with parameter

$$\hat{\xi}_{\hat{\mu}}(x, y) = \sum_{\vec{n}} \hat{\xi}_{\hat{\mu}}(x)\, e^{2\pi i \vec{n}.\vec{y}/L} \qquad (9.37)$$

can be easily shown to lead to the gauge transformations

$$h_{\mu\nu}^{\vec{n}} \;\rightarrow\; h_{\mu\nu}'^{\vec{n}} = h_{\mu\nu}^{\vec{n}} + \partial_\mu \xi_\nu^{\vec{n}} + \partial_\nu \xi_\mu^{\vec{n}} + \frac{2\pi i}{L}n_i \xi_i^{\vec{n}} \eta_{\mu\nu},$$

$$A_{\mu i}^{\vec{n}} \;\rightarrow\; A_{\mu i}'^{\vec{n}} = A_{\mu i}^{\vec{n}} + \partial_\mu \xi_i^{\vec{n}} - \frac{2\pi i}{L}n_i \xi_\mu^{\vec{n}} \qquad (9.38)$$

$$\phi_{ij}^{\vec{n}} \;\rightarrow\; \phi_{ij}'^{\vec{n}} = \phi_{ij}^{\vec{n}} - \frac{\pi i}{L}n_i \xi_j^{\vec{n}} - \frac{\pi i}{L}n_j \xi_i^{\vec{n}},$$

on the brane. This is a problem, since it means that a different choice of bulk coordinates (different gauge) would lead to coupled equations of motion for the metric fields. We, therefore, wish to work with bulk-gauge invariant objects on the brane, and hence we need to construct such fields out of the $h^{\vec{n}}_{\mu\nu}$, $A^{\vec{n}}_{\mu i}$ and $\phi^{\vec{n}}_{ij}$. This bulk-gauge invariant construction has actually been made by Han, Lykken and Zhang, and we reproduce their results below.

Given the vector $\vec{n} = \{n_1, n_2, \ldots, n_p\}$ we define projection operators in the compact space.

$$\widetilde{P}^{\vec{n}}_{ij} = \frac{n_i n_j}{\vec{n}^2}, \qquad P^{\vec{n}}_{ij} = \delta_{ij} - \widetilde{P}^{\vec{n}}_{ij}, \tag{9.39}$$

so that $P^{\vec{n}}_{ij} + \widetilde{P}^{\vec{n}}_{ij} = \delta_{ij}$, and the projection operators satisfy the following relations:

$$\begin{aligned}
P^{\vec{n}}_{ij} P^{\vec{n}}_{jk} &= P^{\vec{n}}_{ik}, & \widetilde{P}^{\vec{n}}_{ij} \widetilde{P}^{\vec{n}}_{jk} &= \widetilde{P}^{\vec{n}}_{ik}, & P^{\vec{n}}_{ij} \widetilde{P}^{\vec{n}}_{jk} &= 0, \\
P^{\vec{n}}_{ii} &= n-1, & \widetilde{P}^{\vec{n}}_{ii} &= 1, & \widetilde{P}^{\vec{n}}_{ij} n_i &= n_j, \\
P^{\vec{n}}_{ij} n_i &= 0.
\end{aligned} \tag{9.40}$$

We now define the fields

$$\begin{aligned}
\widetilde{h}^{\vec{n}}_{\mu\nu} &= h^{\vec{n}}_{\mu\nu} - \frac{iL}{2\pi}\frac{n_i}{\vec{n}^2}(\partial_\mu A^{\vec{n}}_{\nu i} + \partial_\nu A^{\vec{n}}_{\mu i}) \\
&\quad - \frac{1}{3}(P^{\vec{n}}_{ij} + 3\widetilde{P}^{\vec{n}}_{ij})\left(\frac{2}{M_{\vec{n}}^2}\partial_\mu\partial_\nu - \eta_{\mu\nu}\right)\phi^{\vec{n}}_{ij}, \\
\widetilde{A}^{\vec{n}}_{\mu i} &= P^{\vec{n}}_{ij}\left(A^{\vec{n}}_{\mu j} - \frac{iLn_k}{\pi\vec{n}^2}\partial_\mu\phi^{\vec{n}}_{jk}\right), \\
\widetilde{\phi}^{\vec{n}}_{ij} &= \sqrt{2}\left(P^{\vec{n}}_{ik}P^{\vec{n}}_{jl} + \alpha P^{\vec{n}}_{ij}P^{\vec{n}}_{kl}\right)\phi^{\vec{n}}_{kl},
\end{aligned} \tag{9.41}$$

where α satisfies the equation $3(n-1)\alpha^2 + 6\alpha - 1 = 0$ (see below). Each of these fields is bulk-gauge invariant, as can be verified by direct substitution. On the brane, they transform as the Lorentz indices indicate, viz., as rank-2 tensor, vector and scalar fields respectively. We therefore identify these fields on the brane as

- $\widetilde{h}^{\vec{n}}_{\mu\nu}$: physical massive Kaluza-Klein graviton fields;

- $\widetilde{A}^{\vec{n}}_{\mu i}$: physical massive Kaluza-Klein vector boson fields;

- $\widetilde{\phi}^{\vec{n}}_{ij}$: physical massive Kaluza-Klein scalar fields.

Here $\widetilde{\phi}^{\vec{n}}_{ii} \equiv \widetilde{\phi}^{\vec{n}}$ is the physical massive dilaton. Note that there is a tower of each of these states, corresponding to different values of $\vec{n}$.

In terms of these physical fields, we can now re-write the bulk Fierz-Pauli Lagrangian. This involves three steps, *viz.*

1. Substituting the Kaluza-Klein decompositions into the Lagrangian.

2. Inverting the definitions of $\widetilde{h}^{\vec{n}}_{\mu\nu}$, $\widetilde{A}^{\vec{n}}_{\mu i}$ and $\widetilde{\phi}^{\vec{n}}_{ij}$ to get $h^{\vec{n}}_{\mu\nu}$, $A^{\vec{n}}_{\mu i}$ and $\phi^{\vec{n}}_{ij}$ in terms of these.

3. Substituting these into the decomposed Lagrangian.

After all this (admittedly messy!) algebra is done, one has the decoupled Lagrangian[5]

$$
\begin{aligned}
\mathcal{L} = \sum_{\vec{n}} \frac{1}{4} \Big\{ & \partial^\mu \widetilde{h}^{\vec{n}\alpha\beta} \partial_\mu \widetilde{h}^{-\vec{n}}_{\alpha\beta} - \partial^\mu \widetilde{h}^{\vec{n}} \partial_\mu \widetilde{h}^{-\vec{n}} - 2\partial_\alpha \widetilde{h}^{\vec{n}\mu\alpha} \partial^\beta \widetilde{h}^{-\vec{n}}_{\mu\beta} \\
& + \partial_\alpha \widetilde{h}^{\vec{n}\mu\alpha} \partial_\mu \widetilde{h}^{-\vec{n}} + \partial_\alpha \widetilde{h}^{-\vec{n}\mu\alpha} \partial_\mu \widetilde{h}^{\vec{n}} \\
& - M_{\vec{n}^2} \left(\widetilde{h}^{\vec{n}\alpha\beta} \widetilde{h}^{-\vec{n}}_{\alpha\beta} + \widetilde{h}^{\vec{n}} \widetilde{h}^{-\vec{n}} \right) \Big\} \\
& + \sum_{i=1}^{p} \sum_{\vec{n}} \left(-\frac{1}{4} \widetilde{F}^{\vec{n}\mu\nu}_i \widetilde{F}^{-\vec{n}}_{i\mu\nu} + \frac{1}{2} M^2_{\vec{n}} \widetilde{A}^{\vec{n}\mu}_i \widetilde{A}^{-\vec{n}}_{i\mu} \right) \\
& + \sum_{i=1}^{p} \sum_{j=1}^{p} \sum_{\vec{n}} \left(\frac{1}{2} \partial_\mu \widetilde{\phi}^{\vec{n}}_{ij} \partial^\mu \widetilde{\phi}^{-\vec{n}}_{ij} - \frac{1}{2} M_{\vec{n}^2} \widetilde{\phi}^{\vec{n}}_{ij} \widetilde{\phi}^{-\vec{n}}_{ij} \right), \quad (9.42)
\end{aligned}
$$

where $\widetilde{F}^{\vec{n}}_{i\mu\nu} = \partial_\mu \widetilde{A}^{\vec{n}}_{i\nu} - \partial_\nu \widetilde{A}^{\vec{n}}_{i\mu}$. This immediately leads to Euler-Lagrange equations of motion corresponding to massive spin-0, spin-1 and spin-2 particles :

$$
\left(\Box + M^2_{\vec{n}} \right) \widetilde{\phi}^{\vec{n}}_{ij} = 0,
$$

---

[5] One can also start from this Lagrangian and work back to find the precise form of the $h \to \widetilde{h}$ etc. transformations. This is done for example, to get α, which is chosen to get decoupled equations of motion.

$$\partial_\mu \widetilde{F}^{\vec{n}}_{i\mu\nu} + M_{\vec{n}}^2 \widetilde{A}^{\vec{n}}_{i\nu} = 0,$$

$$\Box \left(\widetilde{h}^{\vec{n}}_{\mu\nu} - \frac{1}{2}\widetilde{h}^{\vec{n}}\eta_{\mu\nu} \right) - \partial_\mu\partial^\alpha \left(\widetilde{h}^{\vec{n}}_{\nu\alpha} - \frac{1}{2}\widetilde{h}^{\vec{n}}\eta_{\nu\alpha} \right) \qquad (9.43)$$

$$-\partial_\nu\partial^\alpha \left(\widetilde{h}^{\vec{n}}_{\mu\alpha} - \frac{1}{2}h^{\vec{n}}\eta_{\mu\alpha} \right) + \partial^\alpha\partial^\beta \left(\widetilde{h}^{\vec{n}}_{\alpha\beta} \right.$$

$$\left. -\frac{1}{2}\widetilde{h}^{\vec{n}}\eta_{\alpha\beta} \right)\eta_{\mu\nu} + M_{\vec{n}}^2 \left(\widetilde{h}^{\vec{n}}_{\mu\nu} - \widetilde{h}^{\vec{n}}\eta_{\mu\nu} \right) = 0.$$

This Lagrangian can now be quantized for each $\vec{n}$ by defining suitable path integrals with the $\widetilde{h}^{\vec{n}}_{\mu\nu}$, $\widetilde{A}^{\vec{n}}_{i\mu}$ and $\widetilde{\phi}^{\vec{n}}_{ij}$ as dynamical variables. The propagators for these fields can be determined from the free action written above. We also need to expand the spin-1 and spin-2 fields in terms of polarization vectors/tensors, which is a standard procedure.

Let us go over the motivation for introducing the $\widetilde{h}^{\vec{n}}_{\mu\nu}$, $\widetilde{A}^{\vec{n}}_{i\mu}$ and $\widetilde{\phi}^{\vec{n}}_{ij}$ fields again. The basic reason was that they are bulk-gauge invariant. Another way to look at it is that with these fields, the bulk gauge conditions reduce to

- $\partial_\mu \widetilde{h}^{\vec{n}}_{\mu\nu} = 0$ and $\widetilde{h}^{\vec{n}} = 0$, *i.e.*, the de Donder gauge on the brane, which means the graviton is traceless and transverse:

- $\partial_\mu \widetilde{A}^{\vec{n}}_{i\mu} = 0$ and $n_i \widetilde{A}^{\vec{n}}_{i\mu} = 0$, *i.e.*, a Lorentz gauge on the brane, which means the gauge bosons are also transverse.

- $n_i \widetilde{\phi}^{\vec{n}}_{ij} = 0$, which is an orthogonality condition.

The Lagrangian density is also manifestly diagonal in these fields. The degrees of freedom are, therefore,

- $\widetilde{h}^{\vec{n}}_{\mu\nu}$: starting from $\vec{n} = 0$, *i.e.*, massless graviton plus massive gravitons (spin-2);

- $\widetilde{A}^{\vec{n}}_{i\mu}$: no zero mode but $(p-1)$ massive spin-1 fields, $\widetilde{P}_{ki}\widetilde{A}^{\vec{n}}_{i\mu} = 0$ removes one field;

- $\widetilde{\phi}^{\vec{n}}_{ij}$: again no zero mode, but $p(p-1)/2$ massive spin-0 fields.

Another way of stating this is to say that the de Donder gauge in $(4+p)$-dimensions leads to mixed states on the brane, which are difficult to quantise. We thus choose a different gauge in which the graviton Kaluza-Klein modes obey a four-dimensional de Donder gauge and the spin-1 fields obey a Lorentz gauge. We can also look upon the vacuum configuration $\hat{g}_{\hat{\mu}\hat{\nu}} = \eta_{\hat{\mu}\hat{\nu}}$ as breaking the symmetry of translation around the p-circles spontaneously. As in the Higgs mechanism, the $\widetilde{h}^{\vec{n}}_{\mu\nu}$ absorb one of the $\widetilde{A}^{\vec{n}}_{i\mu}$ and $\hat{\phi}^{\vec{n}}_{ij}$ to become massive. This is a spacetime analogue of the Higgs mechanism.

Given the free-field equations of motion, we can find their Green's functions, *i.e.*, the propagators. For the gravitons $\widetilde{h}^{\vec{n}}_{\mu\nu}$, the propagator has the form:

$$i\Delta^{\vec{n},\vec{n}'}_{\mu\nu,\rho\sigma} = \frac{i\delta_{\vec{n},-\vec{n}'}\, P_{\mu\nu,\rho\sigma}(k)}{k^2 - M^2_{\vec{n}} + i\epsilon}, \tag{9.44}$$

where $P_{\mu\nu,\rho\sigma}(k)$ represents the sum over polarizations:

$$\begin{aligned} P_{\mu\nu,\rho\sigma}(k) &= Q_{\mu\rho}Q_{\nu\sigma} + Q_{\mu\sigma}Q_{\nu\rho} - \frac{2}{3}Q_{\mu\nu}Q_{\rho\sigma}, \\ Q_{\mu\nu} &= \eta_{\mu\nu} - \frac{k_\mu k_\nu}{M^2_{\vec{n}}}, \end{aligned} \tag{9.45}$$

etc. Similarly, for $\widetilde{A}^{\vec{n}}_{i\mu}$, the propagator has the form:

$$i\Delta^{\vec{n},\vec{n}'}_{i\mu,j\nu} = \frac{i}{2}\,\frac{\delta_{\vec{n},-\vec{n}'}\, P^{\vec{n}}_{ij}\, Q_{\mu\nu}}{k^2 - M^2_{\vec{n}} + i\epsilon}, \tag{9.46}$$

and for $\widetilde{\phi}^{\vec{n}}_{ij}$, the propagator has the form:

$$i\Delta^{\vec{n},\vec{n}'}_{ij,kl} = \frac{i}{2}\,\frac{\delta_{\vec{n},-\vec{n}'}\left(P^{\vec{n}}_{ik}P^{\vec{n}}_{jl} + P^{\vec{n}}_{il}P^{\vec{n}}_{jk}\right)}{k^2 - M^2_{\vec{n}} + i\epsilon}. \tag{9.47}$$

For the graviton propagator $P^\mu{}_{\mu,\rho\sigma} = 0$ if $\widetilde{h}^{\vec{n}}_{\mu\nu}$ is on-shell, *i.e.*, $k^2 = M^2_{\vec{n}}$.

9.4 ADD Feynman rules

In order to study the gravitational interactions of these Kaluza-Klein states, we need to couple them to matter. We have already seen that the prescription for this is to write an extra piece in the action: $\hat{S} = \hat{S}_g + \hat{S}_m$, where

$$\hat{S}_m = \int d^4x \, d^p\vec{y} \, \sqrt{\hat{g}} \, \mathcal{L}_m(\hat{g}, \Phi, V_\mu, \Psi) \, \delta^p(\vec{y}). \qquad (9.48)$$

The delta-function in the integrand ensures that the matter fields remain confined on the brane, *i.e.*, at $\vec{y} = \vec{0}$. Now, integrating over the y-coordinates (which corresponds to compactification) and with the substitution

$$\hat{g}_{\mu\nu} = \eta_{\mu\nu} + \kappa(h_{\mu\nu} + \phi\eta_{\mu\nu}), \qquad (9.49)$$

($\phi = \phi_{ii}$ and $\kappa = \hat{\kappa}/\sqrt{V_p}$) we arrive at the well-known result

$$S_m = -\frac{\kappa}{2} \sum_{\vec{n}} \int d^4x \, (h_{\vec{n}}^{\mu\nu} \, T_{\mu\nu} + \phi_{\vec{n}} \, T^\mu{}_\mu) + \mathcal{O}(\kappa^2). \qquad (9.50)$$

In the above

$$T_{\mu\nu} = \left(-\eta_{\mu\nu}\mathcal{L}_m + 2\frac{\delta\mathcal{L}_m}{\delta\hat{g}_{\mu\nu}} \right)\bigg|_{\hat{g}_{\mu\nu}=\eta_{\mu\nu}}. \qquad (9.51)$$

Note that the metric fields have been expanded in their Kaluza-Klein modes. Also, in this equation $\hat{g}_{\mu\nu}$ is the induced metric on the brane and not the full $(4 + p)$-dimensional metric, i.e.

$$\begin{aligned}
\sqrt{\hat{g}} &= 1 + \frac{\kappa}{2}h + 2\kappa\phi, \\
\hat{g}_{\mu\nu} &= \eta_{\mu\nu} + \kappa(h_{\mu\nu} + \phi\eta_{\mu\nu}), \\
\hat{g}^{\mu\nu} &= \eta^{\mu\nu} - \kappa(h^{\mu\nu} + \phi\eta^{\mu\nu}).
\end{aligned} \qquad (9.52)$$

We now have to replace the $h_{\mu\nu}^{\vec{n}}$ and $\phi^{\vec{n}}$ by the physical (gauge-invariant) fields $\widetilde{h}_{\mu\nu}^{\vec{n}}$ and $\widetilde{\phi}^{\vec{n}}$. This also brings in the $\widetilde{A}_{i\mu}^{\vec{n}}$. Making the appropriate substitutions, and using the result

$$P_{ij}^{\vec{n}}\phi_{ij}^{\vec{n}} = \sqrt{\frac{3}{2(p+2)}} \; \widetilde{\phi}^{\vec{n}},$$

we get an interaction term

$$\mathcal{S}_m = -\frac{\kappa}{2} \sum_{\vec{n}} \int d^4 x \left(\widetilde{h}^{\vec{n}}_{\mu\nu} T^{\mu\nu} + \sqrt{\frac{2}{3(p+2)}} \, \widetilde{\phi}^{\vec{n}} T^{\mu}{}_{\mu} \right), \qquad (9.53)$$

(plus terms of $\mathcal{O}(\kappa^2)$) where $\widetilde{\phi}^{\vec{n}} = \widetilde{\phi}^{\vec{n}}_{ii}$.

It is interesting that the $\widetilde{A}^{\vec{n}}_{i\mu}$ have disappeared from the interaction Lagrangian; so we have the $\widetilde{\phi}^{\vec{n}}_{ij}$, except for the dilaton field. This is ultimately a consequence of putting matter on the brane, *i.e.*, taking the bulk-energy-momentum tensor as $\hat{T}_{\hat{\mu}\hat{\nu}} = \delta^{\rho}_{\hat{\mu}} \delta^{\sigma}_{\hat{\nu}} T_{\rho\sigma}$. This, in fact, ensures that the vector fields $\widetilde{A}_{\mu i}$ decouple from the energy-momentum tensor even at $\mathcal{O}(\kappa)$.

With this prescription, we now consider various kinds of matter fields coupled to gravity, *i.e.*, different forms of $\mathcal{L}_m(g, \Phi, V_m, \Psi)$.

First, for a scalar field Φ in a gauge theory, the Lagrangian is

$$\mathcal{L}_m = g^{\mu\nu} D_\mu \Phi^\dagger D_\nu \Phi - m_\Phi^2 \Phi^\dagger \Phi - \lambda^2 (\Phi^\dagger \Phi)^2, \qquad (9.54)$$

which leads to an energy-momentum tensor

$$\begin{aligned} T_{\mu\nu}(\Phi) \;=\;& D_\mu \Phi^\dagger D_\nu \Phi + D_\nu \Phi^\dagger D_\mu \Phi - \eta_{\mu\nu} D_\alpha \Phi^\dagger D^\alpha \Phi \\ &+ \eta_{\mu\nu} m_\Phi^2 \Phi^\dagger \Phi + \eta_{\mu\nu} \lambda (\Phi^\dagger \Phi)^2. \end{aligned} \qquad (9.55)$$

Here $D_\mu = \partial_\mu + ie V^a_\mu T^a$ is a gauge-covariant derivative with coupling e to a gauge field V^a_μ, and T^a are the generators of the gauge group. Hence the interaction Lagrangian for gravity coupling to this energy-momentum tensor comes out to be

$$\begin{aligned} \mathcal{L}_I(\Phi) \;=\;& \kappa \Bigg[-\left(\widetilde{h}^{\vec{n}}_{\mu\nu} - \frac{1}{2} \widetilde{h}^{\vec{n}} \eta_{\mu\nu} \right) D^\mu \Phi^\dagger D^\nu \Phi - \frac{1}{2} \widetilde{h}^{\vec{n}} m_\Phi^2 \Phi^\dagger \Phi \\ &+ \sqrt{\frac{2}{3(p+2)}} \, \widetilde{\phi}^{\vec{n}} \left(D_\mu \Phi^\dagger D^\mu \Phi - 2 m_\Phi^2 \Phi^\dagger \Phi \right) \Bigg] \end{aligned} \qquad (9.56)$$

(plus terms of $\mathcal{O}(\kappa^2)$). We thus have vertices of the following types: schematically,

$$\begin{array}{ccc} h\Phi\Phi, & \phi\Phi\Phi, & h\phi\Phi V, \\ h\Phi\Phi V, & \phi\Phi\Phi V, & \phi\Phi\Phi VV. \end{array}$$

If the Φ-potential is of a spontaneously-broken type, then we will have to make a replacement $\Phi \to \Phi + \langle\Phi\rangle$. In this case, we will also generate vertices:

$$h\Phi V, \qquad hVV, \qquad \phi\Phi V, \qquad \phi VV.$$

Secondly, for a vector field, V_μ, the Lagrangian is

$$\begin{aligned}
\mathcal{L}_m &= \frac{1}{4}g_{\mu\alpha}g_{\nu\beta}F^{\mu\nu}F^{\alpha\beta} - \frac{1}{2\xi}(g_{\mu\nu}\partial^\mu V^\nu - \Gamma^{\mu\nu}{}_\nu A_\mu)^2 \\
&\quad + \frac{1}{2}m_V^2 g_{\mu\nu}V^\mu V^\nu,
\end{aligned} \tag{9.57}$$

where $F_{\mu\nu} = \partial_\mu V_\nu - \partial_\nu V_\mu$. This leads to an energy-momentum tensor

$$\begin{aligned}
T_{\mu\nu}(V) &= \eta_{\mu\nu}\left(\frac{1}{4}F_{\rho\sigma}F^{\rho\sigma} - \frac{1}{2}m_V^2 V_\rho V^\rho\right) \\
&\quad - \left(F_\mu^\rho F_{\nu\rho} - m_V^2 V_\mu V_\nu\right) \\
&\quad - \frac{1}{\xi}\left[\eta_{\mu\nu}\left(\partial^\rho\partial^\sigma V_\rho V_\sigma + \frac{1}{2}(\partial_\rho V^\rho)^2\right)\right. \\
&\quad \left. + (\partial_\mu\partial^\rho V_\rho V_\nu + \partial_\nu\partial^\rho V_\rho V_\mu)\right],
\end{aligned} \tag{9.58}$$

which couples to gravity yielding an interaction Lagrangian

$$\begin{aligned}
\mathcal{L}_I(V) &= \kappa\left[-\frac{1}{8}\left(\widetilde{h}^{\vec{n}}\eta^{\mu\nu} - 4\widetilde{h}^{\mu\nu,\vec{n}}\right)F_\mu{}^\rho F_{\nu\rho}\right. \tag{9.59} \\
&\quad + \frac{1}{4}\left(\widetilde{h}^{\vec{n}}\eta^{\mu\nu} - 2\widetilde{h}^{\mu\nu,\vec{n}}\right)m_V^2 V_\mu V_\nu \\
&\quad + \frac{\widetilde{h}^{\vec{n}}}{2\xi}\left(\partial^\rho\partial^\sigma V_\sigma V_\rho + \frac{1}{2}(\partial^\rho V_\rho)^2\right) - \frac{1}{\xi}\widetilde{h}^{\mu\nu,\vec{n}}\partial_\mu\partial^\rho V_\rho V_\nu \\
&\quad \left. + \sqrt{\frac{1}{6(p+2)}}\,m_A^2\widetilde{\phi}^{\vec{n}}V^\mu V_\mu - \frac{1}{\xi}\sqrt{\frac{2}{3(p+2)}}\,\partial^\mu\widetilde{\phi}^{\vec{n}}\partial^\nu V_\nu V_\mu\right],
\end{aligned}$$

(plus terms of $\mathcal{O}(\kappa^2)$). This leads to the vertices, which we write schematically as:

$$\begin{aligned}
hVV, &\qquad hVVV, &\qquad hVVVV, \\
\phi VV, &\qquad \phi VVV, &\qquad \phi VVVV.
\end{aligned}$$

Finally, for fermions, we need to introduce the concept of a *vierbein*. In curved space, we have, at every point, a set of unit vectors $\hat{n}_\mu$ along the coordinate axes. We also have a locally inertial (Minkowski) coordinate system, which should have unit vectors $\hat{e}_n$ along the axes. Obviously, $\hat{n}_\mu.\hat{n}_\nu = g_{\mu\nu}$ while $\hat{e}_n.\hat{e}_m = \eta_{mn}$. If we now expand

$$\hat{e}_n = e_n^\mu \hat{n}_\mu, \quad \text{and} \quad \hat{n}_\mu = \epsilon_\mu^n \hat{e}_n, \tag{9.60}$$

the coefficients ϵ_n^μ are said to form a vierbein. Obviously, $g_{\mu\nu}\epsilon_m^\mu \epsilon_n^\mu = \eta_{mn}$ *i.e.*, ϵ_m^μ factorizes the metric. The inverse vierbein satisfies $\eta_{mn}\epsilon_\mu^m \epsilon_\nu^n = g_{\mu\nu}$.

A fermion field coupled to gravity is now described by the Lagrangian

$$\mathcal{L}_m(\psi) = \epsilon\, \bar{\psi}\, (i\tilde{\gamma}^\mu D_\mu - m_\psi)\, \psi, \tag{9.61}$$

where $\epsilon = \det \epsilon_\mu^n$ and $\tilde{\gamma}^\mu = \epsilon_n^\mu \gamma^n$ (recall that the usual Dirac matrices γ^n are defined in Minkowski space). The covariant derivative on the fermion field is given by

$$\mathcal{D}_\mu \psi = \left(D_\mu + \frac{1}{2}\omega_\mu^{mn}\sigma_{mn} \right) \psi, \tag{9.62}$$

where ω^{mn} is the spin connection and $\sigma_{mn} = \frac{1}{4}[\gamma_m, \gamma_n]$. The spin connection can be written in terms of the vierbein as,

$$\omega_{\mu mn} = \frac{1}{2}\left(\partial_\mu \epsilon_{m\nu} - \partial_\nu \epsilon_{m\mu}\right)\epsilon_n^\nu - \frac{1}{2}\left(\partial_\mu \epsilon_{n\nu} - \partial_\nu \epsilon_{n\mu}\right)\epsilon_m^\nu$$
$$-\frac{1}{2}\epsilon_m^\rho \epsilon_n^\sigma \left(\partial_\rho \epsilon_{q\sigma} - \partial_\sigma \epsilon_{q\rho}\right)\epsilon_\mu^q. \tag{9.63}$$

In the weak-field approximation, the vierbein becomes

$$\epsilon_\mu^n = \delta_\mu^n + \frac{\kappa}{2}(h_\mu^n + \delta_\mu^n \phi). \tag{9.64}$$

One can now show that the energy-momentum tensor for the fermion field is

$$T_{\mu\nu}(\psi) = -\eta_{\mu\nu}\left(\bar{\psi}i\gamma^\rho D_\rho \psi - m_\psi \bar{\psi}\psi\right) + \frac{i}{2}\bar{\psi}\gamma_{(\mu}D_{\nu)}\psi$$
$$+\frac{1}{2}\eta_{\mu\nu}\partial^\rho\left(\bar{\psi}i\gamma_\rho\psi\right) - \frac{1}{4}\partial_{(\mu}\left(\bar{\psi}i\gamma_{\nu)}\psi\right) \tag{9.65}$$

(plus terms of $\mathcal{O}(\kappa)$). In the above, some of the indices have been symmetrized and this is denoted by the parenteses on the subscripts. Now we are in a position to write the interaction Lagrangian for fermions with gravity:

$$
\begin{aligned}
\mathcal{L}_I(\psi) \;=\; & \frac{\kappa}{2}\Bigg[(\widetilde{h}^{\vec{n}}\eta^{\mu\nu} - \widetilde{h}^{\mu\nu,\vec{n}})\bar{\psi}i\gamma_\mu D_\nu\psi - m_\psi \widetilde{h}^{\vec{n}}\bar{\psi}\psi \\
& + \frac{1}{2}\bar{\psi}i\gamma^\mu(\partial_\mu \widetilde{h}^{\vec{n}} - \partial^\nu \widetilde{h}^{\vec{n}}_{\mu\nu})\psi \\
& + \sqrt{\frac{6}{p+2}}\,\widetilde{\phi}^{\vec{n}}\bar{\psi}i\gamma^\mu D_\mu\psi - 4\sqrt{\frac{2}{3(p+2)}}\,m_\psi \widetilde{\phi}^{\vec{n}}\bar{\psi}\psi \\
& + \sqrt{\frac{3}{8(p+2)}}\,\partial_\mu \widetilde{\phi}^{\vec{n}}\bar{\psi}i\gamma^\mu\psi \Bigg]
\end{aligned}
\tag{9.66}
$$

plus terms of $\mathcal{O}(\kappa^2)$, as usual. Thus, schematically, the vertices involving fermions are of the form:

$$
h\bar{\psi}\psi, \quad hV\bar{\psi}\psi, \quad \phi\bar{\psi}\psi, \quad \phi V\bar{\psi}\psi,
$$

where the vector field V_μ comes from the gauge-covariant derivative D_μ.

With all these ingredients, we are now in a position to read-off Feynman rules from the interaction Lagrangian. These can be grouped into three classes, *viz.* three-, four- and five-point vertices. We exhibit the three-point vertices in Fig. 9.1. The four- and five-point vertices can be found in the paper by Han, Lykken and Zhang.

Explicitly, the three-point vertices are

$$
\begin{aligned}
(1a) \quad h\Phi\Phi \;&:\; -i\frac{\kappa}{2}\left(M_\Phi^2 \eta_{\alpha\beta} + C_{\alpha\beta\rho\sigma}p^\rho p'^\sigma \right), \\
\phi\Phi\Phi \;&:\; -i\omega\kappa(p.p' - 2M_\Phi^2), \\
(1b) \quad hVV \;&:\; -i\frac{\kappa}{2}\left[(M_V^2 + p.p')C_{\alpha\beta\mu\nu} + D_{\alpha\beta\mu\nu}(p,p') \right], \\
\phi\Phi\Phi \;&:\; -i\omega\kappa\delta_{ab}M_V^2 \eta_{\mu\nu}, \\
(1c) \quad h\bar{\psi}\psi \;&:\; -i\frac{\kappa}{8}\left[\gamma_\alpha(p_\beta + p'_\beta) + \gamma_\beta(p_\alpha + p'_\alpha) \right.
\end{aligned}
$$

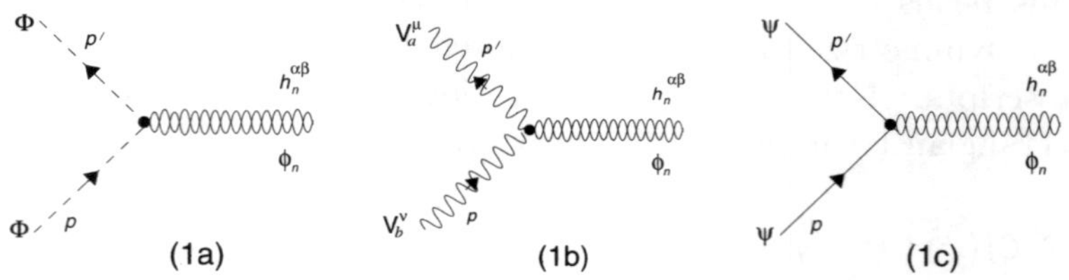

Figure 9.1: Three-point vertices.

$$-2\eta_{\alpha\beta}(\not{p}+\not{p}' - 2m_\psi)\Big],$$

$$\phi\bar{\psi}\psi \quad : \quad -i\omega\kappa\left[\frac{3}{4}(\not{p}+\not{p}') - 2m_\psi\right],$$

where $\omega = \sqrt{\frac{2}{3(p+2)}}$, and

$$C_{\alpha\beta\rho\sigma} = \eta_{\alpha\rho}\eta_{\beta\sigma} + \eta_{\alpha\sigma}\eta_{\beta\rho} - \eta_{\alpha\beta}\eta_{\rho\sigma},$$

$$D_{\alpha\beta\rho\sigma}(p,p') = \eta_{\alpha\beta}p_\sigma p'_\rho - \Big[\eta_{\alpha\sigma}p_\beta p'_\rho \tag{9.67}$$

$$+\eta_{\alpha\rho}p_\sigma p'_\beta - \eta_{\rho\sigma}p_\alpha p'_\beta - (\alpha \leftrightarrow \beta)\Big].$$

For brevity, we choose the unitary gauge for the V-boson, i.e. $\xi \to \infty$. Note that at $\mathcal{O}(\kappa)$ each vertex cannot contain more than one graviton. Multple gravitons at a vertex will arise in higher orders of κ.

Some of the general features of these vertices are:

- Gravity couples to all particles through the energy-momentum tensor. The three-point couplings are blind to flavour and colour, but sensitive to spin. Four- and five-point graviton vertices are sensitive to colour.

- Each vertex is proportional to $\kappa = \sqrt{16\pi G_N} = \sqrt{16\pi}/M_P$, *i.e.*, each graviton KK mode couples very weakly to matter (coupling being $\propto 1/M_P$).

- The dilaton (ϕ) couplings vanish for massless fermions and gauge bosons (in the unitary gauge); but not for massless scalars.

- All the couplings have powers of momentum, *i.e.*, the interaction grows stronger for higher energies. This, of course, is a basic feature of the gravitational interaction, embodied even in Newton's action-at-a-distance law of gravitation.

9.5 ADD phenomenology

We see therefore, that at low energies, *i.e.*, below a cutoff scale $M_S \sim \hat{M}_P$—which we shall call the *string scale*—we have, on the brane, an effective theory of massive graviton (and dilaton) towers interacting with the Standard Model fields. How does this affect processes at low energies? There are two possibilities:

- *Virtual graviton exchange*, in which the initial and final states are Standard Model particles, but there are gravitons in the intermediate states.

- *Real graviton emission*, in which the initial states are Standard Model particles, but the final state contains gravitons in addition to Standard Model particles.

Let us consider virtual exchange[5] first. We choose a typical 4-fermion process

$$f(p_1) + \bar{f}(p_2) \longrightarrow f'(p_3) + \bar{f}'(p_4) \tag{9.68}$$

where f and f' are dissimilar fermions (this choice is made to avoid the complication of t-channel diagrams). For simplicity, we shall also neglect the fermion masses, i.e. $m_f = m_{f'} = 0$.

In the SM, this process would correspond to a diagram of the form given in Fig. 9.2 with interference terms between photon-mediated and Z-mediated diagrams, of course.

We now see that for graviton exchange we have vertices of the form

$$-i\frac{\kappa}{8}\left[\gamma_\mu(p+p')_\nu + \gamma_n u(p+p')_\mu - 2\eta_{\mu\nu}(\not{p}+ \not{p}' - 2m_f)\right]$$

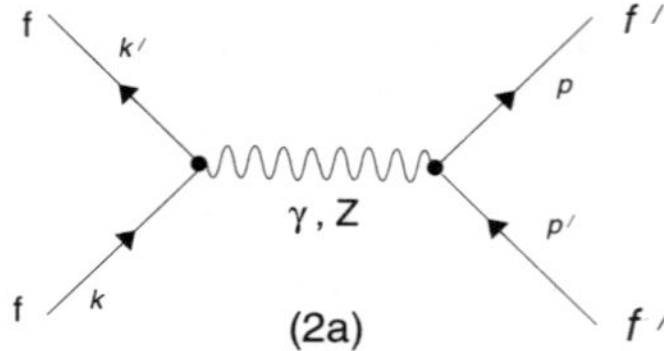

Figure 9.2: Four fermion process in the Standard Model.

which allows us to have a Feynman diagram in Fig. 9.3, where the $\vec{n}$th Kaluza-Klein mode of the graviton (dilaton) is exchanged. If we set $m_f \simeq 0$, the dilaton exchange contribution vanishes. However, there will be a similar diagram for each and every value of $\vec{n}$ and, since the final states are the same for all $\vec{n}$, these will all add *coherently*, i.e., the total amplitude for the $f\bar{f} \to f'\bar{f}'$ transition will be the sum of the amplitudes due to each graviton $h^{\vec{n}}$. Thus the effective matrix element for this transition will be

$$\mathcal{M}(f\bar{f} \to f'\bar{f}') = \sum_{\vec{n}} \mathcal{M}_{\vec{n}}(f\bar{f} \to f'\bar{f}'). \qquad (9.69)$$

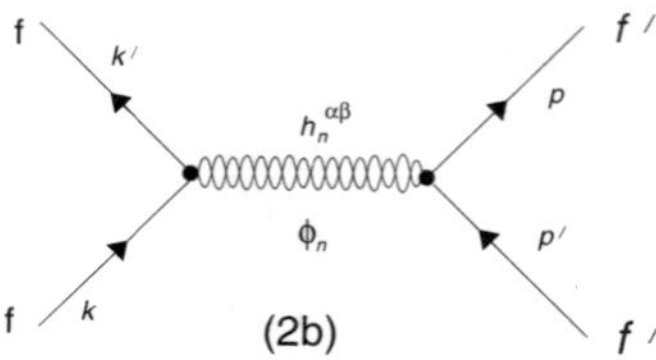

Figure 9.3: Graviton contribution to the four fermion process.

Let us first calculate $\mathcal{M}_{\vec{n}}(f\bar{f} \to f'\bar{f}')$. Using the Feynman rule given above, dropping all terms containing $m_f, m_{f'}$ and using the (on-shell) Dirac equation, we get

$$\mathcal{M}_{\vec{n}}(f\bar{f} \to f'\bar{f}') \;\; = \;\; -i\frac{\kappa^2}{64}\,\frac{1}{q^2 - M_{\vec{n}}^2 + i\epsilon}\,\bar{u}(p)\gamma_{(\rho}(p - p')_{\sigma)}v(p')$$

$$\times \mathcal{P}^{\rho\sigma,\mu\nu}(q) \times \bar{v}(k')\gamma_{(\mu}(k-k')_{\nu)}u(k) \quad (9.70)$$

where, $q = k + k' = p + p'$, we have symmetrized the indices as before, and $\mathcal{P}$ are given in Eq.(9.45).

Consider the last term $\mathcal{Q}_{\mu\nu}\mathcal{Q}_{\rho\sigma}$, which gives a factor

$$\begin{aligned}
\mathcal{Q}^{\mu\nu} \,\bar{v}(k')&\{\gamma_\mu(k-k')_\nu + \gamma_\nu(k-k')_\mu\}u(k) \\
&= \left(\eta^{\mu\nu} - \frac{1}{M_{\bar{n}}^2}q^\mu q^\nu\right)\,\bar{v}(k')\gamma_{(\mu}(k-k')_{\nu)}u(k) \\
&= -2m_f\,\bar{v}(k')u(k) \\
&\simeq 0 \qquad (\text{when } m_f \simeq 0).
\end{aligned} \quad (9.71)$$

We can similarly show that all the $M_{\bar{n}}^{-2}$ terms drop out of $\mathcal{P}^{\rho\sigma,\mu\nu}(q)$ in the limit of massless fermions, leaving us with a simple expression

$$\begin{aligned}
\mathcal{P}^{\rho\sigma,\mu\nu}(q) &= \eta^{\rho\mu}\eta^{\sigma\nu} + \eta^{\rho\nu}\eta^{\sigma\mu} + m_f(\cdots) + m_{f'}(\cdots) \\
&\simeq \eta^{\rho\mu}\eta^{\sigma\nu} + \eta^{\rho\nu}\eta^{\sigma\mu}.
\end{aligned} \quad (9.72)$$

In this approximation, the $\bar{n}$ dependence in $\mathcal{M}_{\bar{n}}(f\bar{f} \to f'\bar{f}')$ is contained in the propagator factor only, *i.e.*,

$$\begin{aligned}
\mathcal{M}_{\bar{n}}(f\bar{f} \to f'\bar{f}') &= -\frac{i}{64}\left(\frac{\kappa^2}{q^2 - M_{\bar{n}}^2 + i\epsilon}\right)\bar{u}(p)\gamma_{(\rho}(p-p')_{\sigma)}v(p') \\
&\times \ (\eta^{\rho\mu}\eta^{\sigma\nu} + \eta^{\rho\nu}\eta^{\sigma\mu})\,\bar{v}(k')\gamma_{(\mu}(k-k')_{\nu)}u(k),
\end{aligned}$$

and hence

$$\begin{aligned}
\mathcal{M}(f\bar{f} \to f'\bar{f}') &= -\frac{1}{64}\left(\sum_{\bar{n}}\frac{i\kappa^2}{q^2 - M_{\bar{n}}^2 + i\epsilon}\right)\bar{u}(p)\gamma_{(\rho}(p-p')_{\sigma)}v(p') \\
&\times \ (\eta^{\rho\mu}\eta^{\sigma\nu} + \eta^{\rho\nu}\eta^{\sigma\mu})\,\bar{v}(k')\gamma_{(\mu}(k-k')_{\nu)}u(k).
\end{aligned}$$

The sum over the graviton tower of states can, therefore, be done separately. Let

$$\mathcal{D}(q^2) = \sum_{\bar{n}}\frac{i}{q^2 - M_{\bar{n}}^2 + i\epsilon}. \quad (9.73)$$

We note that q^2 is typically of the order of tens to hundreds of GeV squared, while the $M_{\vec{n}}^2$ are given by

$$M_{\vec{n}}^2 = \frac{4\pi^2 \vec{n}^2}{L^2} \sim (10^{-4}\ \mathrm{eV})^2 \vec{n}^2. \qquad (9.74)$$

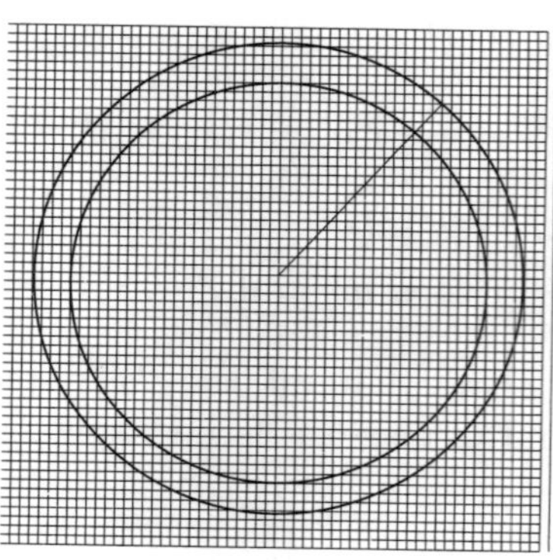

Figure 9.4: Quasicontinuum of Kaluza-Klein states.

The $M_{\vec{n}}$'s are therefore, densely packed at the grid points of a p-dimensional hyperspace, with a lattice size $2\pi/L$ which can be thought of as a quasi-continuum of states in p-dimensions. This is illustrated in Fig. 9.4 which shows the case $p = 2$. For a given value of $M_{\vec{n}}$, therefore, the number of states lying between $M_{\vec{n}}$ and $M_{\vec{n}} + \delta M_{\vec{n}}$ is obtained by calculating the volume of a hyperspherical shell in p-dimensions (of thickness $\delta M_{\vec{n}}$) and dividing by the lattice size. This gives the density of states function

$$\rho(M_{\vec{n}})\delta M_{\vec{n}} = \frac{2\pi^{p/2} M_{\vec{n}}^{p-1}}{\Gamma(p/2)} \delta M_{\vec{n}} \times \frac{1}{(2\pi/L)^p}$$

$$= \frac{2L^p M_{\vec{n}}^{p-1}}{(4\pi)^{p/2}\Gamma(p/2)} \delta M_{\vec{n}}. \qquad (9.75)$$

We can now use this density of states function to write

$$\mathcal{D}(q^2) = \sum_{\vec{n}} \frac{i}{q^2 - M_{\vec{n}}^2 + i\epsilon} \longrightarrow \int_0^\infty dM_{\vec{n}}^2 \frac{i\rho(M_{\vec{n}}^2)}{q^2 - M_{\vec{n}}^2 + i\epsilon}. \qquad (9.76)$$

Since the integrand has poles at $q^2 = M_{\vec{n}}^2$ (*i.e.*, almost everywhere in the $M_{\vec{n}}$ space!), we write

$$\frac{1}{q^2 - M_{\vec{n}}^2 + i\epsilon} = P\left(\frac{1}{q^2 - M_{\vec{n}}^2}\right) - i\pi\,\delta(q^2 - M_{\vec{n}}^2), \qquad (9.77)$$

so that,

$$\begin{aligned}
\mathcal{D}(q^2) &= \int_0^\infty dM_{\vec{n}}^2\, i\rho(M_{\vec{n}}^2) P\left(\frac{1}{q^2 - M_{\vec{n}}^2}\right) \\
&\quad + \pi \int_0^\infty dM_{\vec{n}}^2\, \rho(M_{\vec{n}}^2)\, \delta(q^2 - M_{\vec{n}}^2).
\end{aligned} \qquad (9.78)$$

Making a change of variable $y = M_{\vec{n}}/\sqrt{q^2}$, we get

$$\mathcal{D}(q^2) = 2iP \int_0^\infty \frac{dy\, y\rho(q^2 y^2)}{1 - y^2} + \pi\rho(q^2). \qquad (9.79)$$

The integral can be evaluated to obtain

$$\mathcal{D}(q^2) = \frac{L^p}{(4\pi)^{p/2}\Gamma(p/2)}(q^2)^{p/2-1}\left[\pi + 2i\, I_p(y_0)\right], \qquad (9.80)$$

where y is cut off at $y_0 = M_S/\sqrt{q^2}$ and

$$I_p(y_0) = P \int_0^\infty dy\, \frac{y^{p-1}}{1 - y^2}. \qquad (9.81)$$

When this integral is worked out, we get

$$I_p(y_0) = \begin{cases}
-\frac{1}{2}\ln(y_0^2 - 1) - \sum_{k=1}^{p/2-1} \dfrac{y_0^{2k}}{2k} & \text{for even } p, \\[2ex]
\frac{1}{2}\ln\left(\dfrac{y_0+1}{y_0-1}\right) - \sum_{k=1}^{(p-1)/2} \dfrac{y_0^{2k-1}}{2k-1} & \text{for odd } p.
\end{cases} \qquad (9.82)$$

The overall factor in the amplitude $\mathcal{M}(f\bar{f} \to f'\bar{f}')$ now works out as

$$\kappa^2 \mathcal{D}(k^2) = \frac{\kappa^2 L^p}{(4\pi)^{p/2}\Gamma(p/2)}(q^2)^{p/2-1}\left(\pi + 2I(y_0)\right), \qquad (9.83)$$

where $y_0 = M_S/\sqrt{k^2}$. We now invoke Gauss' law[6] to relate

$$\kappa^2 L^p = 8\pi \ (4\pi)^{p/2}\Gamma(p/2)M_S^{-(p+2)}. \tag{9.84}$$

Writing $M_S \simeq \widehat{M}_P$ and remembering that $\kappa^2 = 16\pi G_N = 16\pi/M_P^2$, we get

$$\frac{\kappa^2 L^p}{(4\pi)^{p/2}\Gamma(p/2)} = \frac{8\pi}{M_S^{p+2}}, \tag{9.85}$$

which immediately leads to

$$\begin{aligned}
\kappa^2 \mathcal{D}(k^2) &= \frac{8\pi}{M_S^{p+2}} \, (k^2)^{p/2-1} \ (\pi + 2iI_p(y_0)) \\
&= \frac{8\pi}{M_S^4} \, y_0^{2-p} \ (\pi - 2iI_p(y_0)) \tag{9.86} \\
&\equiv \frac{\lambda(y_0)}{M_S^4}. \tag{9.87}
\end{aligned}$$

In the low energy limit $k^2 \ll M_S^2$, *i.e.*, $y_0 \gg 1$, it can be easily shown that

$$\lambda(y_0) \simeq \begin{cases} -\,8\pi i \log y_0^2 & \text{for } p = 2, \\ -\,\frac{16\pi i}{p-2} & \text{for } p > 2. \end{cases} \tag{9.88}$$

Two interesting features come out of this. One is the fact that at low energies, the graviton effective coupling is almost a constant ($\ln y_0$ is a slowly varying function). The other is that the Planck scale κ has cancelled out of the final result, leaving a suppression by M_S^4, which is not so severe.

We can now write

$$\begin{aligned}
\mathcal{M}_{\bar{n}}(f\bar{f} \to f'\bar{f}') &= -\frac{\lambda(y_0)}{64 M_S^4} \, \bar{u}(p_3)\gamma_{(\rho}(p_3 - p_4)_{\sigma)}v(p_4) \\
&\quad \times \ (\eta^{\rho\mu}\eta^{\sigma\nu} + \eta^{\rho\nu}\eta^{\sigma\mu}) \ \bar{v}(p_2)\gamma_{(\mu}(p_1 - p_2)_{\nu)}u(p_1),
\end{aligned}$$

---

[6]The details of this are explained in the lectures of Sandip Trivedi, in this volume.

and go on to calculate $\sum_{\text{spins}} |\mathcal{M}|^2$ and the cross-section

$$\sigma = \frac{1}{\text{Flux}} \int \overline{|\mathcal{M}|^2} \, d\,\text{Lips}(f\bar{f} \to f'\bar{f'}). \tag{9.89}$$

Note that the only new parameters in σ are the string scale M_S and the number of extra dimensions p, apart from the energy $s = q^2$ and the angular dependences.

We also note that

$$\mathcal{M}_{\bar{n}}(f\bar{f} \to f'\bar{f'}) \propto \frac{1}{M_S^4} \tag{9.90}$$

implies that

$$\sigma \propto \frac{1}{M_S^8} \sim \left(\frac{s}{M_S^2}\right)^4. \tag{9.91}$$

For energies $q^2 \ll M_S^2$ (as we have taken) this suppression is very severe, *e.g.* for $\sqrt{q^2} = \frac{1}{2} M_S$, this gives a factor $\sim 4 \times 10^{-3}$. Does this mean that gravity effects, even with large extra dimensions, are still too small to be seen? The answer is no, because there is also a SM contribution to the cross-section.

$$\mathcal{M}(f\bar{f} \to f'\bar{f'}) = \mathcal{M}_{SM}(f\bar{f} \to f'\bar{f'}) + \mathcal{M}_{grav}(f\bar{f} \to f'\bar{f'})$$

and hence

$$\sigma \propto |\mathcal{M}|^2 = \sigma_{SM} + \sigma_{grav} + \sigma_{int}, \tag{9.92}$$

where σ_{int} is an interference term. Now,

$$\sigma_{grav} \sim \frac{k^8}{M_S^8}$$

but

$$\sigma_{int} \sim \frac{q^4}{M_S^4} \sim 10^{-1} \quad \text{for} \quad \frac{\sqrt{q^2}}{M_S} \simeq \frac{1}{2}.$$

We see that the interference term is not so badly suppressed, and for reasonably large energies, it is generally larger than the radiative corrections to the tree-level SM cross-sections. We can, therefore use experiments of reasonably high sensitivity to constrain the ADD model (i.e. constrain M_S, given p). The higher the energy, the larger the gravitational effect.

For higher energies closer to M_S, the variation of $\lambda(y_0)$ with $y_0 = M_S/\sqrt{k^2}$ has to be taken into account. This has the form illustrated in Fig. 9.5 below. It is clear from the figure that the constant-λ approximation has its limitations. Nevertheless, it has been widely used in the literature, and, in fact, absorbed into M_S to write

$$\frac{\lambda(y_0)}{M_S^4} = \frac{1}{\widetilde{M}_S^4}.$$

In this case, the graviton interaction behaves as a contact interaction, with the $\widetilde{M}_S$ acting like a compositeness scale.

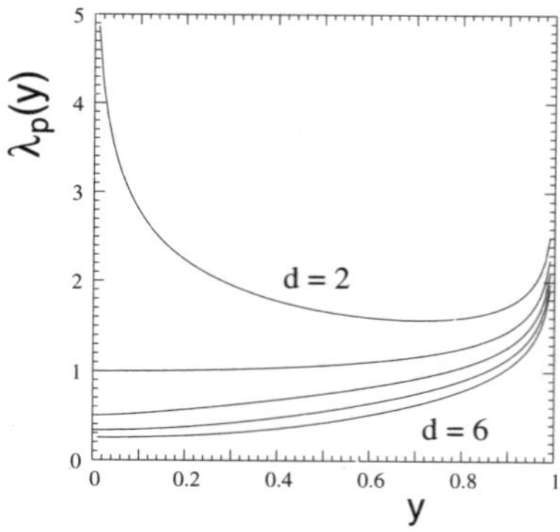

Figure 9.5: The function $\lambda(y_0)$.

If we now consider an actual process like, for example, $e^+e^- \to \mu^+\mu^-$, we can easily compute the cross-sections. The theoretical predictions should then be compared with data. The principal areas where the gravitational interaction will show up are as follows.

- The total cross-section would show an excess over the Standard Model prediction; this excess is basically from the interference term and increases with energy as s^2/M_S^4.

- The angular distribution $d\sigma/d\cos\theta$ would show deviations from the Standard Model prediction because of the spin-2 exchange — this is obviously a result of having spherical harmonics of order 2 in the distribution.

- The invariant mass distribution of the final-state muons would show deviation for high values of invariant mass — this is completely analogous to the case of contact interactions.

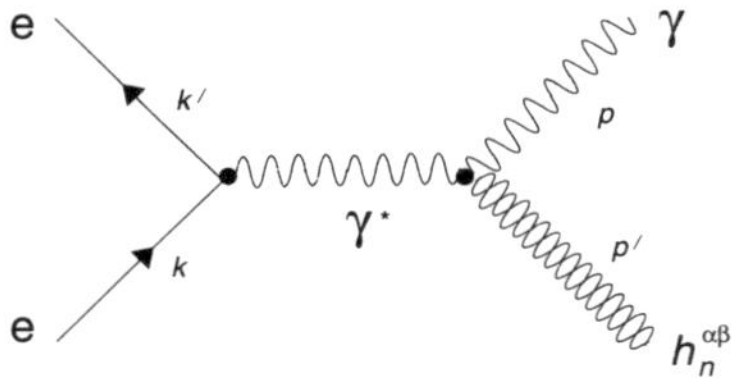

Figure 9.6: Process producing a real graviton.

We now turn to real graviton production[6]. The prototype process for this is the one first studied by Mirabelli, Peskin and Perelstein, *viz.*

$$e^+e^- \to \gamma^* \to \gamma + G_{\vec{n}}.$$

Unlike the case with virtual graviton exchange, every diagram with a different $h_{\vec{n}}^{\alpha\beta}$ is a different diagram, with a cross-section

$$\sigma_{\vec{n}} = \frac{1}{\text{Flux}} \int \overline{|\mathcal{M}|_{\vec{n}}^2} \, d\,\text{Lips}\,(e^+e^- \to \gamma G_{\vec{n}}). \qquad (9.93)$$

Note that not only the matrix element, but the two-body phase space for this is different for each graviton Kaluza-Klein mode because of the different mass. However, we now consider the experimental situation. Every graviton produced escapes detection since its coupling with the matter in the detector(s) is very weak ($\propto \kappa \sim M_P^{-1}$)—far weaker than the couping of neutrinos. For every graviton mode produced, this leads to a final state with a hard photon recoiling against apparently nothing. This is the well-known photon plus missing energy signal, which has been studied in LEP, for example, as a neutrino-counting experiment. Now, every graviton mode contributes to the same signal, which means the contributions must be added *incoherently*

$$\sigma = \sum_{\vec{n}} \sigma_{\vec{n}} \rightarrow \int dM_{\vec{n}}^2 \, \rho(M_{\vec{n}}^2) \, \sigma_{\vec{n}}. \tag{9.94}$$

Without going into the details of the calculation, it is easy to see that the amplitude for each mode $\sim \kappa$, the cross-sections $\sigma_{\vec{n}} \sim \kappa^2$ and hence, on integration with the density of states in Eq.(9.75), we finally get a suppression factor of s^2/M_S^4. This is an effect of the same size as the one obtained in virtual graviton production.

The above ideas have been used in a variety of ways to obtain bounds on the parameters of the ADD model, *viz.* the string scale M_S and the number of extra dimensions p. A detailed discussion of these is beyond the scope of the present lectures. Some of the principal results are described below.

- For $p = 1$, in order to avoid deviations from Newtonian gravity at observable scales, it is necessary for $M_S > 10^{13}$ TeV. This removes the primary motivation for the ADD models, which is to bring down the string scale close to the electroweak scale (~ 0.1 TeV).

- For $p = 2, 3$ the most stringent bounds come from a consideration of graviton radiation in supernovae. If the cross-section for graviton radiation off nucleons inside a supernova (which is extremely sensitive to the supernova temperature) is too large, then the supernova would cool down very fast due to graviton radiation and would not produced the observed optical and neutrino fluxes. This lead to an upper bound on the graviton emission cross-sections and hence constrains $M_S > 30$ TeV for $p = 2$ and $M_S > 4$ TeV for $p = 3$.

- For $p \geq 4$, the best bounds come from collider studies (of many different kinds) which generically predict $M_S > 1$ TeV (give or take a couple of hundred GeV).

- Extra dimensions have been used to explain the smallness of the neutrino masses by invoking a hgher-dimensional seesaw mechanism with a sterile neutrino in the bulk. This has the nice feature that it explains large mixing angles as the sum of a large number of small mixing angles. However, since the bulk neutrino is sterile, it cannot account for the positive results of recent appearance experiments. However, many variations exist, in which the large size of the extra dimensions is used to generate the seesaw mass scale.

- There also exists a variety of grand unification models which invoke large extra dimensions to lower the grand unification scale and hence solve the gauge hierarchy problem.

- Brane world ideas have also been used to explain the supersymmetric flavour problem by placing the hideen sector of supersymmetry on a distant brane away from the brane which constitutes the observable Universe. These are the so-called anomaly-mediated supersymmetry-breaking (AMSB) models.

Before ending this section, it is important to note that the string scale M_S in the ADD model is in the nature of a cut-off. It is therefore, not of very great importance to determine exact numerical bounds on it, as has been done by many authors. Another point to note is that all the current studies have been done for the toroidal compactifiction introduced above. It is, in fact, fair to say that the phenomenological study of brane worlds and extra compact dimensions is still in its infancy. There is a wealth of brilliant ideas and ingenious constructions using these ingrediatents, which address various current problems in the study of fundamantal particles and their interactions. Our purpose was to provide a pedagogic introduction to some of the first of these ideas. The interested reader is invited to go on and discover for herself a whole world of ideas just waiting there in the literature.

Bibliography

[1] See, for example, the well-known textbook *Gravitation and Cosmology*, by S. Weinberg (John Wiley, 1972).

[2] N. Arkani-Hamed, S. Dimopoulos and G. Dvali, *Phys. Lett.* **B429** 263 (1998);
I. Antoniadis, N. Arkani-Hamed, S. Dimopoulos and G. Dvali, *Phys. Lett.* **B436**, 257 (1998).

[3] T. Han, J. Lykken and R.-J. Zhang, *Phys. Rev.* **D59**, 105006 (1999).

[4] G.F Giudice, R. Rattazzi and J. Wells, *Nucl. Phys.* **B544**, 3 (1999).

[5] Some of the basic phenomenological ideas are discussed in the early papers. See, for example,
J.L. Hewett, *Phys. Rev. Lett.* **82**, 4765 (1999);
P. Mathews, S. Raychaudhuri and K. Sridhar, *Phys. Lett.* **B450**, 343 (1999), *ibid.* **B455**, 115 (1999);
T.G. Rizzo, *Phys. Rev.* **D59**, 115010 (1999);
K. Agashe and N.G. Deshpande, *Phys. Lett.* **B456**, 60 (1999);
K. Cheung and W.-Y. Keung,. *Phys. Rev.* **D60**, 112003 (1999).

[6] The single-photon process was first studied by
E.A. Mirabelli, M. Perelstein and M.E. Peskin, *Phys. Rev. Lett.* **82**, 2236 (1999);
constraints from the supernova SN1987A are discussed by
S. Cullen and M. Perelstein (SLAC), *Phys. Rev. Lett.* **83**, 268 (1999).

Contributors

Rajiv. V. Gavai
Tata Institute of Fundamental Research
Homi Bhabha Road
Mumbai 400 005
India
gavai@theory.tifr.res.in

Debashis Ghoshal
Harish-Chandra Research Institute
Chhatnag Road, Jhusi
Allahabad 211 019
India
ghoshal@mri.ernet.in

Dileep P. Jatkar
Harish-Chandra Research Institute
Chhatnag Road, Jhusi
Allahabad 211 019
India
dileep@mri.ernet.in

Anjan S. Joshipura
Theoretical Physics Division
Physical Research Laboratory
Navarangpura
Ahmedabad 380 009
India
anjan@prl.ernet.in

Biswarup Mukhopadhyaya
Harish-Chandra Research Institute
Chhatnag Road, Jhusi
Allahabad 211 019
India
biswarup@mri.ernet.in

V. Ravindran
Harish-Chandra Research Institute
Chhatnag Road, Jhusi
Allahabad 211 019
India
ravindra@mri.ernet.in

Sreerup Raychaudhuri
Department of Physics
Indian Institute of Technology (Kanpur)
Kanpur 208 016
India
sreerup@iitk.ac.in

Saurabh D. Rindani
Theoretical Physics Division
Physical Research Laboratory
Navarangpura
Ahmedabad 380 009
India
saurabh@prl.ernet.in

Sandip P. Trivedi
Tata Institute of Fundamental Research
Homi Bhabha Road
Mumbai 400 005
India
sandip@theory.tifr.res.in

Index